Springer Series in Geomechanics and Geoengineering

Geomechanics deals with the application of the principle of mechanics to geomaterials including experimental, analytical and numerical investigations into the mechanical, physical, hydraulic and thermal properties of geomaterials as multiphase media. Geoengineering covers a wide range of engineering disciplines related to geomaterials from traditional to emerging areas.

The objective of the book series is to publish monographs, handbooks, workshop proceedings and textbooks. The book series is intended to cover both the state-of-the-art and the recent developments in geomechanics and geoengineering. Besides researchers, the series provides valuable references for engineering practitioners and graduate students.

Indexed by SCOPUS, EI Compendex, INSPEC, SCImago.

Jafar Hassanpour · Jamal Rostami
Editors

Performance of Tunnel Boring Machines (TBM) in Rock

Springer

Editors
Jafar Hassanpour
University of Tehran
Tehran, Iran

Jamal Rostami
Colorado School of Mines
Golden, CO, USA

ISSN 1866-8755 ISSN 1866-8763 (electronic)
Springer Series in Geomechanics and Geoengineering
ISBN 978-3-032-20423-3 ISBN 978-3-032-20424-0 (eBook)
https://doi.org/10.1007/978-3-032-20424-0

This Springer imprint is published by the registered company Springer Nature Switzerland AG
The registered company address is: Gewerbestrasse 11, 6330 Cham, Switzerland

Preface

The field of tunnel construction has seen remarkable advancements in recent decades, particularly with the evolution of Tunnel Boring Machines (TBMs). At this time, TBM tunneling is the dominant method of tunnel construction in the world, as compared to conventional tunneling. However, accurate prediction of TBM performance in varying rock conditions remains a challenge. The motivation behind writing this book is to consolidate contemporary knowledge, best practices, and practical insights on hard rock TBM performance prediction. Despite advancements in technology, a significant gap remains in comprehensive, integrated resources that address the intricacies of TBM operation in rock environments. This book is timely, given the increasing demand for infrastructure development, including transport networks and underground facilities, as well as the renewed interest in mining industry to use TBM as the mines go deeper. Proper selection, operation, and successful application of these machines necessitate optimized TBM performance and cost-effective solutions.

The primary aim of this book is to provide a thorough examination of how TBMs operate in diverse rock formations and to present strategies for examining the machine performance and identification of the geological and operational risks in using rock TBMs. This book will serve as a valuable reference for engineers, researchers, and industry professionals seeking to understand the nuances of TBM performance, including its limitations and opportunities for improvement in their operations.

The book is prepared to be easy to read and follow, yet comprehensive, designed for both experienced professionals and newcomers in the field. It will employ a novel approach by integrating theoretical concepts with practical case studies and technological insights. This combination will provide a well-rounded perspective that bridges the gap between academic research and real-world application.

A distinguishing feature of this book will be its inclusion of a global range of case studies, reflecting diverse rock conditions and TBM technologies. These case studies will offer readers a broad perspective on TBM performance and practical solutions across different ground conditions. Additionally, the book will feature contributions from leading experts in the field, ensuring authoritative and up-to-date content.

In summary, this book intends to provide a solid foundation of understanding TBM operation that addresses current challenges, explores innovative solutions, and supports the advancement of tunneling technology.

We sincerely thank all the authors for their invaluable contributions to this book. Their expertise and dedication have significantly enriched the content and broadened our understanding of underground infrastructures. We appreciate their effort in sharing their knowledge and insights across different chapters. This collaborative work would not have been possible without their commitment. Our heartfelt gratitude goes to each of them for making this publication a useful resource in this field. We hope to update the content of this book frequently to reflect the advances in the field as tunneling technology enters a new era of innovations. This includes more intensive use of computational power in our industry, in the form of additional sensory systems, automation, and application of the artificial intelligence (AI) and Machine Learning (ML) as emerging horizons in tunneling practice to help the engineers/operators uncover some potentials of TBM tunneling that have not been reachable before.

Tehran, Iran
Golden, USA
July 2025

Jafar Hassanpour
Jamal Rostami

About This Book

This book offers an in-depth examination of TBM performance prediction models, specifically designed for use in rock formations. It effectively bridges theoretical concepts with practical applications, offering engineers, geologists, and tunneling professionals the essential knowledge and analytical tools to evaluate and predict TBM performance during rock excavation. The content covers a wide range of topics, including fundamental principles of TBM performance prediction—encompassing theoretical, semi-theoretical, and empirical models—along with different factors that influence model accuracy. Additionally, it presents real-world case studies demonstrating how these models are applied in practice. The increasing demand for mechanized tunneling makes such knowledge more critical than ever, as efficient and cost-effective tunneling is crucial for various infrastructure projects.

The book also addresses the inherent complexities of tunneling in difficult rock conditions, which often lead to delays, cost overruns, and safety hazards. By providing a systematic approach to TBM performance prediction, it aims to enable professionals with the insights needed to make informed decisions, thereby reducing risks and enhancing project outcomes. Its comprehensive approach helps address the challenges posed by unpredictable geological conditions and machine responses, ultimately contributing to more reliable and efficient tunneling operations. The targeted audience includes engineers, geologists, construction managers, researchers, and students specializing in tunneling, mining, engineering geology, and geotechnical engineering. Whether experienced practitioners seeking to deepen their understanding of performance models or students seeking entry level knowledge of rock tunnel excavation, this book serves as a valuable resource for anyone involved in underground construction projects, offering practical guidance and foundational understanding to advance their expertise in the field.

Contents

Editors and Contributors

About the Editors

Jafar Hassanpour is a distinguished expert in Engineering Geology and Tunnelling, with a career spanning academia, research, and practical engineering. He earned his Bachelor's degree in Geology with a focus on Hydrogeology from Shiraz University in 1993, followed by a Master's in Engineering Geology in 1997 and a Ph.D. in the same field in 2010 from Tarbiat Modares University (TMU). His doctoral dissertation focused on the impact of engineering geology parameters on tunnel boring machine (TBM) performance and developing new empirical prediction models. He completed postdoctoral studies at EPFL in Switzerland in 2011–2012, specializing in Tunnelling in Difficult Grounds. Professionally, he gained extensive experience working on large civil projects, including water conveyance and metro tunnels, from 1996 to 2013 with various consulting firms. Since 2013, he has served as a faculty member at the University of Tehran's School of Geology, supervising theses and courses in rock mechanics, engineering geology, tunnelling, and geotechnical hazards. His research centers on TBM performance in various rock formations, mechanized tunnelling under challenging conditions, and geological hazards. Dr. Hassanpour has significantly contributed to professional bodies; he has been on the board of the Iranian Tunnelling Association (IRTA) since 2013, chairing scientific committees for regional and national tunnel conferences, and organizing workshops and lectures. He is also an active participant in international conferences related to tunnel construction. His scholarly output includes over 50 publications in reputable journals and over 150 conference presentations, establishing him as a leading figure in tunnelling and geological engineering.

Jamal Rostami holds a Bachelor's degree in Mining Engineering from the University of Tehran, and Master's and Ph.D. degrees in Mining Engineering from the Colorado School of Mines (CSM), earned in 1992 and 1997 respectively. He has a diverse background, including academia as a faculty member at the University of Tehran, and professional experience as a full-time consultant with international companies from 2002 to 2007. In 2007, he joined Pennsylvania State University (PSU), and in 2016, he returned to CSM, where he currently serves as the director of the Earth Mechanics Institute (EMI). With over 30 years of experience, Dr. Rostami has contributed significantly to design, management, research, and teaching in mining, tunneling, and underground construction sectors. He is actively involved in scholarly publishing, serving as an editor for *Tunnelling and Underground Space Technology* and on the editorial board of *Rock Mechanics and Rock Engineering*. His leadership roles include vice president of the International Tunnelling Association (ITA), member of the Society for Mining, Metallurgy, and Exploration (SME), and chair of the 2019 International TBM Digs Conference. He has delivered numerous training courses globally on tunneling and shaft construction. His prolific publication record boasts over 120 research articles in peer-reviewed journals and more than 150 conference papers. Beyond academia, Dr. Rostami is a founding member of the Iranian American Academicians and Professionals (IAAP), a board member of the Child Foundation, and founder of Professors Without Borders (PWOB). His multifaceted contributions exemplify his dedication to advancing research, education, and professional development in mining and tunneling.

Contributors

Nick Barton NB&A, Oslo, Norway

Nuh Bilgin Istanbul Technical University (ITU), Istanbul, Turkey

Jafar Hassanpour School of Geology, College of Science, University of Tehran, Tehran, Iran

Morteza Karami Technital S.p.A, Verona, Italy

Javier Macias JMConsulting-Rock Engineering AS, Oslo, Norway

Arsham Moayedi Far Subsurface Engineering, Technical University of Leoben, Leoben, Austria

Jamal Rostami Excavation Engineering and Earth Mechanics Institute (EMI), Colorado School of Mines, Golden, CO, USA

Hanan Samadi University of Tehran, Tehran, Iran;
R&D Department, SYSTRA Co., Trento, Italy

Marlene C. Villeneuve Subsurface Engineering, Technical University of Leoben, Leoben, Austria

List of Parameters

(Cutterhead) penetration	P, ROP, i_p (mm/rev)
Penetration rate	PR, ROP, i_0 (m/h, mm/min)
Advance rate	AR (m/day)
Utilization factor	U (%)
Boring time	t_B (min, h)
Thrust force	T_h (kN, ton)
Cutter load, Normal force	F_n (kN, ton)
Rolling force	F_r (kN, ton)
Cutterhead torque	T_q (kN.m, ton.m)
Power	P (kW)
Cutterhead rotation speed	RPM (rev/min)
Rolling velocity	V_D (m/min)
Machine diameter	d_{TBM}, D_{TBM} (m)
Cutter tip width	T (mm)
Cutters spacing	ac, S (mm)
Number of cutters	N_{TBM}
Number of cylinders	N_c
Shield friction force	T_f (kN, ton)
Gripper friction force	F_G (kN, ton)
Cutter coefficient	k_c
Cutter constant	C_c
Field penetration index	FPI (kN/c/mm/rev)
Specific energy	SE (Mj/m^3)
Conveyor belt capacity	C_{CB} (m^3/h, ton/h)
Equivalent thrust	M_{ekv} (kN, ton)
Uniaxial compressive strength	UCS, σ_c (MPa)
Brittleness index	BI
Tensile strength	σ_t (MPa)
Elastic Modulus	E (GPa)
Point load index	I_s (MPa)
Schmith number	N

Vicker's Hardness Number	VHNR
Drilling rate index	DRI
Siever's J value	S_j (0.1 mm)
Brittleness index	S_{20}(%)
Discontinuity class or NTH class	NTH
Fracturing factor	K_s
Total fracturing factor	$K_{s\text{-}tot}$
Equivalent fracturing factor	K_{ekv}
Rock Fracture Index	RFI
Rock mass boreability index	RMBI
Rock mass cuttability index	RMCI (MPa)
Rock quality designation	RQD (%)
Joint count number	J_v
Rock mass rating	RMR
Basic RMR	BRMR
Rock Condition Rating	RCR
Rock structure rating	RSR
Q system	Q
Rock mass number	N
Number of joint sets index	J_n
Joint roughness index	J_r
Joint alteration index	J_a
Stress reduction factor	SRF
Groundwater condition index	J_w
Geological strength index	GSI
Rock mass index	RMi (MPa)
Rock mass strength	UCS_{rm}, σ_{crm} (MPa)
Disc cutter life	H_f (m^3/cutter)
Disc cutter life	H_m (m/cutter)
Disc cutter wear	W_m (Cutter/m)
P-wave velocity	VP, V_P (km/s, m/s)

Introduction

Jafar Hassanpour

1 Introduction

Mechanized tunneling has emerged as the dominant mode of tunneling in the past couple of decades due to the efficiency, safety, and high rate of advance it offers. At this stage, there are thousands of TBMs working across the world, with a smaller fraction in rock and by the open mode TBMs. Continued advancements in TBM technology have revolutionized tunneling practices and significantly accelerated project timelines. Today, Tunnel Boring Machines (TBMs) are available in various diameters, ranging from less than 1.0 m to over 15 m for rock excavation. These machines are successfully employed in various geological conditions, including soft soils, hard rock, fault zones, and mixed ground conditions. This versatility highlights the remarkable progress in mechanized tunneling, allowing for the efficient and effective execution of tunneling projects across diverse settings. Yet, one of the challenges in selection and application of rock TBMs is the capabilities of the machines to mine various formation at reasonable rate and their effectiveness across different geological conditions.

Ongoing technological advancements necessitate the updating of performance prediction models to align with the specifications and capabilities of modern TBMs. This chapter serves as an introduction to rock TBMs as a prelude to the other chapters in this book. It discusses the latest developments in hard rock TBMs, explains the rock cutting process and the factors that influence mining rate, and provides details on the parameters used to estimate machine performance.

J. Hassanpour (✉)
School of Geology, College of Science, University of Tehran, Tehran, Iran
e-mail: hassanpour@ut.ac.ir

J. Hassanpour and J. Rostami (eds.), *Performance of Tunnel Boring Machines (TBM) in Rock*, Springer Series in Geomechanics and Geoengineering,
https://doi.org/10.1007/978-3-032-20424-0_1

2 Tunnel Boring Machines

Tunnel boring machines (TBMs), tunnel excavation machines, and tunneling machines are terms commonly used in the mechanized tunneling industry, referring to a variety of machines designed for tunnel excavation and completion. In this book, we specifically define tunnel boring machines as those equipped with a rotating cutter head and cutting tools that excavate the tunnel in a full-face manner. These machines can be categorized into two main types based on their application: those designed for rock and those suited for soft ground, each exhibiting distinct characteristics. Given the focus of this book on examining the performance of machines in rock, we will exclusively consider tunnel boring machines used in rock excavation, excluding other types of tunnel boring machines.

One should note that the reference to soft ground machines often refers to single shield TBMs, including Earth Pressure Balance (EPB) and Slurry machines used to counter the face pressure and prevent water discharge into the tunnel. However, it is very common that along the tunnel alignment, the machine needs to transition from soil to rock or in rock formations under groundwater table where there is need to prevent water ingress into the tunnel, hence single shield machines (EPB or Slurry) are used with the rock cutterheads featuring disc cutters, hence the models discussed in this book can be used to estimate penetration rate of the machine. In general, soft ground machines operate under constant volume, meaning that to control face pressure in soil, the amount of muck mined should be equal the bulk material mined, which is the product of face area and rate of penetration. This in practice means that most of these machines are set to go at constant rate of penetration to maintain this balance. However, the rock TBMs are operated under constant force to avoid overloading the disc cutters. This means that the rate of thrust is set and the penetration is variable in different rock types. Obviously, the thrust is set to a level where there is sufficient torque to rotate the cutterhead, which means that in softer rock, the machine operates at lower than nominal cutterhead thrust, yet it is constant force setting.

To effectively understand how these machines operate and the methods used to predict their performance, it is essential to first describe their specifications. This section will outline the types of machines used in hard rock tunnelling projects and highlight their most important components.

3 Classification of Tunnel Boring Machines

The highly variable ground conditions that TBMs may encounter during excavation, as well as the practical considerations for tunnel construction, have necessitated the design and manufacturing of a range of machines with distinct specifications. Consequently, machines can be classified based on various parameters. For instance, as illustrated in Fig. 1, Rostami (2016) classified tunnel boring machines according to their application range across various geological conditions.

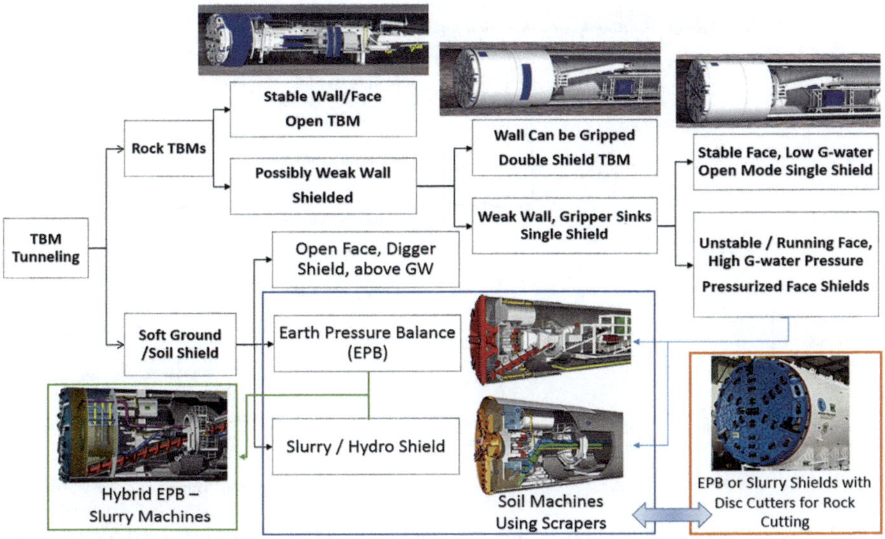

Fig. 1 General classification of TBMs for various ground condition (Rostami 2016)

One of the most famous TBM classifications is proposed by Working Group No. 14 (WG14) of the International Tunneling Association (ITA), as presented in Fig. 2. This classification categorizes tunnel excavation machines according to excavation methods (full face or partial face), types of cutterhead (rotation or non-rotation), and methods of applying reaction force (by grippers or erected concrete segments) (ITA 2000). This classification does not include specific machines with limited use, such as reamer type machines designed for widening tunnel diameter.

The ITA classification (Fig. 2) categorizes tunnel boring machines (TBMs) into two primary groups:

- Hard rock machines (Open type I and closed type I);
- Soft ground machines.

Hard rock machines (marked by * in Fig. 2) are typically utilized for excavating tunnels in stable conditions (medium to high strength rocks, without wide crushed zones). These machines may also be equipped with shields for occasional unstable and difficult ground conditions.

On the other hand, soft ground machines, always equipped with shields, are employed in soft grounds such as alluvial deposits, loose soils, very soft rocks such as marls and mudstones, water bearing zones, and severely crushed zones and faults, with the main purpose of ensuring stability of the tunnel walls and face. They employ open mode when the tunnel face is stable and closed mode when the face is unstable, during which a special mechanism is utilized to apply pressure to the face for balancing ground and water pressures and preventing face movement or collapse.

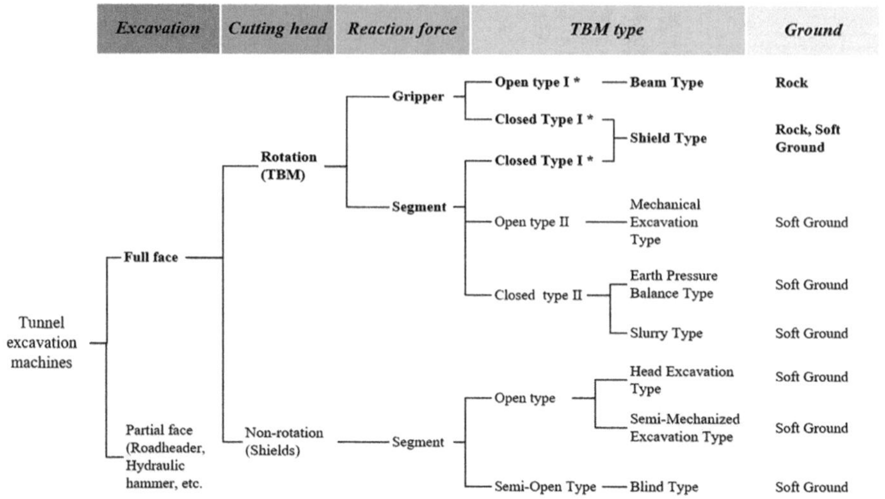

Fig. 2 Classification of tunnel excavation machines (ITA 2000) (*Machines used for excavating rock formations)

A more comprehensive classification of tunneling machines, proposed by the German tunnel committee (DAUB[1]), is available to those who want to examine different classifications. This classification, divides tunneling machines into four main categories: tunnel boring machines (TBM; including open/gripper machines), shield machines (SM), double shield machines (DSM), and convertible shield machines (KSM). Within the shield machines category, various types of full-face and partial-face machines employing different methods of supporting the tunnel face are included.

3.1 Hard Rock Tunnel Boring Machines

According to DAUB classification, the most commonly utilized machines for the excavation of rock tunnels are TBM (open/gripper machines), SM-V1 (single shield machines), and DSM (double shield machines). Therefore, the focus of this section is primarily on the details of these three types of machines, in line with the objectives of this book. The selection of the most suitable type of machine for hard rock geological conditions depends on factors such as the quality of the rock mass, its behavior within the tunnel, tunnel diameter and length, and various construction phase considerations.

Schematic representations of the components and diverse elements of full-face tunnel boring machines operating in hard rock geological conditions are depicted in Fig. 3. As shown, key components include the boring system (cutterhead), clamping

[1] Deutscher Ausschuss für unterirdisches Bauen.

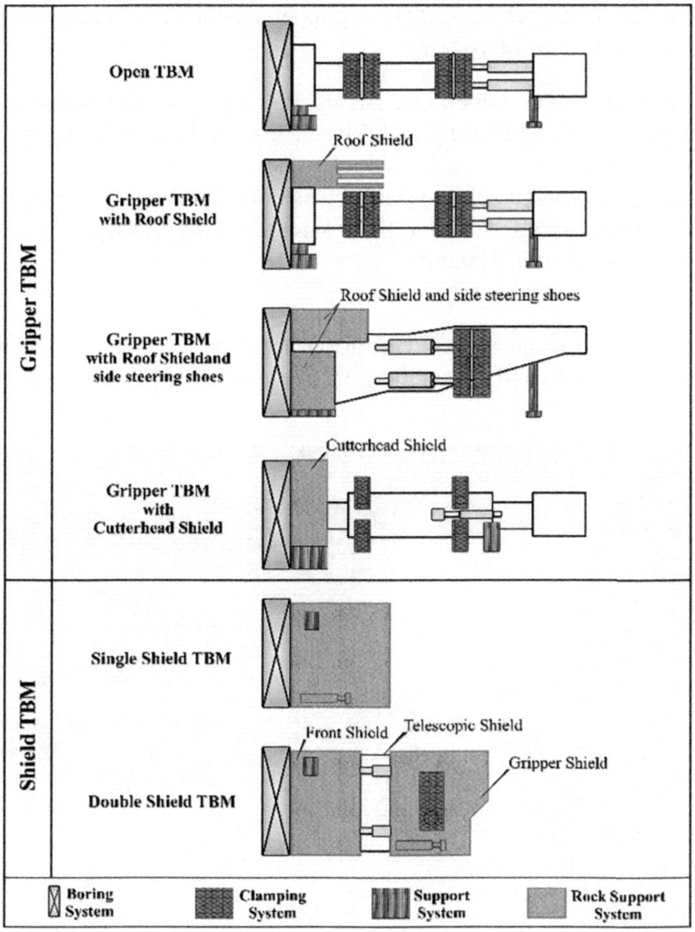

Fig. 3 Overview of various machine systems of tunnel boring machines with full-face excavation (Maidl et al. 2008)

system (grippers), support system and rock support systems (shields). Subsequent sections will provide an explanation of the specific characteristics of each of these machines and their components.

3.1.1 Open/Gripper-Type Machines

In stable rock formations with low water pressure/flow, open or gripper-type machines are commonly used. The International Tunneling Association (ITA) recommends using these machines in rocks with a uniaxial compressive strength (UCS) >40 MPa.

According to Wittke et al. (2007) a gripper machine, as shown in Fig. 4, is composed of four groups of systems:

- Cutting/boring system, which consists of the cutterhead, the cutting tools (disc cutters), the cutterhead bearing and drive. The cutterhead is driven by hydraulic or electric motors;
- The thrust, gripper and bracing system consists of the jacking cylinders, the gripper jacks, the gripper shoes as well as the front leg and the rear leg;
- The support installation system;
- The mucking and conveyor system.

Open machines are manufactured in various types based on the design and size of the cutting head support and the small shield used mainly to protect the cutting head.

Fig. 4 Main components of an open type machine

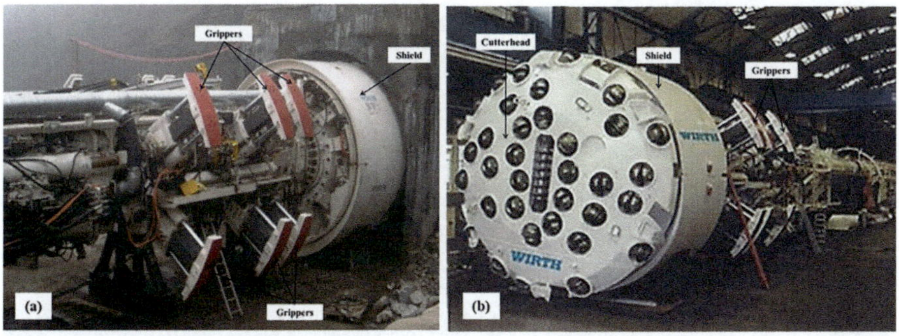

Fig. 5 An example of an open TBM (double-gripper type) manufactured by Wirth Co.: **a** in the project site and **b** in the factory; (employed for completing a service tunnel in Alborz road tunnel project, Iran)

A key feature of these machines is the use of grippers, which create friction against the tunnel wall. This friction is crucial for providing the support needed to apply thrust force to the rock face. These machines can be categorized into two groups, single and double gripper machines, based on the type of grippers they utilize. An example of a double-gripper machine is shown in Fig. 5.

Additionally, the maximum pressure applied by the gripper pads (shoes) depends on the strength of the rock mass surrounding the tunnel. Consequently, the size of the gripper pads is determined by the force/pressure they exert and the maximum allowable bearing pressure of the rock mass. Typically, the gripper force is considered to be approximately twice the machine's thrust force. If the rock mass strength is insufficient, the machine cannot fully utilize its thrust force for advancement.

A potential issue when using open/gripper machines is the insufficient stand-up time of the rock mass. In such cases, the machine is designed to facilitate the installation of a support system immediately behind the cutterhead. The temporary support system typically includes rock bolts, wire mesh, steel frames, and shotcrete. In modern machines, the support system can be installed simultaneously with the advance.

The advance rate of gripper machines is dependent upon the quantity, type, and timing of the installation of the necessary support system, taking into account the rock conditions. In these machines, the installation of the support system can be initiated immediately behind the front shield. In single-gripper machines, the motors and main drive system of the cutterhead are positioned in the front sections of the machine, behind the cutterhead, while in double-gripper machines, they are located in the rear section of the machine, behind the grippers. Additionally, to prevent the rotation of the double-gripper machines, the first gripper must be as close to the cutterhead as possible. This characteristic results in limited available space for the installation of support systems in these machines compared to single-gripper machines, which is considered a shortcoming for this type of machine.

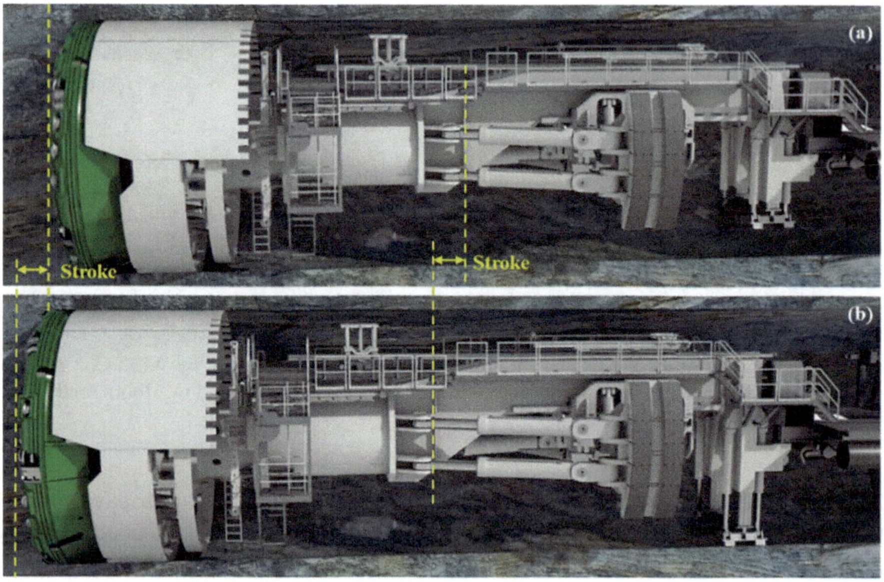

Fig. 6 Boring cycle of an open type machine (single-gripper); **a** advance phase: grippers extended, rear leg retracted, **b** repositioning of the gripper: grippers retracted, rear leg extended (Herrenknecht website[2])

Figure 6 illustrates a single-gripper machine and its boring cycle. Thrust cylinders between the gripper unit and cutterhead push the cutterhead against the rock face while the machine slides on an invert shield or sliding shoe (Fig. 6a). Hydraulic-jacked gripper pads, pressing against sidewalls, support the thrust force and cutterhead torque during advancement. The rear leg lifts during this phase. At the end of the stroke, the rear leg extends and the grippers retract (Fig. 6b). For the next cycle, the thrust cylinders retract, the gripper unit slides forward, the rear leg lifts, and a new stroke begins.

The machine's rotational axis deviation caused by gripper slippage can be corrected during re-gripping by pressing the grippers against the wall and adjusting the cutterhead and beam accordingly.

3.1.2 Shielded Hard Rock Machines

Shielded machines are the most commonly used machines in hard rock mining due to their high flexibility for operating in various geological conditions. They have a shield skin that serves as rock support until the tunnel lining is installed.

The design and manufacturing of shielded machines are tailored for projects situated in challenging and unstable conditions, often associated with the risk of

[2] https://www.herrenknecht.com/de/produkte/productdetail/gripper-tbm/.

rock collapses. Primarily, they serve to safeguard workers and equipment from such collapses, while also facilitating the installation of support systems. This type of machines are the ideal machines for tunneling in both unstable weak rocks and hard rocks.

These machines are categorized into single-shield and double-shield types. The basic mode of operation of both machine types is described in the following, and the differences in individual functions explained.

(I) Single shield machines

Single shield machines have a cylindrical design and feature a cutterhead located at the front, a segment erection system, and thrust cylinders positioned at the rear. Additional details and various components of these machines are illustrated in Fig. 7. An example of a single shield machine can be seen in Fig. 8.

These machines can be used in highly crushed, fragmented, and fault zones with unstable walls and limited stability where grippers cannot provide thrust. According to ITA, they are best suited for low to medium strength rocks, where rock support is needed during excavation but the tunnel face remains stable.

In single shield machines, as opposed to open type machines, there is no requirement for the installation of a temporary support. The machine's boring cycle involves a sequential process of excavating and installing precast concrete segments as the final lining. The thrust force of the machine is derived from thrust cylinders that apply the required thrust by pushing against the erected segments. The boring cycle in single shield machines is illustrated in Fig. 9. The lining involves precast concrete segments that are picked up with a hydraulic manipulator, the so-called segment erector, and positioned to form a complete ring. For this purpose, the thrust cylinders are retracted at the respective installation point. The segment is set in place and

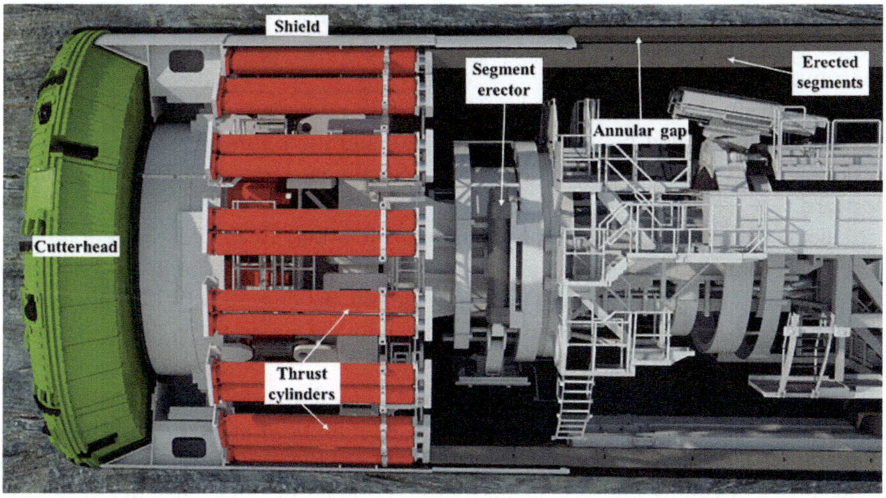

Fig. 7 Different parts of single shielded TBM

Fig. 8 Single shield TBM, Mont Sion road tunnel in France (diameter = 11.835 m)

held in position by extending the thrust cylinders again. The key segment placed last completes the ring building, allowing the next tunneling phase (boring cycle) to begin.

Because segment erection and boring are not simultaneous, these machines have a long advance cycle. Long advance cycles and low advance rates make single shield machines unsuitable for long tunnels. The erection time of concrete segments can range from 12 to 15 min for smaller diameter machines (3–4 m) to 30–40 min for larger machines (about 10 m). The installation time of concrete segments is contingent upon their quantity, which can range from 4 + 1 to 8 + 1, between small and larger diameter machines, respectively. Factors constraining the quantity of concrete segments include the weight of each segment and the machine's maneuvering capability.

As soon as the shield is pushed forward in relation to the tunnel lining, an annular gap is created between the rock mass and the segment extrados. This gap is continuously backfilled with pea gravel and cement grout as the machine moves forward.

Due to the absence of a gripper in single shield machines, the correction of the machine's rotational movement around its axis is facilitated through the rotation of the cutterhead in each boring cycle. Consequently, the cutterhead's buckets must be designed to allow for bidirectional movement, with the cutterhead rotating clockwise in one cycle and counterclockwise in the next.

To reduce the risk of shield jamming, it is essential to minimize the shield length. Swelling and squeezing grounds can cause jamming, which can be mitigated by overcutting. Understanding ground convergence and the forces exerted on the shield, especially in swelling and squeezing rocks, is crucial in shield machine design. In

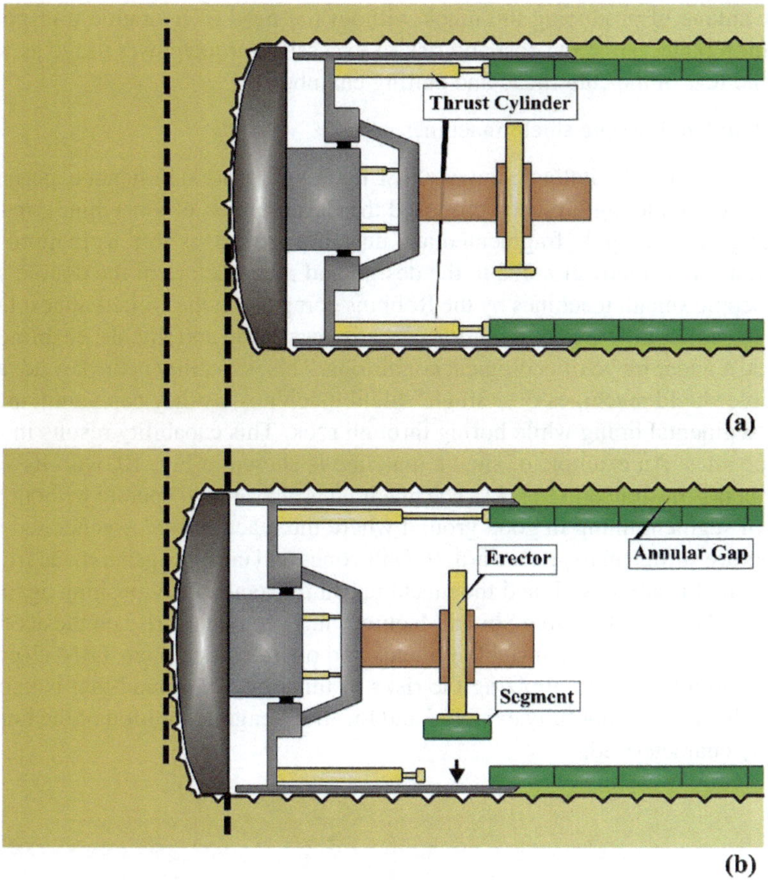

Fig. 9 Boring cycle of a single shield TBM, **a** phase of advance, **b** installation of segmental lining (Modified after Wittke et al. 2007)

such conditions, the thrust force must exceed the combined frictional forces and the required cutterhead's thrust.

The shielded TBMs for rock cutting come in two general categories, (1) open mode, where ground water can flow in during the mining cycle, (2) pressurized mode, where the groundwater is not allowed to flow in and face pressure is maintained at all times. The second group of single shields use either Earth Pressure Balance (EPB) or Slurry machine principals for maintaining face pressure. The choice of EPB or Slurry machine depends on the conditions at the face but most of the pressurized face machines used in rock are Slurry shields, which facilitates transpiration of the excavated materials by slurry. These machines are slightly more expensive to operate but offer better management of muck and face pressure compared to EPB, where the muck has to be conditioned by a combination of foam and injected bentonite to be extracted from the cutting chamber while keeping the torque lower. EPB offer the advantage of producing the muck without the need of hydraulic transport and separation plant, but at the cost and risk of excessive torque/power usage as well as wear and tear of the cutterhead and cutting chamber.

(II) Double/telescopic shield machines

For the first time, the Italian company SELI, known for its significant experience in mechanized tunneling projects, proposed the development of a machine capable of operating in both highly fragmented and unstable ground, as well as in more stable conditions. This innovation led to the design and manufacture of the double shield, or telescopic shield, machines by the Robbins company in the United States. Double shield machines utilize a combination of a gripper system and shields, enabling them to tackle a wider range of geological conditions. This versatility is the key advantage of double shield machines over single shield machines, as they can simultaneously install segmental lining while boring through rock. This capability results in higher advance rates. An example of such a machine is shown in Fig. 10, with its various components illustrated in Fig. 11. The machine is also able to operate without installation of segment lining in good ground where the reaction forces generated by the grippers are sufficient to go forward. In fault zones and unstable ground, the front and gripper shields are locked, and the machine can move ahead by pushing against the segments. The use of Double Shield Machines has been on the rise on the account of its flexibility to deal with unstable grounds and operate as an open TBM elsewhere. This would allow for minimizing the risks of tunneling in ground that is known to have fault zones but not fully explored and location/length/condition of the faults are not fully characterized.

Fig. 10 An example of double shield machine in Lanzhou water source project, China

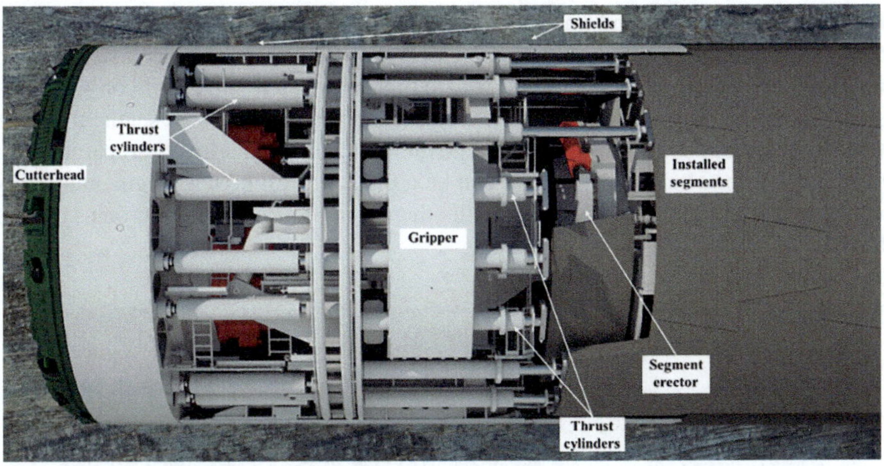

Fig. 11 Different components of a typical double shield TBM

The main components of this machine include the cutterhead (with mounted cutting tools and muck buckets) the front shield which covers the main bearing, motors, and thrust cylinders, the main or gripper shield which protects the auxiliary thrust system, the tail shield, covering the segment erector and the conveyor system (Fig. 11). The gripper shield and the front shield overlap through a telescopic section called the telescopic shield. Additionally, the tail shield is sometimes considered part of or an extension of the gripper shield.

The primary function of the thrust cylinders is to advance the machine during normal operation or in double-mode (when the rock quality is good and gripping is possible), by relying on the gripper jacks. The auxiliary thrust system, consisting of a number of hydraulic cylinders, can also propel the machine forward under specific conditions (when the rock mass is fractured or unstable and gripping is not possible) by relying on the last erected ring of segments (single shield-mode).

Double shield machines are longer than single shield machines, increasing the risk of getting trapped in fault zones and weak ground conditions. Consequently, they are not recommended for projects with high ground squeezing potential, which could occur at higher depth and weaker formations with high ground stresses. To mitigate the risk of tunnel convergence and machine trapping, the shield diameter gradually decreases from the cutterhead toward the rear of the machine, which is referred to as "Conicity".

When the tunnel wall can withstand the load imposed by the grippers, concrete segments can be installed simultaneously with the boring process. In cases where the rock mass quality is favorable and support installation is not necessary, the front shield and cutterhead are advanced using main thrust jacks, with the support provided by the grippers. Under such conditions, it may even be possible to construct the tunnel without installing full ring of concrete segments (Fig. 12), requiring only a limited number of rock bolts for local support behind the tail shield if needed.

For smaller diameters and special cases, steel sets can be used instead of precast concrete segmental lining, and in some cases, they can serve as the primary support for machine driving. It is important to note that this system is suitable only for dry, stable ground conditions. However, it is very common that an invert segment is used with Double Shield TBMs to speed up the tunneling and avoid invert preparation for rail installation. The invert segment can be a part of a full precast segmental lining or just as a stand-alone invert piece, with other form of ground support on as needed basis.

Fig. 12 Images of the Uma-oya Water Conveyance Tunnel project in Sri Lanka, excavated using a double shield machine. As shown, in certain sections of the tunnel, the high quality of the rock mass allowed for the installation of only the invert segment, without completing the full ring of segments

The boring cycle of this type of machine in double mode is shown in Fig. 13. If the rock mass is unfavorable and cannot support the load imposed by the grippers, the machine operates similarly to a single shield machine, with the boring process temporarily halted during segment erection.

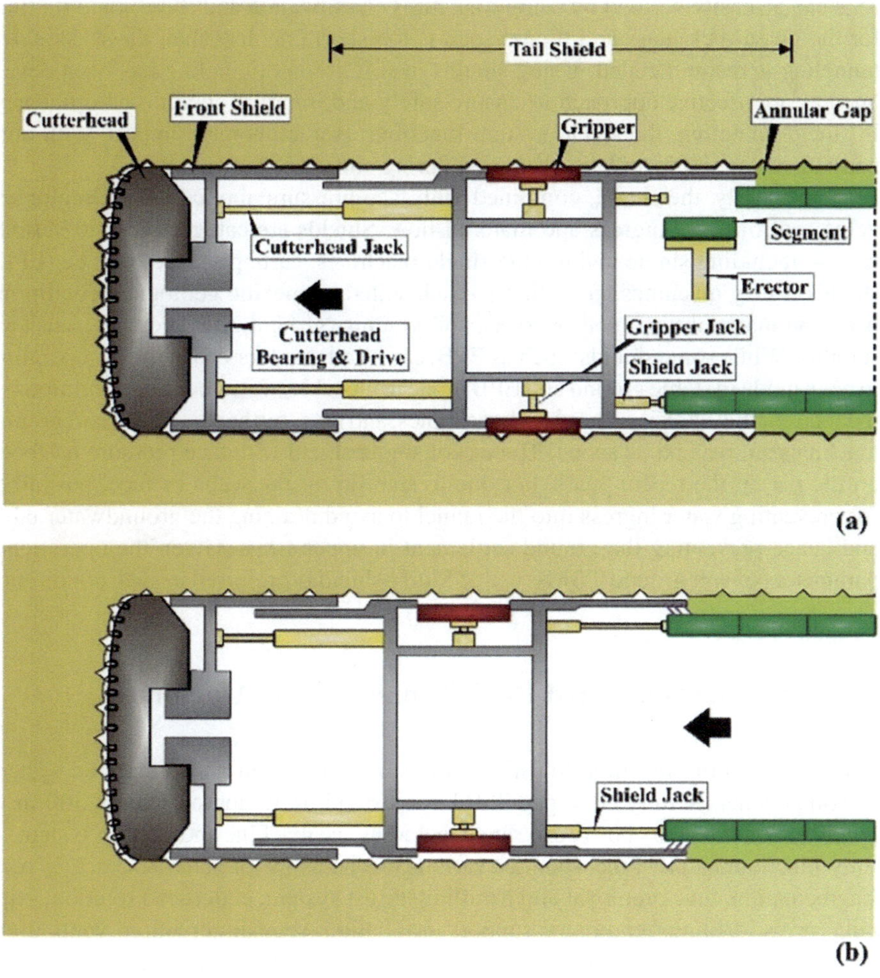

Fig. 13 Boring cycle with double shield machine in double-mode, **a** Simultaneous boring in front shield and segment erection in tail shield; gripper jacks extended, **b** pushing forward of the tail shield and back-up system using auxiliary jacks; regripping (Modified after Wittke et al. 2007)

3.2 *Soft Ground Tunnel Boring Machines*

The immediate collapse of the tunnel face, roof, and walls, along with water and mud ingress into the tunnel in weak/soft rock and Quaternary alluvial formations, necessitates a tunneling method that consistently maintains stability during excavation. This can be easily verified by using rock mass classification and if the stand-up time for the given rock mass and tunnel span is too short (say less than days), shielded tunneling is recommended. Using shields in soft ground tunneling has been developed as an effective approach to ensure safety and improve operational efficiency. In shield tunneling, the shield system functions as a temporary support, with only permanent support installation required during and after excavation.

Additionally, the shield, combined with face pressure application technologies, helps control water ingress and material flow. Shields are categorized into various types—including single and double shield machines, earth pressure balance (EPB) shields, slurry machines, and others—each suited to specific geological conditions based on material type (soil or rock), soil grading, groundwater pressure, and face stability. While some shields, such as EPB, can be adapted for use in rock—operating in open mode in stable ground and EPB mode in unstable, water-bearing conditions—the focus of this book is on hard rock machines, and thus, further details on soft ground machines are beyond its scope. The use of single shield with face pressure has been on the rise in the recent years, not due to stability of the walls or face, but rather for preventing water ingress into the tunnel to avoid draining the groundwater table and hence preventing the ground settlement in upper strata. Given the operational parameters of soft ground TBMs, use of Slurry shield is preferred in such conditions.

4 Components of Hard Rock Tunnel Boring Machines

The tunnel boring machine (TBM) features a complicated and integrated system of components. A fully equipped TBM can extend up to approximately 300 m in length but often in the 100–120 m range including the machine and backup system. A fully functioning TBM incorporates various mechanisms for activities such as rock fragmentation, muck removal and handling, thrust system, cutterhead rotation, gripping, probe drilling, temporary support installation, segment erection, ventilation, grouting, dewatering, and more. All these mechanisms must operate in a seamless setting with the machine's advance, along with essential services like power cables, compressed air and water pipes, and ventilation ducts, which are designed to be installed at certain intervals and move simultaneously with the TBM.

The term "integrated" is particularly significant, as all these systems must function at appropriately coordinated rates to match the machine's overall progress. Any malfunction or delay in one of these mechanisms can lead to a slowdown in tunneling process, potentially impacting the project schedule and increasing costs.

This section of the chapter provides a brief description of the operation of various components in hard rock tunnel boring machines. As mentioned previously, a hard rock machine consists of several key components, including the cutterhead, cutting tools, thrust system, torque system (or cutterhead rotation system), gripping system, muck removal system, probe drilling system, installation system for concrete segments or temporary supports, and a backup system that includes hydraulic, electrical, dewatering, ventilation, injection, and other subsystems. The pictures a to j in Fig. 14 illustrate the key components of a hard rock tunnel boring machine, which are briefly described in the following sections.

4.1 Cutterhead

The cutterhead, located at the front of the TBM, is the most critical component of the machine. Its primary functions include:

- Supporting cutting tools and maintaining contact with the rock face;
- Transferring thrust and rotational forces to the cutting tools and tunnel face;
- Collecting and directing excavated materials toward the conveyor system;
- Supporting the stability of the tunnel face;
- Reducing dust through water spray application on the face.

The cutterhead can be either convex / domed (Fig. 15a) or flat in shape (Fig. 15b and c), with the face partially or fully open (Fig. 15d). Dome-shaped cutterheads were typically designed for machines with diameters of 4–5 m but have been gradually phased out in favor of flat cutterhead design. For larger diameter machines, flat cutterheads are preferred due to better performance, stress distribution on the tunnel face, and reduced risk of fracturing and wear. Flat cutterheads have gained popularity for their ease of construction, higher efficiency, maintenance convenience, and operational benefits such as improved machine guidance, leading to a significant increase in their usage.

Machines with small diameters featuring a fully open cutterhead are ideal for hard rock excavating but unsuitable for fragmented, loose rock masses, and zones with high potential of rock bursting. This configuration may lead to safety issues and hamper machine advance rate. The closed cutterhead, however, presents advantages for excavating in fragmented rocks, as it prevents blocks extruding in the face to block the cutterhead.

The arrangement of disc cutters and muck buckets on the cutterhead is designed to facilitate chipping, collection, and removal of excavated materials from the tunnel face. Manufacturers offer varying configurations for cutterhead based on rock mass conditions and other parameters.

Figure 16 illustrates two types of cutterheads based on how cutting tools are replaced. In Fig. 16a, cutters are replaced from the front while the cutterhead is retracted, which may pose risks due to ground instability. In contrast, Fig. 16b shows back loading cutters, which is a design where cutters are replaced from inside the

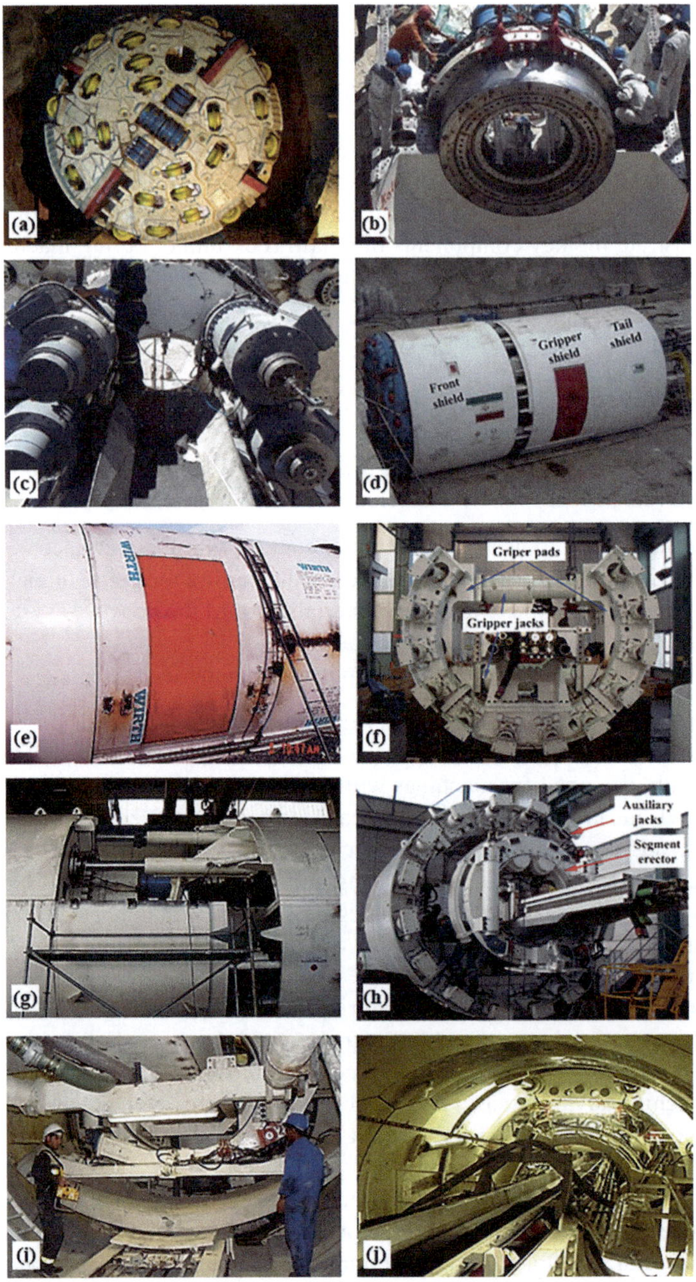

Fig. 14 Pictures of the main components of a double shield machine; **a** cutterhead; **b** main bearing; **c** motors; **d** shields in a double shield machine; **e, f** Gripper jacks and pads; **g** thrust jacks; **h** auxiliary jacks; **i** segment erector; **j** conveyor belt

Fig. 15 Different types of cutterhead: **a** dome shape cutterhead (courtesy of the Maidl et al. 2008), **b** and **c** closed cutterhead with side muck buckets, **d** open cutterhead

cutterhead, eliminating the need to retract it to access the work face. A comprehensive discussion of the cutterhead design and configuration can be found in Rostami and Chang (2017).

4.2 Cutting Tools

The cutting tools, mounted on the cutterhead, are used to excavate the ground and break down rock. The type of cutting tool utilized varies based on the specific ground being excavated, and includes different types such as picks, drag bits, as well as disc cutters and roller cutters (Fig. 17). Picks and drag bits are suitable for soft ground, such as alluvial materials and soft rocks, hence rarely used on rock TBMs, whereas disc cutters and rollers are used for hard rock excavation. While machines for soft ground or hard rock usually use a single type of cutting tool, those operating in inhomogeneous conditions, like EPB-hard rock machines, may employ multiple tools.

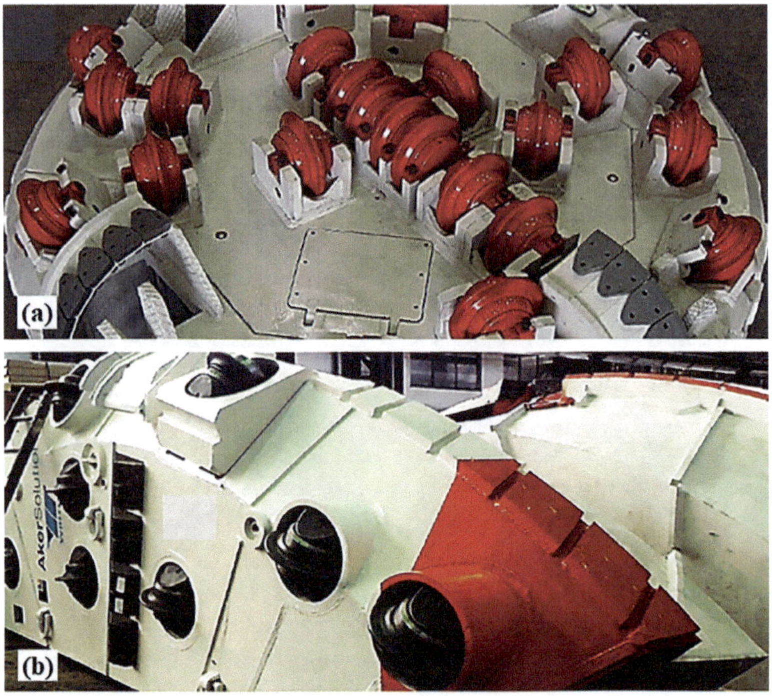

Fig. 16 Examples of cutterhead, demonstrating the approach to replace the cutting tools, **a** front of the cutterhead, **b** within the cutterhead

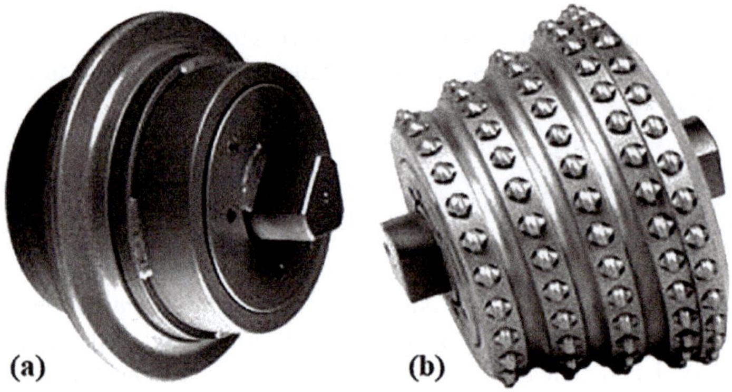

Fig. 17 The view of disc cutter (**a**) and multi row TCI roller cutter (**b**)

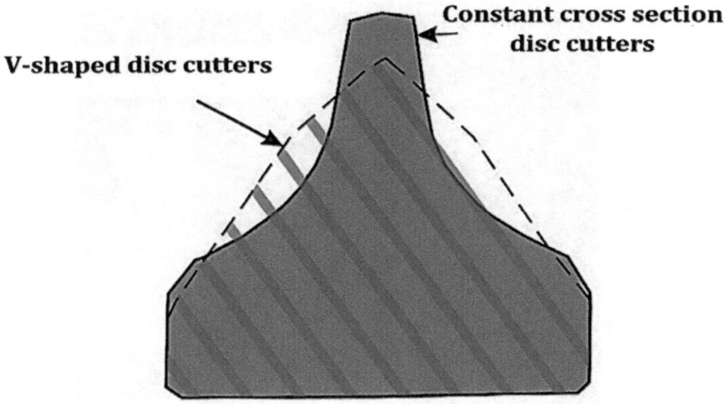

Fig. 18 Comparison between V-shaped disc cutters with constant cross section disc cutters (Rostami 1997)

In recent years, disc cutters have been recognized for their superior effectiveness compared to other cutting tools. Initially, the first disc cutters had a V-shaped cross-section (Fig. 18), which caused them to experience rapid performance deterioration due to widening of the tip caused by wear. As a result, in recent years, constant cross-section disc cutters have been adopted to address wear issues and maintain efficiency, replacing the older V-shaped cutter profiles.

Recently, through the utilization of novel designs and the advancement of wear-resistant materials and alloys, disc cutters have experienced a substantial increase in efficiency and load-bearing capacity.

Typically, the diameter of disc cutters ranges from 380 to 480 mm (15–19 in), with the 432 mm (17 in) disc cutters being the most prevalent. However, in recent years, there has been a growing use of disc cutters with a diameter exceeding 19 in (483 mm) for extremely hard and abrasive rocks. Notably, the production and development of 17-in disc cutters with load capacity of around 25 ton (~250 kN) in the 1980s marked a significant milestone in the advancement of TBM manufacturing technology, as the use of these tools enabled TBMs to operate more efficiently and economically in tunneling projects.

Moreover, high power machines equipped with 19-in disc cutters, which have a load capacity exceeding 300 kN (30 tons), have been available in the market for over two decades. These machines are particularly effective in extremely strong rocks, and their favorable penetration rates have been observed and documented in various countries. Due to their high load-bearing capacity, 19-in disc cutters have the potential to achieve very high penetration rates. Additionally, their larger diameter contributes to a longer lifespan. However, in normal conditions and when working with moderately strong to strong rocks, these cutters consume more energy than their 17-in counterparts, resulting in less frequent use. On the other hand, advancements in the metallurgical industry have enhanced the load capacity of 17-in disc cutters, allowing them to exceed a load-bearing capacity of 25 tons and increase linear speed limitations from 150 to 175 m/min.

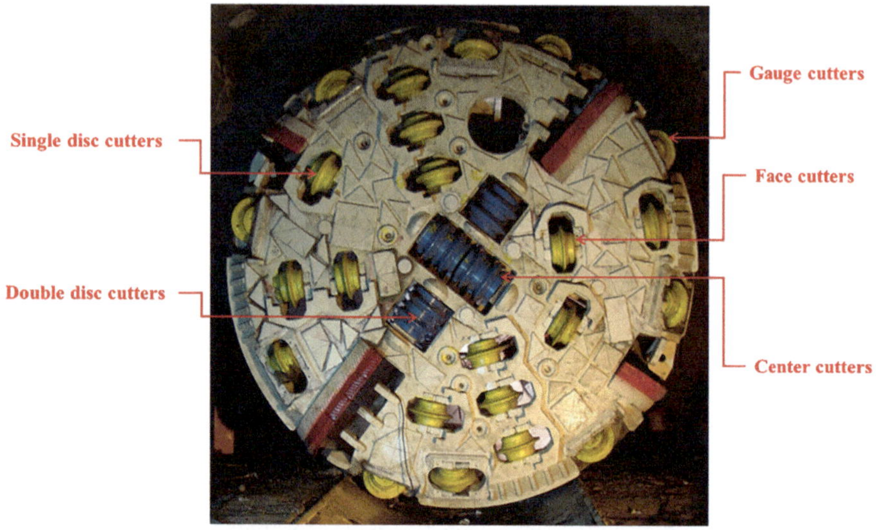

Fig. 19 Classification of disc cutters on the cutterhead based on their position

The recent application of larger rock TBMs has also been possible by development and use of 20-in (~500 mm) disc cutters that allow for higher velocity limit. These cutters have been installed on cutterhead of the larger machines in the 14–15 m range and are anticipated to be used on such large diameter machines in the future. The reluctance to use cutters over 17-in stems in their excessive weight and difficulty of handling through the tunnel as well as installation and removal from the cutterhead after their useful life is over.

The disc cutters, positioned on the cutterhead as shown in Fig. 19, are categorized into center, face, and gauge cutters. Figure 20 shows a general cutterhead profile with various disc cutter groups. Additionally, disc cutters can be further categorized by shape into single, double, or multiple disc cutters.

The center cutters are responsible for excavating the central part of the tunnel face. They typically consist of one or more double or multiple disc cutters, as there is limited space in the central area of the cutterhead. The most common setting at the center is to use a quad (4 cutters combined) or a set of 6 cutters.

The face cutters are installed individually on the cutterhead and form the main part of the excavation system, tasked with cutting the tunnel face.

Toward the periphery of the cutterhead, the spacing of cutters reduces gradually, and their tilt angle increases. The gauge cutters are located in the outer region of the cutterhead. Their primary function is to cut the tunnel's periphery and maintain the minimum required boring diameter. To ensure the required distance between the machine and the tunnel wall is maintained, these gauge cutters are installed at a slight angle relative to the tunnel axis and often perpendicular to the rock face at the location of the installation. This tilt angle results in greater wear on these cutters compared to other cutters mounted on the face area of the cutterhead. Consequently, they are

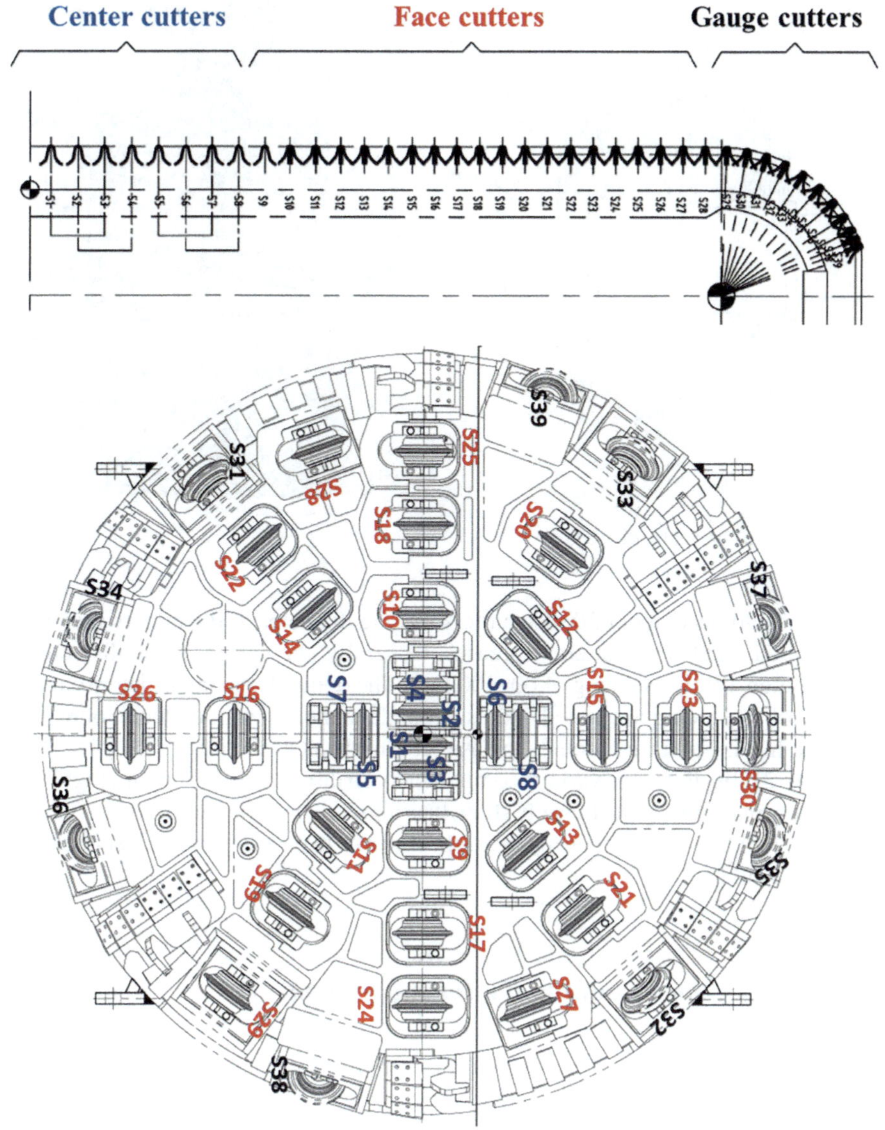

Fig. 20 Typical profile of the cutterhead with various cutting tools

made from special alloy steels or more durable tungsten carbide materials. Moreover, these cutters are spaced closer to reduce cutter load and extend their lifespan.

The total number of disc cutters needed for excavating a tunnel face depends on the diameter of the machine, increasing proportionally with size. For example, a tunnel boring machine with a diameter of 4 m ($d_{TBM} = 4$ m) typically requires 25–28 disc cutters (17 in in diameter), and a machine with an 8-m diameter (d_{TBM}

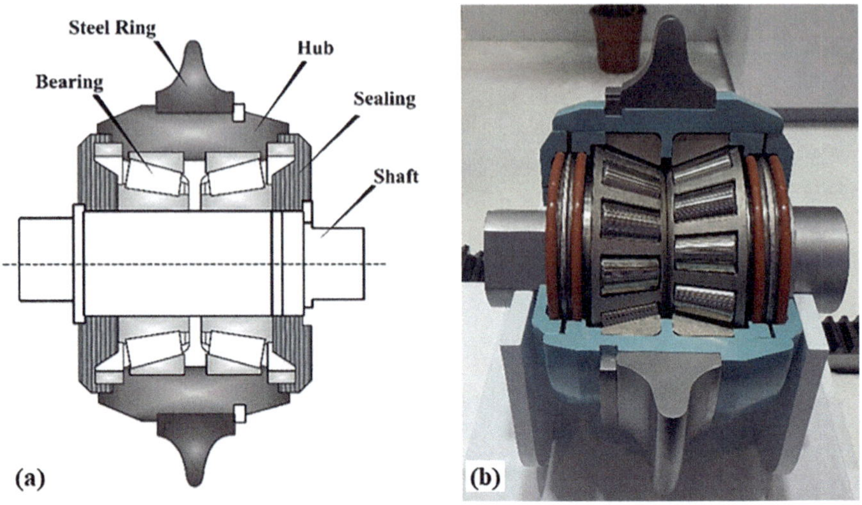

Fig. 21 Various parts of a disc cutter

= 8 m) would need about 50 disc cutters. As the diameter of the cutters increases, the required number of cutters decreases. Generally, the design of the disc cutters allows for an average spacing of 60–100 mm. The optimal spacing is determined by the type of rock being excavated and its strength, and the size of cutters mounted on the cutterhead. For further details, please refer to the next sections.

Figure 21 illustrates the various components of a disc cutter. This figure highlights several key parts, including the hub, shaft, bearings, replaceable steel ring, and sealing rings. Each of these components plays a specific role in ensuring the proper operation of the disc cutter.

4.3 Cutterhead Rotation and Torque System

The cutterhead torque system of a machine comprises motors and the driving train, which includes a set of bull gears and the main bearing. The main bearing (Fig. 14b) is a crucial component responsible for transferring the thrust delivered to the head by the thrust cylinders and torque generated by the rotation of the motors to the cutterhead. The motors (Fig. 14c) can be either hydraulic or electric. Hydraulic systems operate with an efficiency of approximately 75% and can have variable RPM, but electric motors, with their higher efficiency (85–90%), are typically preferred in machine designs and have become the dominant type of TBM drive systems. These electric motor systems can operate at two or three specified and constant RPM. The most notable recent change in the drive system of tunnel boring machines is the adoption of variable frequency drive (VFD) motors. These VFD systems enable the operator to have unlimited control over the cutterhead speed. Furthermore, there has been a significant rise in high voltage and high-power motors in recent years. Most machines have the capability to offer 2–2.5 times operating torque momentarily or for a short time to break the cutterhead jamming. This allows the machine to deal with higher torque needed in the case of a face collapse and start the operation of the next stroke.

4.4 Shields

The primary function of shields in shielded tunneling boring machines is to create a secure working environment for personnel responsible for machine guidance, installation of concrete segments, and other tunneling activities. With the presence of shields, TBM can continue to advance even in unstable and challenging geological conditions. Key functions of the shields include preventing tunnel collapse in soft/weak materials such as fault zones and minimizing issues caused by groundwater flow into the tunnel.

Among hard rock TBMs, the most advanced type features double shields. As shown in Fig. 14d, these machines are equipped with four shields: the front shield, the telescopic shield, the gripper shield, and the tail/rear shield.

In this type of machine, grippers are located on the gripper shield, securing the gripper and tail shields in place against the tunnel wall. This setup allows the front shield, which is connected to the cutterhead, to move forward when the thrust jacks are opened. The telescopic shield then covers the gap between the front and gripper shields as the thrust jacks are activated.

4.5 Grippers

To move a boring machine forward using thrust cylinders, it must have adequate support to transfer the loads to tunnel wall. For machines that have a precast concrete segmental lining installed directly behind them, several jacks can be used at the tail shield to push the machine forward against these segments. However, in open-type machines without a segmental lining, or in double shield machines that do not allow for the use of auxiliary jacks, an alternative support mechanism is required for advancing the machine.

To address this issue, pairs of jacks, known as grippers, are installed on both sides of the machine, oriented perpendicular to its axis. These jacks stabilize the machine by pressing steel plates against the tunnel wall, providing the necessary support for its movement. Once the grippers are secured to the tunnel wall, the cutterhead is gradually advanced using the thrust cylinders.

At the ends of the gripper jacks, steel pads are installed to reduce pressure on the tunnel wall, thereby preventing instability or failure in rock mass. Figure 14e and f illustrates the gripper pads and gripper jacks, while Fig. 22 shows the operation of the gripper pads against the tunnel wall.

Fig. 22 Two examples of pressing gripper pads against tunnel walls to support the force exerted by thrust cylinders while pushing the cutterhead forward

4.6 Thrust Jacks

Cutterhead thrust jacks are a key component of the steering and forward driving system of TBMs. This thrust system operates exclusively on hydraulics, functioning within a pressure range of 35–50 MPa. The thrust cylinders typically have a stroke length between 1.2 and 1.6 m. One end of these hydraulic jacks is connected to the grippers, while the other end is attached to the cutterhead, and their primary function is to propel the cutterhead forward (see Fig. 14g). The number of thrust jacks varies among different TBM models. For open type TBM a set of two or four thrust jacks are used.

4.7 Auxiliary Jacks

The advancement of shielded machines is facilitated by the use of jacks that are supported by the last erected ring of concrete segments at the tail shield. The number of thrust jacks or hydraulic cylinders installed on the shielded machines is proportional to machine circumference/diameter. The number of these jacks typically ranges from 8 to 20, depending on the machine size and the type of concrete segments used (Fig. 14h). These jacks apply pressure on the last erected ring, thereby moving the shield(s), cutterhead, and back-up system forward.

As mentioned earlier, double shield machines utilize two distinct systems for propelling the machine forward. The main system, situated in the front shield, facilitates movement of the front shield relative to the grippers. The auxiliary system can be employed in unstable, weak ground conditions, allowing the entire machine to advance by pushing off segments installed at the back. Notably, this design enables a double shield machine to be converted into a single shield machine by locking the tail and front shields together.

4.8 Segment Erection System

The segment erection system, used in shield machines, places segments in the tunnel lining by facilitating pickup, rotation, lifting, and precise installation, as shown in pictures h and i of Fig. 14. There are two main types: vacuum and screw pick-up with locking mechanisms. The vacuum system uses pumps to lift and position segments, while the screw system employs special holes for connection to a displacement arm, allowing up to 360-degree rotation. Recent advancements include fully automatic installation systems, initially developed in Japan, though their widespread industry adoption has yet to occur.

4.9 Muck Removal System

The muck removal system facilitates the transfer of excavated material from the tunnel face to the rear section of the TBM, ensuring uninterrupted excavation. The performance of this system is crucial; deficiencies can hinder operational capacity and accelerate wear on key components such as disc cutters, the cutterhead surface, the muck ring, and associated parts. The system employs muck buckets to lift material from the tunnel bottom, deposit it onto the hopper, and then transfer it via conveyor belts (Fig. 14j) to muck cars for removal. Proper operation of the muck removal system is vital for increasing the cutting rate, optimizing the utilization factor, and improving the durability of the machine.

Figure 23 illustrates the penetration rate of a 12-m diameter tunnel boring machine, which depends on operational performance and the conveyor belt's pick-up capacity. An increase in rock penetration enhances the production rate, thereby necessitating a greater number of buckets on the cutterhead, provided the conveyor system can accommodate the higher throughput. Insufficient bucket capacity for material collection and discharge can result in the accumulation of crushed material between the cutterhead and the tunnel face, leading to reductions in advance rate and accelerated wear of cutting tools and cutterhead components. In extreme cases, the cutterhead may become jammed or cause a substantial rise in torque due to the increased friction from unremoved material.

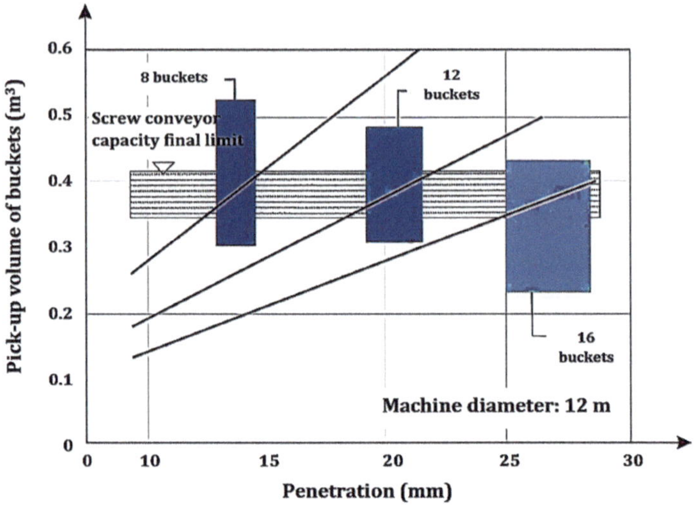

Fig. 23 The relationship between the penetration of cutterhead, the pick-up volume and number of buckets for a 12-m machine diameter (Maidl et al. 2008)

4.10 Probe Drilling System

The prob drilling system includes a drilling hammer mounted on the segment erector arm at the rear shield, capable of rotating around the tunnel perimeter. This system is installed on most machines to identify the rock mass ahead of the machine prior to excavation, to prevent sudden encounters with geological hazards such as water inrush and tunnel collapse in fault zones.

To drill into the rock and pass the drilling rods from inside the shield to the tunnel wall, holes are made on the shield at intervals that vary based on geological conditions. Figure 24a and b illustrate the drilling hammer and the holes embedded in the shield, respectively.

This system typically detects geological and hydrogeological conditions up to 50 m in front of the machine. In open-type machines, it is usually installed behind the cutterhead and can drill at a slight angle in the direction of advance. Some machines use bolt installation drills for this purpose, which can be repositioned on both sides at 3 and 9 o'clock, and, with prior planning, can drill borehole in the upper perimeter if necessary. Many projects now perform probe drilling during maintenance shifts routinely to increase machine utilization.

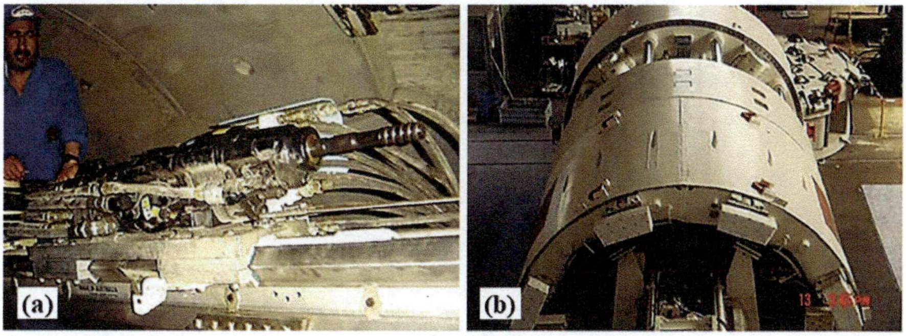

Fig. 24 A view of probe drilling system in TBM used for Zagros water conveyance tunnel, with holes strategically spaced at varying intervals on the shield

5 Cutting Process in Rock

Tunnel excavation with a TBM involves the use of a cutterhead equipped with cutters that apply a significant amount of force to the tunnel face. As the cutterhead rotates, the disc cutters penetrate into the rock, leaving behind concentric circles on the tunnel face (see Fig. 25). Numerous theories have been developed to model and explain the fragmentation mechanism of rock by disc cutters. According to a theory formulated at the Colorado School of Mines (Rostami and Ozdemir 1993), rock cutting by disc cutter encompasses processes such as crushing, chipping, and cracking, which are influenced by the physical and mechanical properties of the rock. These processes collectively define the characteristics of rock mass boreability and cuttability.

Fig. 25 The trace of disc cutters on the tunnel face

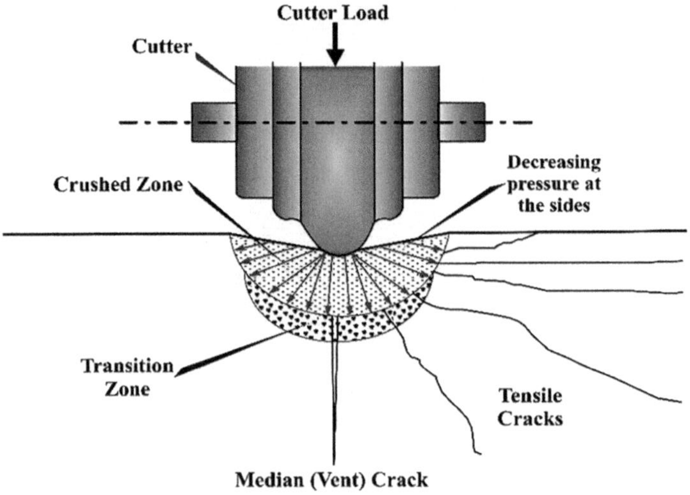

Fig. 26 Stress field and various zones formed beneath the cutter edge. Reproduced from Rostami and Ozdemir (1993)

Laboratory indentation tests and numerical analyses have shown that when the cutter load is applied to the rock, three distinct zones form immediately beneath the cutter edge. These zones include:

- Crushed zone
- Crack zone
- Elastic zone

In terms of the type and mechanism of stresses, it is possible to identify three distinct zones—compression, transition, and tensile—immediately beneath the cutter edge on the rock surface (see Fig. 26).

In summary, the rock fragmentation process, induced by the cutter's penetration into the rock, can be delineated into two stages:

- Formation of stress field and crushed zone;
- Formation of chips and grooves.

In the initial phase, the cutter penetrates the rock under the influence of thrust forces, leading to the formation of a small zone of crushed material under the cutter due to high stresses. This crushed zone expands as pressure increases until it reaches a critical level. At this level, tensile strains in the area surrounding crushed zone result in the formation of various size radial tensile cracks at the transition zone between crushed zone and the rock face. The propagation of these cracks continues as long as the tensile stresses are lower than the rock's tensile strength, marking the penetration or indentation stage of the cutter edge. Subsequently, if the tension stresses developed in the crushed zone are sufficient, cracks formed between two adjacent discs can extend and join together, resulting in the formation of large chips

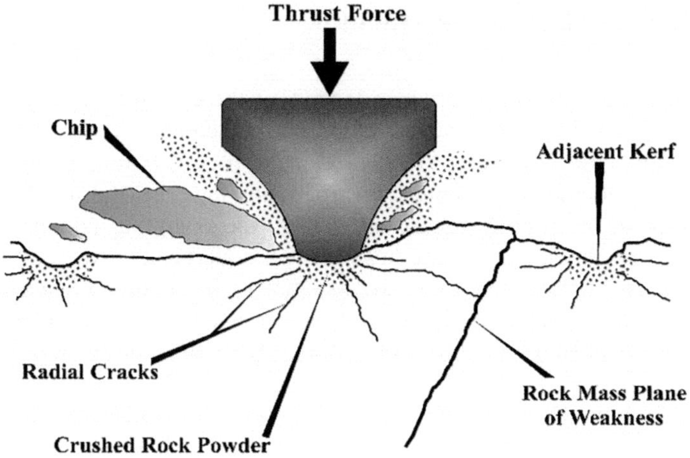

Fig. 27 A schematic model of the cutting process and the formation of a chip under disc cutter (Rostami and Ozdemir 1993)

between the disc cutters (Fig. 27). This, in turn, completes the rock cutting process and allows the cutter to break the rock and cutterhead to advance in the tunnel.

The shape and dimensions of the indenter, as well as confining stresses, significantly influence the formation of crushed zones and crack development. Additionally, the direction and length of radial cracks in the crack zone play a crucial role in the efficiency of rock cutting and chipping by the machine.

In practice, the penetration process during cutterhead rotation at the tunnel face consists of successive cycles, each including crushing and chip formation. The crushed zone continuously expands beneath the cutters while cracks form and expand. Typically, only one or a few of these cracks reach the free surface of the rock or join other cracks created by adjacent cutters, completing the chipping process.

Snowdon et al. (1982) conducted a series of tests to explore the impact of machine specifications on the rock cutting process using a linear cutting machine (LCM). Their findings revealed that cutter spacing and cutter load significantly influence specific energy and its cutting efficiency. Furthermore, Rostami and Ozdemir (1993) utilized similar experiments to examine the effect of cutting geometry on chip formation. Their research demonstrated that the geometry and shape of the cutter edge control the dimension of the crushed zone under the disc cutter. They also revealed that cutter spacing and cutter load affect crack propagation and the formation of chips between two adjacent discs.

Gong and Zhao (2007) employed numerical simulations and established that cutter load, diameter of the disc cutters, and the width of the cutter edge impact the development of stress fields in the rock and the indentation process.

In general, the rock cutting process and the penetration rate in the rock are heavily reliant on machine parameters, such as TBM diameter, cutter spacing, cutter diameter, width of the cutter edge, and cutterhead thrust force and torque.

6 Key Parameters of the Machine

As shown in Fig. 28, TBM performance depends on three main parameters: (1) cutterhead design (including specifications and tool arrangement), (2) operating parameters, and (3) geological features at the face and along the tunnel. This section will introduce the cutterhead's technical specifications (such as the number, spacing, and diameter of the cutting tools), as well as the operating parameters (which include machine thrust force, cutterhead torque, cutterhead power, and cutterhead rotation speed). All of these factors are essential for evaluating the performance of the machine. Additionally, parameters relevant to machine performance, including penetration rate, utilization factor, and advance rate, will be discussed.

6.1 Specification of the TBM and Cutting Tools

These specifications are parameters that play a crucial role in the boring process and the performance of the machine. They are determined during the design of the machine based on geological conditions, rock mass properties, and other project-specific characteristics. Typically, these specifications remain constant throughout the project. However, in the event of significant design issues, they may be subject to change and redesign, albeit at a high cost and potentially leading to project suspension for an extended period. This section briefly outlines some of the technical specifications of the machine, cutterhead, and cutting tools, which impact the machine performance.

6.1.1 Machine Diameter

In modern times, tunnel boring machines (TBMs) with diameters ranging from less than 1.0 to approximately 15 m have been designed and manufactured for excavating

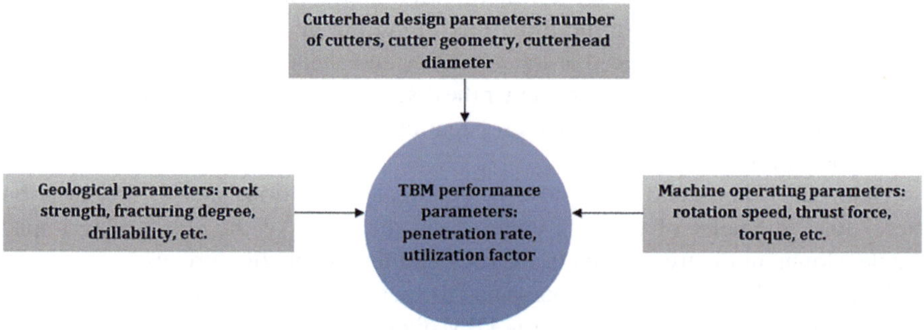

Fig. 28 The relationship between the three sets of parameters and TBM performance

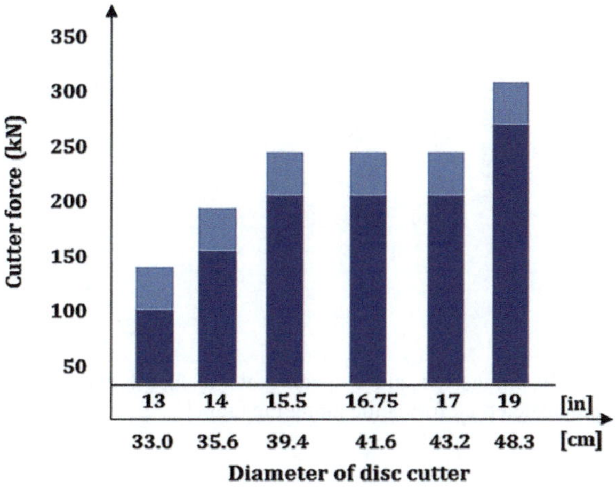

Fig. 29 The cutter load capacity of disc cutters in term of their diameter (Wittke et al. 2007, Wirth Group, Germany, Technical Documents)

tunnels in hard rock formations. The diameter of the machine is a crucial parameter for evaluating its performance, as many of the machine's specifications and performance metrics are directly related to this factor. For open machines, the diameter can be adjusted within a range of 10–20%, depending on the manufacturer. This means increasing or decreasing the diameter of the head, and some minor modifications in the machine structure and gripper shoes, while the main bearing and thrust system can be reused.

6.1.2 Disc Cutter Diameter

As noted before, the diameter of disc cutters is widely recognized as one of the most influential parameters affecting machine performance. Disc cutters used between the 1950s and 1980s typically had smaller diameters, ranging from 13 to 15 in (356–394 mm). In recent years, especially for tunnels with larger diameters, disc cutters with diameters of 17–19 in (432–483 mm) and up to 20 in (500 mm) have been employed. For these larger disc cutters, the cutter load (F_n) generally ranges from 200 to 300 kN (as illustrated in Fig. 29). As Fig. 29 shows, an increase in the diameter of disc cutters is associated with a higher load-bearing capacity. Currently, 17-in (432 mm) disc cutters are primarily used in hard rock machines.

6.1.3 Spacing of Disc Cutters

The spacing of disc cutters in various machines is a direct function of the rock strength and can vary based on the configuration and number of disc cutters mounted on the cutterhead. Furthermore, cutter spacing can differ across different parts of the cutterhead. For example, in a cutterhead equipped with 19-in (483 mm) disc cutters,

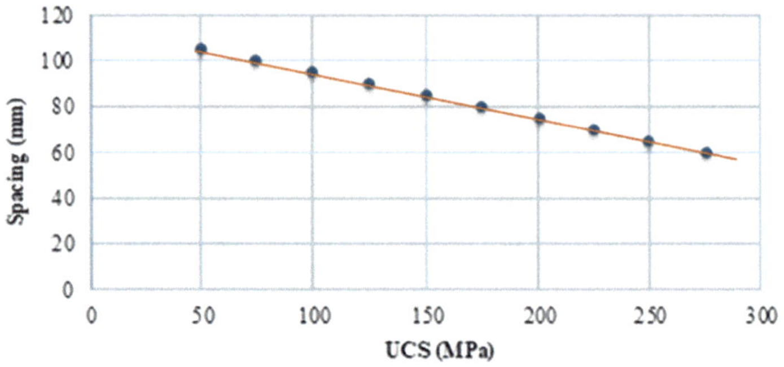

Fig. 30 Suggested cutter spacing at the face based on strength of intact rock in the hardest anticipated rock formation along the tunnel

the average spacing usually ranges from 90 to 100 mm. In contrast, the effective spacing for gauge cutters typically falls between 10 and 20 mm. Figure 30 can be used to calculate an average value for cutter spacing within the cutterhead (for d_c = 17 in).

6.1.4 Number of Disc Cutters

The number of disc cutters is directly related to the diameter of the cutterhead as well as the engineering geological characteristics of the rock mass along the tunnel route. Figure 31 illustrates the correlation between the typical number of cutters on the cutterhead and the machine's diameter. It is evident that as the machine diameter increases from 3.5 to 9 m, the number of disc cutters rises from approximately 25–65. Additionally, the diameter of the disc cutters plays a significant role in determining their quantity; when smaller diameter disc cutters are used, a greater number are typically installed on the cutterhead.

An average number of cutters based on cutter spacing within the cutterhead, measured in mm, can typically be derived using Eq. 1:

$$N_{TBM} = \frac{d_{TBM} \cdot 1000}{2 \cdot S} \cdot k \tag{1}$$

In this equation, S (in mm) represent cutter spacing at the face area, d_{TBM} (m) is the diameter of the machine in m, and N_{TBM} is the number of disc cutters. The coefficient k accounts for the additional cutters in the gage area. It varies from 1.1 on smaller machines (<5 m) to 1.05 for larger TBMs.

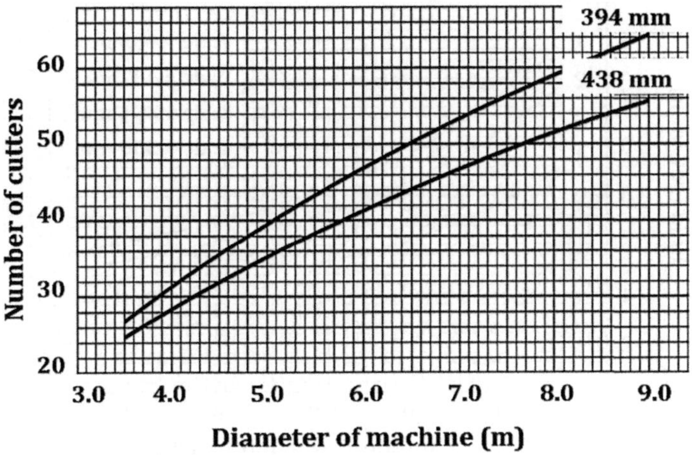

Fig. 31 The relationship between the typical number of disc cutters mounted on the cutterhead and the diameter of the machine (Bruland 1998)

6.2 Operating Parameters

The machine operating parameters are adjusted by the operator during tunnel excavation to ensure optimal penetration rates and safe tunnel stability. An experienced operator considers the geological conditions of the tunnel when setting machine parameters to achieve the best performance and minimize instability. The key parameters include cutterhead rotation speed, thrust force, and once these parameters are selected in a given formation, cutterhead torque and power will follow. Below is a brief description of each of these parameters.

6.2.1 Cutterhead Rotation Speed (RPM)

The rotation speed of the cutterhead, RPM, is dependent on the rolling velocity of the disc cutters, V_D. The rolling velocity is highest for the gauge disc cutters, which are located on the outer edge of the cutterhead. This velocity can be calculated using Eq. 2.

$$V_D = RPM \cdot \pi \cdot d_{TBM} \tag{2}$$

Furthermore, Eq. 2 can be expressed in the following manner:

$$RPM = \frac{V_D}{\pi \cdot d_{TBM}} = f_m \cdot \frac{1}{d_{TBM}} \tag{3}$$

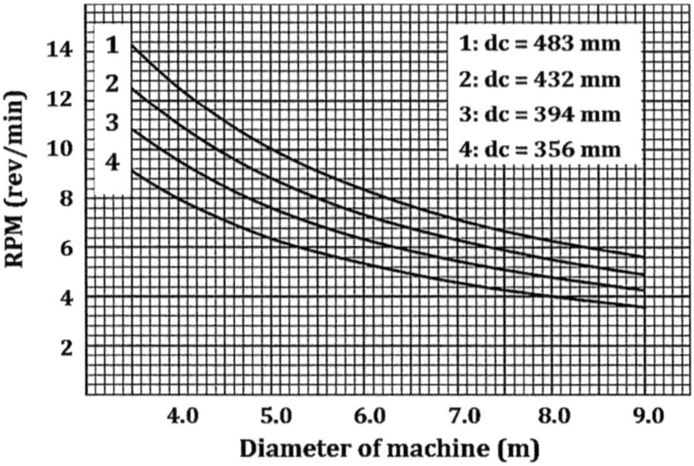

Fig. 32 Variations of the rotation speed (RPM) of the cutterhead vs variations of the machine diameter for different diameters of disc cutters (Bruland 1998)

Equation 3 shows that a constant velocity of the gauge cutter decreases RPM as the tunnel diameter increases. In this formula, V_D is in m/min and d_{TBM} is in m. Consequently, as shown in Fig. 32, cutterhead RPM inversely correlates with machine diameter.

Due to the limitations of the rolling velocity of gauge cutters, the coefficient f_m in Eq. 3 is generally considered to range between 45 and 55. The rolling velocity of the disc cutter depends on both the diameter of the disc cutter and the type and quality of main bearing used. For example, a 17-in disc cutter typically has a rolling velocity (V_D) ranging from 165 to 175 m/min, while a 19-in disc cutter can achieve a higher velocity of around 200 m/min (Wittke et al. 2007).

It is clear that higher RPMs result in increased penetration rates. However, there are limitations to consider, such as the maximum allowable rolling velocity of gauge cutters, the condition of the ground, tunnel face stability, and the capacity of collecting buckets and conveyor belts. These factors prevent the possibility of excessively increasing RPMs.

6.2.2 Thrust

The thrust generated by the machine is a crucial parameter in its operation. Gross thrust is determined by summing the forces of the thrust cylinders and includes the total loads on cutters (net thrust), the friction of the cutterhead and shields with tunnel walls, as well as hydraulic pressure losses in the hydraulic system during excavation. To calculate net thrust, these frictional and hydraulic losses must be measured and subtracted from the gross thrust. Additionally, the high pressure exerted by the cutterhead on the tunnel walls can significantly increase friction. This increase

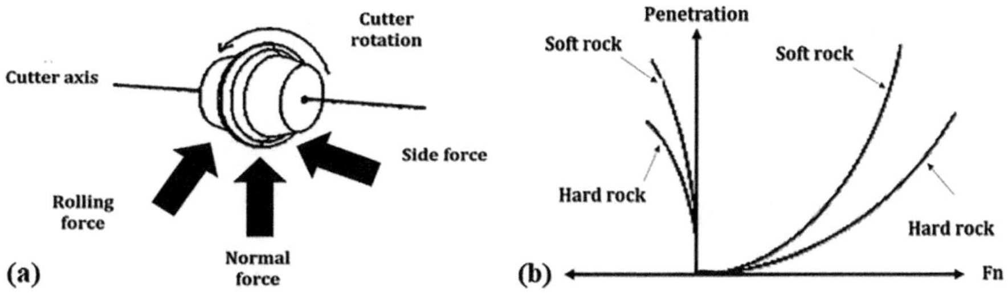

Fig. 33 **a** The active forces on a disc cutter (Rostami and Ozdemir 1993; US Army Corps of Engineers 1997); **b** Interaction of normal and rolling forces and penetration in rock (Nelson 1993)

in friction often occurs when machine advances in sharp curves or when machine operators are still in training. Furthermore, the towing force that pulls the machine's backup system during excavation also consumes part of the total thrust.

The maximum cutter load is typically limited by factors such as the lifespan of the disc cutter's bearing, the quality of the steel used in the cutter ring and its connection to the body. The total thrust is also restricted by the ability of the cutterhead's main bearing to handle the load and torque caused by the machine getting stuck and boring through layers with different strengths.

The forces acting on a disc cutter while applying thrust and rotating the cutterhead on the rock face include (Fig. 33a):

- Normal force (F_n) required to penetrate the cutter disc into the rock;
- Rolling force (F_r) necessary to overcome resistance against the disc rolling;
- Side force (F_s) caused by non-symmetrical loading on the disc cutter by the rock due to the release of pressure on one side of the cutter ring as a chip is formed.

The interaction between these forces is demonstrated in Fig. 33b. The inflection points of the curves indicate the critical thrust at which the gradual transition from grinding to chip formation occurs in the rock. Chips will only form if this specific amount of cutter load is applied to the rock. The critical thrust is directly influenced by the rock mass strength and boreability, and it increases with larger cutter spacing and a wider cutter tip.

The machine's performance is significantly affected by the thrust force and the load applied by the disc cutters to the rock. Equations 4 and 5 can be used to calculate these parameters.

$$T_h = \frac{9.81 \cdot N_c \cdot \pi \cdot d_c^2}{4} \cdot P_c \tag{4}$$

$$F_n = \frac{T_h - T_f - T_{tow}}{N_{TBM}} \tag{5}$$

In Eqs. 4 and 5, Th (in kN) is the total thrust force, P_c (in MPa) is the oil pressure in the thrust cylinders, d_c (in mm) is the diameter of the cylinders, N_c is the number of cylinders, and F_n (in kN) is the cutter load. Additionally, T_f is the friction force between the shields and tunnel wall, and T_{tow} is the towing force of the back-up system, both measured in kN.

6.2.3 Cutterhead Torque

The torque of a machine can be determined by utilizing the specific equations provided by the machine manufacturer, which depend on the type of driving system being used. For example, Eq. 6, as demonstrated by Robbins' Company for open type machines, offers a clear example of the relationships involved in this calculation.

$$T_q = \frac{U_T \cdot \sqrt{3} \cdot 60 \cdot n_m \cdot \cos\varphi \cdot \eta}{2 \cdot \pi \cdot 1000 \cdot RPM} \cdot I_T \tag{6}$$

In the Eq. 6, T_q (in kN m) is the cutterhead torque, n_m is the number of motors, U_T (in V) is the motor voltage, I_T is the amperage (in Amp), and Cos $\varphi \cdot \eta$ represents the efficiency coefficient. This formula calculates the torque value based on the power generated by commonly used three-phase electric motors.

Once the machine's torque is determined, the rolling force can be calculated using a simple equilibrium equation (Eq. 7).

$$F_r = \frac{2 \cdot T_q}{r_{mc} \cdot d_{TBM} \cdot N_{TBM}} \tag{7}$$

$$r_{mc} = \frac{\sum_{i=1}^{N_{TBM}} r_i / N_{TBM}}{0.5 \cdot d_{TBM}} \tag{8}$$

In these equations, d_{TBM} (in m) is the diameter of the machine, F_r (in kN) is the cutter rolling force, r_i (in m) is the position radius of the ith disc cutter, r_{mc} (in m) is the mean position radius of disc cutters on the cutterhead, and RPM (in rev/min) represents the cutterhead rotation speed.

6.2.4 Cutterhead Power

The cutterhead power is responsible for generating the necessary torque to facilitate the rotation of the machine. Typically, the power of the machine is a function of the diameter of the machine and the diameter of the disc cutters mounted on the cutterhead. This relationship is illustrated in Fig. 34.

In general, the cutterhead power is a function of the torque and RPM of the head as reflected in Eq. 9:

$$Po = T_q \cdot RPM \cdot K^* \tag{9}$$

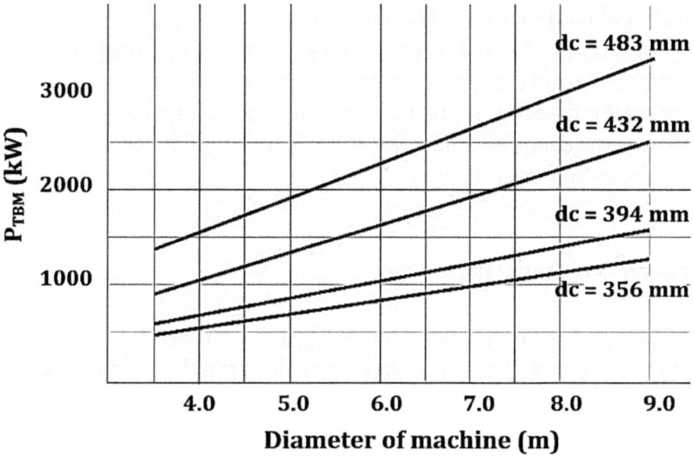

Fig. 34 The relationship between installed cutterhead power with diameter of machine and disc cutters (Bruland 1998)

where power (Po) is in kW or HP, RPM in rev/min, and Torque (T_q) in ft-lb or kN m and K* is the coefficient to convert ft-lb to HP or kN m to kW. This is the cutterhead required power but the installed power is higher and should be determined by dividing the cutterhead power by the mechanical efficiency, often 90%. This means that installed power is higher than the power available on the head to use for rotation of the cutterhead.

6.3 TBM Performance Parameters

The advance rate (AR) of a tunnel boring machine is influenced by two key tunneling activities: the boring of the rock and the installation of support systems and other time-consuming activities (i.e. transport, repair, etc.). The rate at which the machine bores into the rock is known as the penetration rate (PR) or rate of penetration (ROP) which is expressed in m/h. By adjusting PR based on the machine utilization factor (U, %), the AR (m/day) can be calculated. Thus, the advance rate of machine is obtained by multiplying the penetration rate in the utilization factor of the machine and number of working hours per day, as depicted in Eq. 10:

$$AR = PR \cdot U \cdot n_{shifts} \cdot n_{hrs} \tag{10}$$

The TBM utilization factor takes into account various factors such as the installation of support systems, repair and maintenance of the machine, replacement of cutting tools, removal of excavated materials, power supply, dewatering, etc. which is collectively called downtime. The number of shifts (n_{shifts}) is often 3 per day and number of hours (n_{hrs}) is 8 h/shift for 24-h operation of the machine.

In summary, the penetration rate and utilization factor are critical parameters for quantifying the machine's advance rate and overall tunneling efficiency, thereby influencing project scheduling and cost. Both are contingent upon various geological, operational, and managerial factors.

6.3.1 Penetration Rate (PR)

The penetration rate of a machine is defined as the ratio of the boring length (L, in m) to the time it takes to bore that section of the tunnel (t_B, in hr), as presented in Eq. 11.

$$PR = L/t_B \tag{11}$$

The penetration depth of the cutter in each rotation of the cutterhead (P, in mm/rev) can also be used to measure the amount of penetration. The cutterhead penetration (P) is related to penetration rate (PR) as shown in Eqs. 12 and 13.

$$P = \frac{1000 \cdot PR}{60 \cdot RPM} \tag{12}$$

Or:

$$PR = \frac{60 \cdot P \cdot RPM}{1000} \tag{13}$$

where RPM is rotation speed of cutterhead (in rotation per minute).

It is important to distinguish between the terms "penetration" and "indentation" when discussing the performance of a cutterhead. "Penetration" refers to the rate of progress of the cutterhead through the rock, while "indentation" pertains to the depth of the groove formed by the cutter's edge in the rock due to the applied cutter load. The former can be expressed as the average penetration of the disc cutters in the rock over multiple rotations of the cutterhead, while the latter is not practically measurable and can only be determined through laboratory tests and chip analyses. It should be noted that estimating the average penetration is crucial for predicting project schedule and cost.

By substituting the RPM from Eq. 14 into Eq. 13, the penetration rate (PR) of the machine can be calculated using the TBM diameter (d_{TBM}) and P(mm/rev) as shown in Eq. 15.

$$RPM = \frac{V_D}{\pi \cdot d_{TBM}} = f_m \cdot \frac{1}{d_{TBM}} \tag{14}$$

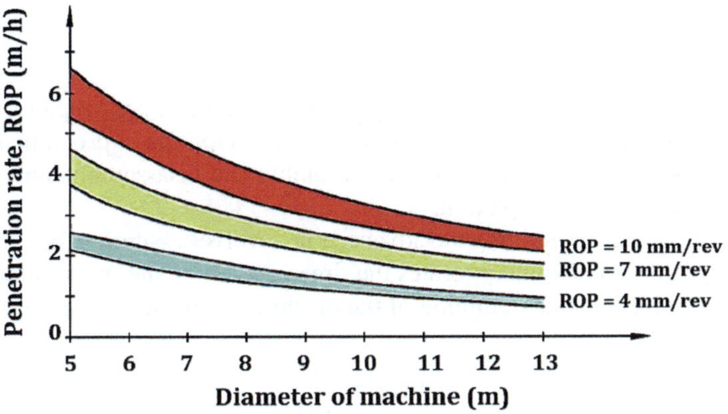

Fig. 35 Changes in the penetration rate of the machine based on the diameter of the machine and the amount of penetration in the rock (Wittke et al. 2007)

$$PR = \frac{60 \cdot P \cdot f_m}{1000 \cdot d_{TBM}} \tag{15}$$

As noted before, V_D is the maximum allowable rolling velocity of the outermost disc cutter on the cutterhead, approximately 165–175 m/min for a 17-in disc cutter. Consequently, f_m ranges between 45 and 55.

Figure 35 depicts the relationship between PR and TBM diameter for *P* values of 4, 7, and 10 mm/rev. As noted earlier, P depends on rock mass properties and the machine's operating parameters. As shown in Fig. 35, larger machines exhibit a decreased PR, assuming constant rock properties, due to limits in gauge cutters rolling velocity and hence lower RPM.

6.3.2 Advance Rate (AR)

The machine's advance rate (AR) at each tunnel section is calculated by:

$$AR_w = \frac{L}{t_w} \tag{16}$$

where AR_w is the advance rate (in m per day, week, month, etc.), L is the length of the mined section (meters), and t_w represents the working times (days, weeks, months, etc.) for activities such as boring, re-gripping, support installation (in open-type machines), and segment ring assembly that are required to complete an advance cycle. Working times also include maintenance and light repairs of the machine and its backup, disc cutter replacement, probe drilling, and other tunneling activities.

The advance rate of the machine can also be calculated based on operating times (as presented in Eq. 17).

$$AR_o = \frac{L}{t_o} \tag{17}$$

Operating times (t_o) include working times, as well as major repairs, extended stops due to difficult geology, installation of special supports in problematic zones, and holiday periods. It excludes machine assembly and disassembly, only covering from the tunnel start to completion.

Another approach to defining the advance rate involves assuming that, in addition to operating times, it also incorporates the time needed for site preparation, as well as the assembling and disassembling of the machine (as presented in Eq. 18).

$$AR_{HB} = \frac{L}{t_{HB}} \tag{18}$$

where, t_{HB} is the commissioning time, and AR_{HB} is the corresponding advance rate derived from it.

6.3.3 Utilization Factor (U)

The preceding discussion underscored the impact of various factors on the average advance rate of tunnel boring machines. Activities such as re-gripping, installation of temporary support systems or segmental lining, and routine maintenance, along with unforeseen delays (like the occurrence of geological/hydrogeological problems, power outages, delays in muck transportation hiccups, machine downtimes, etc.), significantly reduce the utilization factor, potentially to zero in rare cases. Consequently, the utilization factor of a machine is mainly determined by routine tunneling activities, type/specifications of the selected machine, geological conditions of the tunnel route, and site management strategies. The utilization factor of a tunnel boring machine is defined as the ratio of net boring time to a specific time interval (t), as shown in Eq. 19:

$$U = \frac{t_B}{t} \cdot 100 \tag{19}$$

Similar to the advance rate, the utilization factor can be calculated using metrics like working, operating, or commissioning times and can be analyzed for specific tunnel segments, such as geological zones/units or the entire tunnel.

In most tunneling projects, the machine's performance and tunneling activities are documented daily in shift logs by the operator. These logs serve as the primary source for determining the actual machine utilization during a tunneling project. However, in certain instances, identifying the specific cause of stops or delays in tunneling operations can be challenging due to inadequate record-keeping by the operator.

In Table 1, the times spent on different activities in a mechanized tunneling project are divided into three categories: times required for routine activities, delay times due to adverse geological conditions, and delay times due to unforeseen events. Of these three categories, only the first can be predicted and planned based on geological conditions and machine specifications. In all tunneling projects, efforts aim to maximize net boring time while maintaining quality standards and protecting workers and equipment.

Due to complex geology and uncertainties about rock mass properties and geological phenomena at depth, geological hazards and related delays in long, deep tunnels are hard to predict. Most models for estimating the utilization factor in geologically challenging projects lack accuracy because they overlook this key factor. However, thorough geological studies and accurate predictions of geological phenomena are essential for reliable tunneling activities forecasts under difficult conditions. Some of the geological difficulties that cause delays in the TBM advance include:

- Collapse of tunnel face and jamming cutter head in crushed zones;
- Excessive tunnel convergence and machine entrapment;
- Groundwater inflow and related difficulties for tunneling;
- Increasing of cohesive fine-grained particles in muck materials and consuming more time to clean disc cutters;
- Increasing disc cutter changes due to mixed face condition.

The third category of delays results from poor project site set up and management of the logistics, which are often unpredictable and can only be addressed through better management of tunneling sites.

The classification of tunneling activities can vary from one project to another. The following comprehensive classification, which encompasses all major activities in mechanized tunneling, is recommended:

1. TBM advance
    - Boring
    - Re-gripping

Table 1 Categorizing activity and delay times in mechanized tunneling projects

(1) Normal or routine tunneling activities	(2) Delay times due to adverse geological conditions (water inflow, tunnel collapse, squeezing, etc.)	(3) Unforeseen delays caused by inefficient site set up or management
Net boring Muck transportation Cutters inspection Cutter change Re-gripping Ring building/support installation Routine machine repairments and maintenance Routine back-up repairment and maintenance Others	Prob drilling Pumping and dewatering Pre-grouting Extra disc cutter replacements and cutterhead inspection Installing steel segments Excavating the materials behind the shield in case of machine entrapment Extra repairs or modifications in machine and its back-up	Lack of supplies i.e. precast segments Power outage Lack of construction materials (cement, gravel, water, etc.) Inexperienced operators and unskilled workers Other management weaknesses affecting the site/project

- Segment erection or installing the support system

2. Delays due to geological problems

- Cutters inspection
- Replacement of cutters
- Delays due to adverse geological conditions (water, gas, face or wall instability)

3. Repair and service of TBM
4. Repair and service of backup systems
5. Logistics

- Transportation of muck materials
- Transportation of personnel, materials and equipment
- Supply of electricity, water, compressed air, ventilation, etc.
- Surveying
- Shift change, lunch or dinner, etc.

6. Other unplanned delays.

Some of these activities can be subdivided into smaller components. For instance, the process of installing a support system in a segmental lined tunnel within a weak and water-bearing rock mass can be broken down into smaller tasks such as probe drilling, grout injection, segment erection, segment back-filling, and installation of other necessary support systems.

Table 2 and Fig. 36 is an example comparing the performance and activity percentages of a double-shield machine ($d_{TBM} = 5.37$ m) over one year and four 3-month periods in the 1st lot of Zagros Water conveyance tunnel (ZWCT-1A) in western Iran. Built in stable carbonate-argillaceous sedimentary rocks with consistent geology and few fault zones, notable variations in parameters like average advance rate and utilization factors are evident. These changes are attributed to fault zone encounters, increased tunnel length, unforeseen events, and management issues. As shown, despite identical timeframes and using the same machine, the average values of advance length, net boring time, daily advance rate, utilization factor, and activity distribution differ significantly across periods. In this project, 37–50% of operational time was spent on TBM advance (boring, segment assembly and re-gripping). As shown, despite using a double shield machine, there was minimal overlap between segment erection and boring stages.

Disc cutter replacements accounted for 5–8%, while about 20% was for routine repairs of the machine and its backup system. Utility installation (water, air, electricity, etc.) and logistic activities consumed 9–18, and 4–16% of time was lost to other delays, mostly unknown.

Geological conditions were generally favorable, with delays mainly limited to short stops in water-bearing and squeezing faulted zones.

In Fig. 37, the influence of installing a support system on the advance rate of open machines is compared to the advance rate of shielded machines. It is evident that a decrease in the quality of the rock mass necessitates the installation of a heavier

Table 2 The average values of performance parameters in different construction periods of the ZWCT-1A project

No	Period	Section length/ advance (m)	Number of operating days	Net boring time (h)	Average daily advance rate (m/ day)	Operating utilization factor (%)
1	Spring	2012.4	93	527.25	21.64	23.6
2	Summer	2556.1	93	673.17	27.48	30.1
3	Automn	1408.1	90	497.25	15.65	23.0
4	Winter	2480.2	89	738.85	27.86	34.6
5	Total year	8456.8	365	2436.52	23.17	27.8

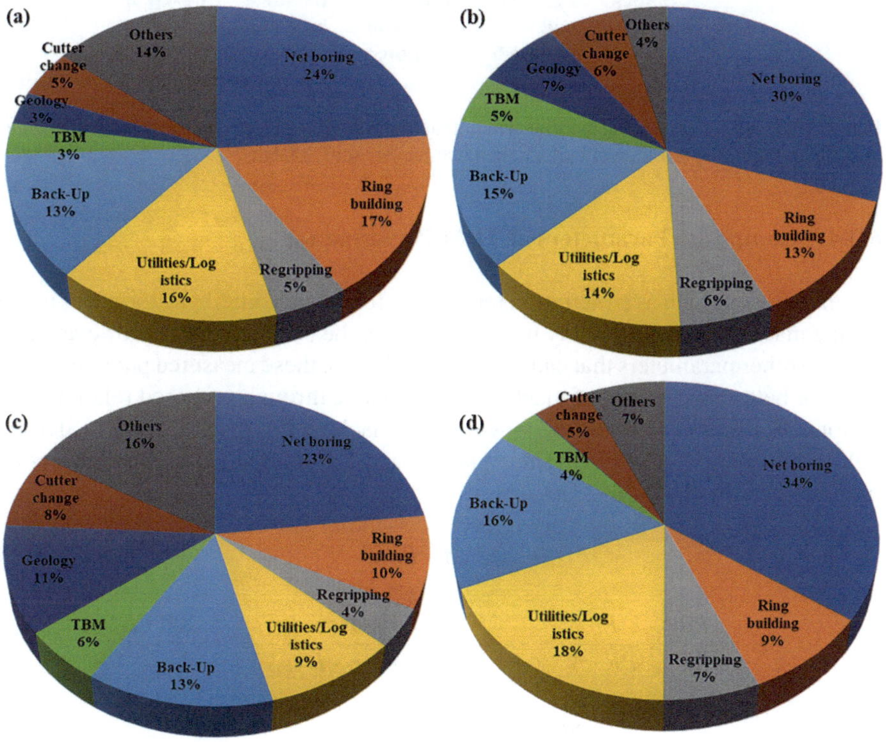

Fig. 36 Examples of required times for different activities in ZWCT-1A project for four different periods; **a** spring months, **b** summer months, **c** automn months and **d** winter months

support system to maintain tunnel stability, resulting in a reduction in the advance rate of open machines. A higher advance rate in open machines (compared to shield machines) is only achievable when, based on the rock mass conditions, a lighter maintenance system is required.

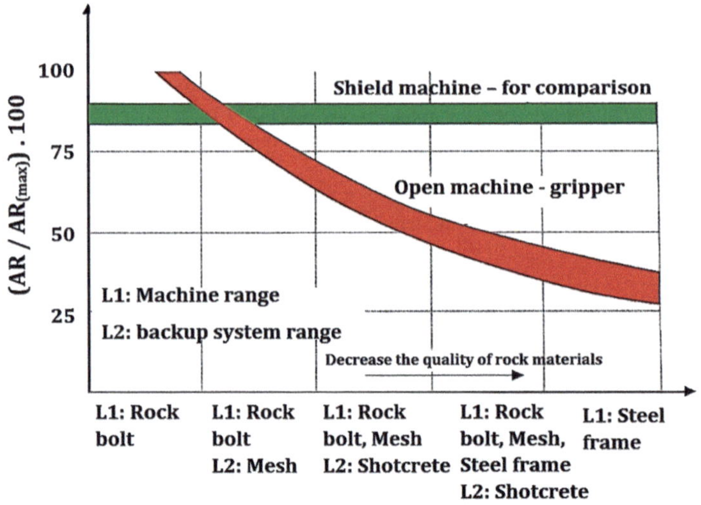

Fig. 37 The influence of implementing a maintenance system on the advance rate of open machine in comparison to the advance rate of shield machine (Schmid 2004)

6.3.4 Composite Parameters of TBM Performance

The parameters such as penetration rate, thrust force, and cutterhead torque of tunnel boring machines can be directly measured during the boring process. However, there are also other parameters that can be calculated using these measured parameters and used for better assessment of machine performance through proposed relationships. Among these parameters are the power of the machine Po, specific energy (SE), and field penetration index (FPI). The equations used to compute these parameters are presented below (Eqs. 20–23):

$$Po = 2 \cdot \pi \cdot RPM \cdot T_q \quad (20)$$

$$SE = \frac{Po}{V} = \frac{Po}{PR \cdot A} = \frac{4 \cdot 1000 \cdot Po}{P \cdot RPM \cdot 60 \cdot \pi \cdot D_{TBM}^2} = \frac{400 \cdot T_q}{3 \cdot D_{TBM}^2 \cdot P} \quad (21)$$

$$SE = \frac{200 \cdot N_{TBM} \cdot r_{mc}}{3 \cdot D_{TBM}} \cdot \frac{F_r}{P} \quad (22)$$

$$FPI = \frac{F_n}{P} \quad (23)$$

In these equations, Po is the power of the machine (kW), SE is the specific energy (kWh/m^3), PR is penetration rate in m/h, t is time for a given stroke or boring time, V is production rate (m^3/h), A = area of tunnel in m^2 and the FPI is the field penetration index (kN/cutter/mm/rev). Other parameters have already been introduced.

6.4 TBM Performance Limitations

The optimization of rock fragmentation and machine performance is limited by certain constraints, which must be considered in machine design. Figure 38 shows a graph of the performance limits of a machine with 17-in disc cutters, illustrating the relationship between cutterhead penetration and cutter load for rocks of different strengths. It highlights that a key limitation is the maximum achievable penetration in the rock, which typically does not exceed 15 mm per revolution of the cutterhead. The other limitation is the maximum load a disc cutter can withstand, typically below 267 kN. The third is the torque required to rotate the cutterhead, which increases with thrust force and penetration due to higher friction during boring and muck removal. Figure 38 shows five curves representing maximum penetration in rocks of different strengths. The chart indicates that in low-strength rocks, maximum penetration is achievable, while in strong rocks, limitations in cutter load and machine torque restrict penetration.

These findings highlight the importance of accounting for these constraints in the design and operation of machinery for efficient rock fragmentation and excavation.

Cutterhead torque is essential for overcoming rolling resistance, which, assuming fixed design parameters like TBM diameter and disc cutter number, depends on operational factors such as:

- The strength of the intact rock or rock mass and the extent of penetration within the rock;
- The condition of the disc cutters, including wear and possible bearing failures;

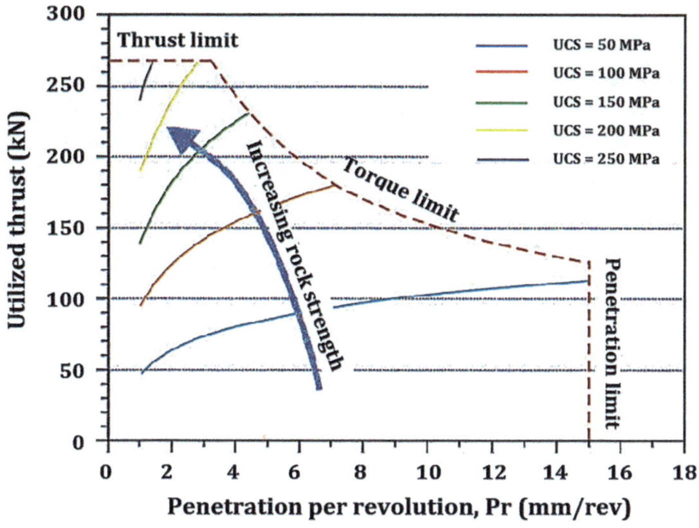

Fig. 38 Limitations for application of a TBM with 17-in diameter cutters in rocks with different uniaxial compressive strength (Frenzel et al. 2008)

- Resistance resulting from incomplete removal of excavated material and accumulation of muck between the rock face and the cutterhead;
- Cutterhead jamming at the start of boring cycle after extended pauses in highly deformable, convergent rocks.

Penetration is the main source of rolling resistance, increasing the cutterhead torque. In soft or fractured rocks, high penetration can exceed the machine's torque capacity, requiring a reduction in thrust force to lower the torque demand. This situation is known as "limited torque".

The machine's performance is also limited by the conveyor belt's capacity to transport muck. Excessive penetration and cutterhead rotational speed can produce more material than the conveyor can handle, disrupting continuous operation. To address this, parameters like rotational speed and penetration depth (via thrust force reduction) must be adjusted. Equation 24 calculates conveyor capacity, where C_{CB} (m^3/h) is the conveyor belt's capacity, PR (m/h) is penetration rate and A (m^2) is the tunnel's cross-sectional area.

$$C_{CB} = A \cdot PR \tag{24}$$

Another operational constraint in machine performance is the gripping force, F_G, which must be sufficient for the machine to achieve optimal performance. In gripper-type machines, F_G depends on the applied thrust force, T_h, and the friction between the gripper pads and the tunnel wall, as shown in Eq. 25, where μ (about 0.5) is the coefficient of friction between rock and steel (gripper pads).

$$F_G = \frac{T_h}{\mu} = 2 \cdot T_h \tag{25}$$

If the gripping force is inadequate, the thrust force cannot be fully utilized. A decrease in thrust force results in reduced penetration capabilities and shorter cutting tool life. The contact pressure beneath the gripper pads is determined by the surface area of the gripper pads and the quality and strength of the rock mass. For instance, when a machine traverses fault damage zones, it is essential to reduce the force applied to the grippers in order to maintain stability. Consequently, this also necessitates a significant reduction in thrust force, which negatively impacts cutterhead penetration and increases tool wear.

Double shield machines can reduce the impact of challenging conditions by changing their operational mode. In favorable rock conditions, the grippers are utilized for load transfer, while the thrust force is generated by cylinders positioned between the front and gripper shields. This setup allows only the front shield and the cutterhead to advance during the boring phase. In cases where the rock is weaker, auxiliary jacks located in the rear shield provide the necessary thrust, enabling the entire shield to move forward during the boring phase. In this case, the negative conditions of the rock mass do not directly affect the machine's thrust force, as they play no role in the TBM advance.

References

Bruland A (1998) Hard rock tunnel boring, vol 1–10. PhD thesis. Norwegian university of Science and Technology (NTNU), Trondheim

Frenzel C, Kasling H, Thuro K (2008) Factors influencing disc cutter wear. Geomechanik und Tunnelbau 1(1):55–60

Gong QM, Zhao J (2007) Influence of rock brittleness on TBM penetration rate in Singapore granite. Tunnel Underground Space Technol 22:317–324

ITA (2000) Recommendations and guidelines for tunnel boring machines. International Tunneling Association, Working group No 14. www.ITA-AITES.org

Maidl B, Schmid L, Ritz W, Herrenknecht M (2008) Hard rock tunnel boring machines. Ernst & Sohn, A Wiley company, Berlin

Nelson PP (1993) TBM performance analysis with reference to rock properties. In: Hudson JA (ed) Comprehensive rock engineering, excavation, support and monitoring, Chapter 10. Pergamon Press, Oxford, pp 261–291

Rostami J (1997) Development of a force estimation model for rock fragmentation with disc cutters through theoretical modeling and physical measurement of crushed zone pressure. PhD thesis. Golden, Co. (USA): Colorado School of Mines

Rostami J (2016) Performance prediction of hard rock tunnel boring machines (TBMs) in difficult ground. Tunnel Undergr Space Technol 57:173–182

Rostami J, Chang SH (2017) A closer look at the design of cutterheads for hard rock tunnel-boring machines (TBMs). Engineering 3(6):892–904. https://doi.org/10.1016/j.eng.2017.12.009

Rostami J, Ozdemir L (1993) A new model for performance prediction of hard rock TBM. In: Bownerman S, Monsees L (eds) Proceedings of rapid excavation and tunneling conference (RETC). Littleton, CO: Soc. for Mining, Metallurgy, and Exploration Inc, Boston, pp 793–809

Schmid L (2004) Die Entwicklung der Methodik des Schildvortriebs in der Schweiz. Geotechnik 27(2):193–200

Snowdon AR, Ryley DM, Temporal J (1982) Study of disc cutting in selected British rocks. Int J Rock Mech Min Sci Geomech 19:107–121

US Army Corps of Engineers (1997) Engineering and design tunnels and shafts in rock: Appendix C: tunnel boring machine performance-concepts and performance prediction. Washington

Wittke W, Druffel R, Erichsen C, Gattermann J, Kiehl L, Schmitt D, Tegelkamp M, Wittke M, Wittke-Gattermann P, Wittke-Schmitt B (2007) Stability analysis and design for mechanized tunneling, geotechnical engineering in research and practice, Aachen: WBI-PRINT 6

Cutterhead Modeling of Hard Rock TBM

Jamal Rostami

1 Introduction

The interaction of disc cutters and the face, as well as the whole TBM and the rock mass is a complex phenomenon that was difficult to capture in models till recently. The early attempts to estimate machine performance in a given rock mass, along with successful operation of a rock TBM was considered an art, and over the past three decades, it has become an engineering problem with possible solutions through the advanced computational tools that has become available to engineers. The introduction of personal computers (PC) in the 1990's opened the door for use of more computational based modeling in evaluating cutterhead design and TBM performance prediction. This was through the use of early codes in Fortran and C programming languages and use of spreadsheets. The evolution of computers and their speed in the past few decades, along with the increased abilities of available Apps and Software in analyzing the laboratory and field-based data has enabled development of new modeling tools for variety of uses in TBM tunneling. These models use computational power to analyze cutting forces from laboratory scale testing on disc cutters to offer a better understanding of the relation between rock parameters and forces, while the complex field data has been digested by a variety of programs to offer models for estimation of rock TBMs in variety of rock masses. The most recent development is the use of Artificial Intelligence techniques in this field that started in late 1990's and early 2000's and with the explosion of these methods and introduction of the machine learning (ML) systems, it has exponentially increased in 2010's and surely has been dominant in the research circles since 2020 in variety of activities.

J. Rostami (✉)
Excavation Engineering and Earth Mechanics Institute (EMI), Colorado School of Mines, Golden, CO, USA
e-mail: rostami@mines.edu

J. Hassanpour and J. Rostami (eds.), *Performance of Tunnel Boring Machines (TBM) in Rock*, Springer Series in Geomechanics and Geoengineering,
https://doi.org/10.1007/978-3-032-20424-0_2

The advances in recent years have propelled the TBM tunneling in rock into the twenty-first century. This is evident through extensive use of computers in variety of stages from preliminary design of the tunnels, to ground modeling during Geotech investigations, to prediction of machine performance in design and bidding stage, monitoring of machine operational data during the construction, and analysis of machine performance in postconstruction stage. Yet, the role of computer models for cutterhead design and performance prediction has remained a useful tool in our disposal.

These models are used in three different levels:

1. Overall models of the TBM cutterhead without details for performance prediction;
2. TBM performance prediction models with more details of the cutterhead design;
3. Detailed cutterhead design models for design evaluation, optimization, forensic analysis, and for performance prediction, especially in mixed ground conditions.

This chapter offers a quick overview of these models and their uses in various stages of TBM tunneling. It offers screenshots of such models as a guide to readers to allow them generate similar computational models for their related projects. It should be noted that there are TBM performance prediction models that are based on field data and are among the empirical models. These models often do not use the cutterhead information in their estimates. However, there are models that are called semi-theoretical [i.e. CSM, Sanio (1985) and Sato et al. (1991) models] that are based on the estimation of the cutting forces. These models have been developed based on full scale cutting tests using disc cutters in different rock types. Such models offer estimates of cutting forces from rock properties, cutter geometry, and disc spacing and penetration. Any of the TBM performance and design models that use cutting forces as a basis of their calculations fall in this category and will be discussed in more details. These models rely heavily on the rock properties for estimation of the cutting forces and do not take joints and rock mass characteristics into account for estimation of such forces. The impacts of rock mass or jointing are often added to the calculations as an adjustment factor to the base calculations as will be discussed later.

2 General TBM Performance Prediction Models

These models are used for estimating the overall performance of a rock TBM in a project as it goes through different formations. They use machine specifications, including thrust, torque, power, cutterhead speed (RPM), and disc cutter geometry to estimate cutting forces and hence potential performance of the machine in terms of penetration per revolution and overall rate of penetration per hour in a given rock type. The general models are based on force equilibrium, meaning the cumulative estimated normal forces are used to estimate machine thrust, cumulative rolling forces estimate torque, and comparing these parameters for a given depth of penetration

EXAMPLE
Screenshot

TBM Performance Prediction/Analysis & Scheduling.

GENERAL INFORMATION									
Project:	Example			Location:	Unknown		Date:	07/20/25	
Owner:	Example			Contractor:	TBD		Machine Manufacturer:	TBD	
				Designer:	TBD		Report Prepared for:	TBD	
TUNNEL INFORMATION									
	(ft)	(m)			(ft)	(m)		(yd^3)	(m^3)
Diameter :	21.5	6.56		Total Length (ft):	21,277	6,489	Total Bank Volume:	286,000	219,300
Area :	363	33.8		Volume (cyd/ft, m^3/m) :	13.44	33.8	Total Excavated Volume:	457,600	350,880
							Swell Factor:	60%	
MACHINE SPECIFICATIONS:									
CUTTERS									
Type:	Disc			Cutter Diameter (in, mm):	19	483	Cost/Hub $:	$3,500	
Face Spacing (in, mm):	3.55	90		Tip Width (in, mm):	0.625	16	Cost/Ring $:	$800	
Suggested No. of Cutters	43			Max. Load Capacity (lbs, kN):	70000	308	Misc. Cost %:	15	
Number of Cutters Used:	39			Max. Linear Speed (ft/min):	620		Avg. Ring/Hub Ratio	7.5	
Performance of Avg. Worn Cutters [% of Max.(Sharp Cutters)]:			75%				Avg. Cost/Change:	$1,456	
MACHINE DATA:									
Max. Thrust (lbs, kN):	3,003,000	13,213		Operating thrust:	2,702,700	(lbs)	RPM (Max. Allowable Calc.):	9.18	
Max. Torque (ft-lbs,kNM,MNM)	2,100,000	2818	2.82	Operating Torque):	1,890,000	2.561	Actual Cutterhead RPM:	9	
Installed Power (hp, kW):	3600	2700		Cutterhead Power (hp):	3,240				
Max Propel Rate (Pump limit):	23	7.015	(ft/hr, m/hr)	Mechanical Efficiency:	90%		Max Average Cutter Load:	69300	
OPERATION INFORMATION:									
Shifts per day:	3			Hours per Shift:	7.5		Working Days per Week:	5	
Hours per day:	22.5			Total Hours per Week:	112.5				

Fig. 1 Screenshot of a general TBM performance prediction and cutterhead model—module 1 and 2 on general project and machine information

with that of machine specification allows one to calculate what is the maximum depth of penetration that can be achieved by the machine in a given formation. The estimated rates are often adjusted to account for the wear of disc cutters and system inefficiencies using a global adjustment factor, often 75%. This reflects the changes in cutter tip width and radius and degradation of the machine performance due to cutter wear.

The layout of an example of these programs is shown in Figs. 1 and 2. This screen shots show the spreadsheet that was developed for performance prediction of a TBM for a known tunnel alignment including rock types, rock mass conditions, and changes in the alignment including slopes, turns, and start up learning curve section.

The model has the following components:

- Heading, which is the general information about the project including name, location, excavation diameter, length, volume of rock, etc.;
- Cutter specifications such as diameter and tip width;
- Machine specifications such as diameter, number of cutters, spacing of the cutters at the face, total installed and available thrust, total and available power and torque, RPM;
- Formation information, including rock types, compressive and tensile strength (UCS and BTS) of the rock, Cerchar Abrasivity Index (CAI) to represent wear and abrasion, density, porosity, rock origin (sedimentary, igneous, metamorphic) and grain size distribution;
- Alignment information including the length of tunnel in each formation, peculiarity of the rock mass in each formation (i.e. joint sets, orientation, joint spacing, etc.), sections of the tunnel in a formation where turns and change in grade happens, which impacts the anticipated machine utilization;

GROUND CONDITIONS:

Lithology #	Rock Type		Origin	Density (gr/cc)	UCS (PSI)	Tensile (PSI)	Grain Size	Porosity (%)	CAI	Ring/Hub	Swell Factor %	Penetration Rate (mm/rev)	(m/hr)	Load (kN)	Torque (kNm)	Power (kw)
1	HC1	Formation 1	2	2.5	11600	1160	2	2	4	7.3	60	16.51	6.69	171	2,475	2,372
2	HC2	Formation 2	2	2.5	14500	1450	2	2	4.5	8.1	60	12.7	5.15	197	2,485	2,380
3	HD1	Formation 3	2	2.5	17400	1740	2	2	4	7.3	60	10.16	4.12	220	2,474	2,370
4	HD2	Formation 4	2	2.5	20300	2030	2	2	4.5	8.1	60	8.382	3.40	241	2,458	2,355
5	TC1	Formation 5	2	2.5	23200	2320	2	2	4	7.3	60	7.112	2.88	261	2,449	2,346
6	TC2	Formation 6	2	2.5	26100	2610	2	2	4.5	8.1	60	6.35	2.57	283	2,506	2,401
7	LOW	Formation 7	2	2.3	29000	2900	2	2	3.5	6.5	60	5.588	2.26	301	2,503	2,398
TOTAL NUMBER OF ROCK TYPES:				7					Average:	8	60					

SEGMENTS OF WORK: Scenario #1, Most likely scenario with an Open Type TBM

Segments #	Length (FT)	Length (m)	Slope %	Rock Type or #	Joint Spacing (in)	Angle (deg)	Ajustment Factor	Adjusted ROP (Ft/hr)	Machine Utilization	Advence Rate (ft/day)	Time of Completion Boring Hours	Working Shifts	Weeks	Cutter Life (hr)	Cutter Costs @ 20% Contingency ($/cyd)	($/ft)
1	5,028	1,534	0.1	1			1.00	6.7	28%	42	751	358	23.9	163	$4.65	$62.45
2	3,964	1,209	0.1	3			1.00	4.1	25%	23	963	513	34.2	163	$7.55	$101.49
3	510	156	0.1	2			1.00	5.1	30%	35	99	44	2.9	145	$6.79	$91.34
4	2,460	750	0.1	4			1.00	3.4	30%	23	724	322	21.5	145	$10.29	$138.39
5	1,210	369	0.1	3			1.00	4.1	30%	28	294	131	8.7	163	$7.55	$101.49
6	5,213	1,590	0.1	5			1.00	2.9	30%	19	1809	804	53.6	163	$10.78	$144.98
7	2,892	882	0.1	6			1.00	2.6	25%	14	1124	599	40.0	145	$13.59	$182.68
											5764					
7	Segments With		Total Footage	21,277			Weighted Average:	3.7	Ft/hr or	23.0	Ft/day	Total Time	185		Total Cutter Cost:	########
			Meters:	6489.485				1.13	m/hr or	7	m/day					
			Maximum Production Rate Bank:	90	Cyd/hr		Average Utilization		28%						Average Cutter life (hrs):	158
			Max. Prod. Rate, Excavated:	144	Cyd/hr											

Fig. 2 Screenshot of a general TBM performance prediction and cutterhead model—module 3 and 4, on rock formation and tunnel section information and summary results

- Summary box that shows the anticipated peak performance and production rate (to allow for sizing of the backup and transport system), average penetration rate, average utilization, average and section by section daily advance rate, average cutter life and number of cutter changes.

The program includes an engine that allows for estimation of the normal and rolling forces for cutting each formation at a given depth of penetration. These forces are used to estimate TBM thrust and torque for the selected depth of penetration. For performance estimate in a formation, penetration per revolution is increased till one of the limits of the TBM is hit and that would be the anticipated depth of penetration that can be achieved in the target formation. The limits include thrust, torque, power, and cutter load limit. From the penetration per revolution, rate of penetration (ROP) in m/h is estimated, along with the pertinent thrust and torque/power level of the machine. Once the anticipated performance of the machine in a given formation is estimated, the same steps are taken to estimate the anticipated ROP in each formation. The rates are adjusted according to the past experience of TBMs to account for average cutter wear conditions. To estimate the cutting forces any of the available models can be used. This includes CSM model (Rostami and Ozdemir 1993; Rostami 1997, 2013; Entacher and Rostami 2019), or similar formulas by others (Sanio 1985; Sato et al. 1991; Gehring 1995; Xia et al. 2018). The cutting force estimation is the core and engine of the model and determines its accuracy. The formation parameters could include additional adjustment factors for cutting forces based on one's experience (i.e. larger grain sizes in igneous rock results in higher forces).

Once the anticipated performance of the machine in a formation is estimated (assuming intact rock), the adjustments for joints and their impact on machine performance can be added. This is done in the next module of the model. In this module, the length of tunnel in a given section is used and a formation is assigned to the selected section. This allows the program to use the pertinent information for that formation to add the adjustment factors for joint orientation and spacing. Once these adjustments are made, the anticipated utilization is added to estimate daily advance rate. This part assumes a three shift 8 h per shift or 24-h operation (and can be adjusted as needed for site settings). The end result is a daily advance rate (AR) for a given section. The adjustment of the jointing in the formation can be done by using one of the available models such as $K_{S\text{-}tot}$ chart by Bruland (1998), Macias et al. (2014) and Macias (2016) or Hassanpour (2009) and Hassanpour et al. (2011) models and adjustment for RQD, etc.

The output of this model can be used to estimate tunnel completion time, daily advance rate, peak production for muck transport design, estimated number of cutters and related cutter cost, and overall, the general information for bidding purposes. These models can be customized to meet the needs of the designer or contractor in their engagement with the project and offers the flexibility for limited changes in the settings to see the overall impact on the project. However, these models are not detailed enough to allow for assessment of machine performance and cutterhead design.

3 Cutterhead Based TBM Performance Prediction Models

The general models use general machine specification to estimate the anticipated TBM performance in a project. However, they do not take into account the cutterhead design intricacies and cut spacing and other details of the design to offer their estimates. The next level of cutterhead modeling programs are the ones that allows for programming each cutter and hence facilitates use of more detailed cutterhead information in the process of estimating TBM performance. Figure 3 shows a typical rock TBM cutterhead profile and Fig. 4 is a screenshot of such models based on the profile.

In this approach, every cutter is individually programmed into a spreadsheet (or a code), and the cutting forces, Normal and Rolling, are estimated based on the formulas and models discussed earlier in previous section.

The detail of the individual cutter includes the cutterhead profile info such as location of the cutter, and tilt angle. The tilt angle starts from 0° (parallel to the tunnel/TBM axis or perpendicular to the face) at the center and face of the cutterhead and gradually goes to around 65° at the gage. The cutter information can include diameter and tip width (and can be adjusted one by one) and in rare occasions where tungsten carbide inset discs used in the center or gage area, these cutters can be programmed separately and cutting forces estimated accordingly (Copur et al. 2024). In this model, the penetration rate is indicated by the user in mm/rev (or in/rev), but while it is the same for all center and face cutters installed perpendicular to the face, it gradually drops as it goes to the cutterhead gage area. This means that penetration of each individual cutter is estimated based on its location on the head and on the cutterhead profile. In the case of double/triple tracking, meaning

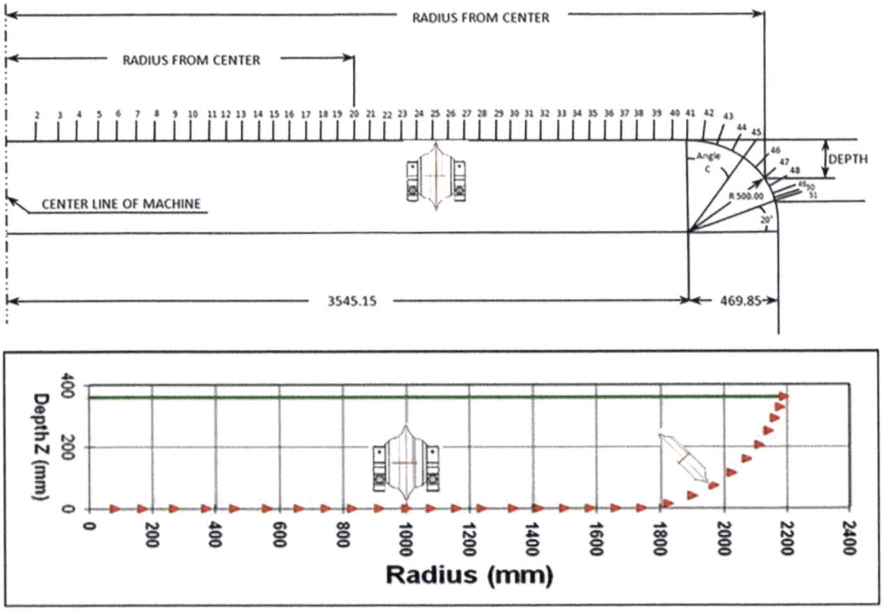

Fig. 3 (Top) A typical cutterhead profile for a hard rock TBM with flat profile (bottom), cutterhead profile as depicted in modelling program for different cutterhead (Rostami and Chang 2017)

more than one cutter per position, the penetration per cutter is adjusted based on the number of cutters per position. This is in case copy cutters are installed in the gage area and hence the cutting forces reflect such reduced workload. Spacing between the cutters are applied individually, and similarly, the estimated cutting forces per each cutter reflects the true spacing between the cuts that each cutter has to excavate. These adjustments allow for more detailed estimation of the cutting forces for each cutter, allowing for more accurate estimation of the performance of the machine.

The normal forces are summed up to estimate the cutterhead thrust. This value is then compared to the available thrust on the cutterhead, which is the sum of forces delivered by thrust cylinders minus friction (which could be substantial for open mode single or double shield TBMs). The thrust limit of the cutterhead is set based on the machine specifications and will be subsequently used in the calculation of the maximum penetration by the disc cutters. Similarly, the rolling forces and actual position of the cutters on the head and radius from the center is used to estimate the cutterhead torque from rolling forces. With selected RPM and cutterhead power, the maximum torque available on the cutterhead can be calculated and used in the performance estimates. In case any of the motors are out of service (gearbox issues, motor failures, etc.), the power can be adjusted and available torque estimated for reduced machine performance.

This model allows the users to check various scenarios with different rock types and develop machine performance charts such as the one shown in Fig. 5. In these

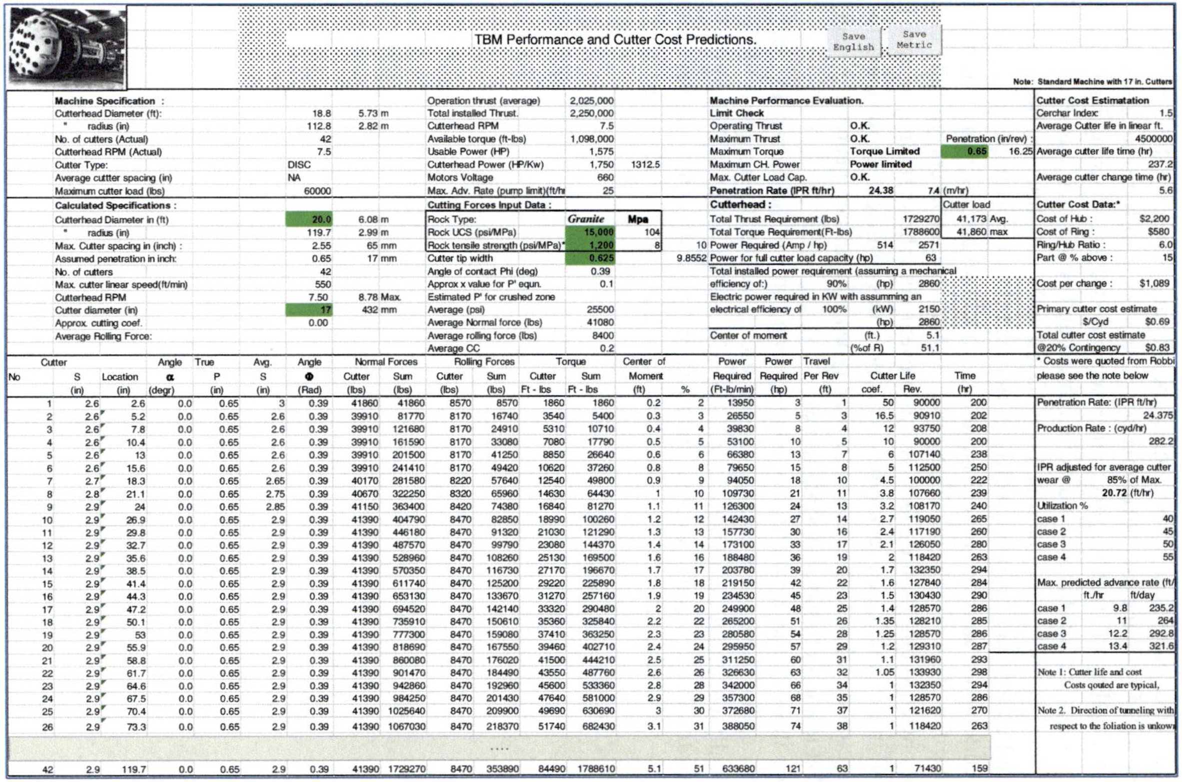

TBM Performance and Cutter Cost Predictions.

Save English | Save Metric

Note: Standard Machine with 17 in. Cutters

Machine Specification :			Operation thrust (average)	2,025,000		
Cutterhead Diameter (ft):	18.8	5.73 m	Total installed Thrust.	2,250,000		
" radius (in)	112.8	2.82 m	Cutterhead RPM	7.5		
No. of cutters (Actual)	42		Available torque (ft-lbs)	1,098,000		
Cutterhead RPM (Actual)	7.5		Usable Power (HP)	1,575		
Cutter Type:	DISC		Cutterhead Power (HP/Kw)	1,750	1312.5	
Average cutter spacing (in)	NA		Motors Voltage	660		
Maximum cutter load (lbs)	60000		Max. Adv. Rate (pump limit)(ft/hr	25		
Calculated Specifications :			**Cutting Forces Input Data :**			
Cutterhead Diameter in (ft)	20.0	6.08 m	Rock Type:	*Granite*	**Mpa**	
" radius (in)	119.7	2.99 m	Rock UCS (psi/MPa)	15,000	104	
Max. Cutter spacing in (inch) :	2.55	65 mm	Rock tensile strength (psi/MPa)*	1,200	8	10
Assumed penetration in inch:	0.65	17 mm	Cutter tip width	0.625		9.8552
No. of cutters	42		Angle of contact Phi (deg)	0.39		
Max. cutter linear speed(ft/min)	550		Approx x value for P' equn.	0.1		
Cutterhead RPM	7.50	8.78 Max.	Estimated P' for crushed zone			
Cutter diameter (in)	17	432 mm	Average (psi)	25500		
Approx. cutting coef.	0.00		Average Normal force (lbs)	41080		
Average Rolling Force:			Average rolling force (lbs)	8400		
			Average CC	0.2		

Machine Performance Evaluation.					
Limit Check					
Operating Thrust	O.K.			Penetration (in/rev) :	
Maximum Thrust	O.K.			0.65	16.25
Maximum Torque	**Torque Limited**				
Maximum CH. Power	**Power limited**				
Max. Cutter Load Cap.	O.K.				
Penetration Rate (IPR ft/hr)	**24.38**	7.4	(m/hr)		
Cutterhead :				Cutter load	
Total Thrust Requirement (lbs)		1729270		41,173 Avg.	
Total Torque Requirement(Ft-lbs)		1788600		41,860 max	
Power Required (Amp / hp)	514	2571			
Power for full cutter load capacity (hp)		63			
Total installed power requirement (assuming a mechanical					
efficiency of:)	90%	(hp)	2860		
Electric power required in KW with assumming an					
electrical efficiency of	100%	(kW)	2150		
		(hp)	2860		
Center of moment		(ft.)	5.1		
		(%of R)	51.1		

Cutter Cost Estimatation		
Cerchar Index:		1.5
Average Cutter life in linear ft.		
		4500000
Average cutter life time (hr)		
		237.2
Average cutter change time (hr)		
		5.6
Cutter Cost Data:*		
Cost of Hub :		$2,200
Cost of Ring :		$580
Ring/Hub Ratio :		6.0
Part @ % above :		15
Cost per change :		$1,089
Primary cutter cost estimate		
	$/Cyd	$0.69
Total cutter cost estimate		
@20% Contingency		$0.83
* Costs were quoted from Robbi		
please see the note below		
Penetration Rate: (IPR ft/hr)		
		24.375
Production Rate : (cyd/hr)		
		282.2
IPR adjusted for average cutter		
wear @	85% of Max.	
	20.72 (ft/hr)	
Utilization %		
case 1		40
case 2		45
case 3		50
case 4		55
Max. predicted advance rate (ft/		
	ft./hr	ft/day
case 1	9.8	235.2
case 2	11	264
case 3	12.2	292.8
case 4	13.4	321.6
Note 1: Cutter life and cost		
Costs qouted are typical,		
Note 2. Direction of tunneling with		
respect to the foliation is unkow		

Cutter No	S (in)	Location (in)	Angle α (degr)	True P (in)	Avg. S (in)	Angle Φ (Rad)	Normal Forces Cutter (lbs)	Normal Forces Sum (lbs)	Rolling Forces Cutter (lbs)	Rolling Forces Sum (lbs)	Torque Cutter Ft - lbs	Torque Sum Ft - lbs	Center of Moment (ft)	%	Power Required (Ft-lb/min)	Power Required (hp)	Travel Per Rev (ft)	Cutter Life coef.	Cutter Life Rev.	Time (hr)
1	2.6	2.6	0.0	0.65	3	0.39	41860	41860	8570	8570	1860	1860	0.2	2	13950	3	1	50	90000	200
2	2.6	5.2	0.0	0.65	2.6	0.39	39910	81770	8170	16740	3540	5400	0.3	3	26550	5	3	16.5	90910	202
3	2.6	7.8	0.0	0.65	2.6	0.39	39910	121680	8170	24910	5310	10710	0.4	4	39830	8	4	12	93750	208
4	2.6	10.4	0.0	0.65	2.6	0.39	39910	161590	8170	33080	7080	17790	0.5	5	53100	10	5	10	90000	200
5	2.6	13	0.0	0.65	2.6	0.39	39910	201500	8170	41250	8850	26640	0.6	6	66380	13	7	6	107140	238
6	2.6	15.6	0.0	0.65	2.6	0.39	39910	241410	8170	49420	10620	37260	0.8	8	79650	15	8	5	112500	250
7	2.7	18.3	0.0	0.65	2.65	0.39	40170	281580	8220	57640	12540	49800	0.9	9	94050	18	10	4.5	100000	222
8	2.8	21.1	0.0	0.65	2.75	0.39	40670	322250	8320	65960	14630	64430	1	10	109730	21	11	3.8	107660	239
9	2.9	24	0.0	0.65	2.85	0.39	41150	363400	8420	74380	16840	81270	1.1	11	126300	24	13	3.2	108170	240
10	2.9	26.9	0.0	0.65	2.9	0.39	41390	404790	8470	82850	18990	100260	1.2	12	142430	27	14	2.7	119050	265
11	2.9	29.8	0.0	0.65	2.9	0.39	41390	446180	8470	91320	21030	121290	1.3	13	157730	30	16	2.4	117190	260
12	2.9	32.7	0.0	0.65	2.9	0.39	41390	487570	8470	99790	23080	144370	1.4	14	173100	33	17	2.1	126050	280
13	2.9	35.6	0.0	0.65	2.9	0.39	41390	528960	8470	108260	25130	169500	1.6	16	188480	36	19	2	118420	263
14	2.9	38.5	0.0	0.65	2.9	0.39	41390	570350	8470	116730	27170	196670	1.7	17	203780	39	20	1.7	132350	294
15	2.9	41.4	0.0	0.65	2.9	0.39	41390	611740	8470	125200	29220	225890	1.8	18	219150	42	22	1.6	127840	284
16	2.9	44.3	0.0	0.65	2.9	0.39	41390	653130	8470	133670	31270	257160	1.9	19	234530	45	23	1.5	130430	290
17	2.9	47.2	0.0	0.65	2.9	0.39	41390	694520	8470	142140	33320	290480	2	20	249900	48	25	1.4	128570	286
18	2.9	50.1	0.0	0.65	2.9	0.39	41390	735910	8470	150610	35360	325840	2.2	22	265200	51	26	1.35	128210	285
19	2.9	53	0.0	0.65	2.9	0.39	41390	777300	8470	159080	37410	363250	2.3	23	280580	54	28	1.25	128570	286
20	2.9	55.9	0.0	0.65	2.9	0.39	41390	818690	8470	167550	39460	402710	2.4	24	295950	57	29	1.2	129310	287
21	2.9	58.8	0.0	0.65	2.9	0.39	41390	860080	8470	176020	41500	444210	2.5	25	311250	60	31	1.1	131960	293
22	2.9	61.7	0.0	0.65	2.9	0.39	41390	901470	8470	184490	43550	487760	2.6	26	326630	63	32	1.05	133930	298
23	2.9	64.6	0.0	0.65	2.9	0.39	41390	942860	8470	192960	45600	533360	2.8	28	342000	66	34	1	132350	294
24	2.9	67.5	0.0	0.65	2.9	0.39	41390	984250	8470	201430	47640	581000	2.9	29	357300	68	35	1	128570	286
25	2.9	70.4	0.0	0.65	2.9	0.39	41390	1025640	8470	209900	49690	630690	3	30	372680	71	37	1	121620	270
26	2.9	73.3	0.0	0.65	2.9	0.39	41390	1067030	8470	218370	51740	682430	3.1	31	388050	74	38	1	118420	263
....																				
42	2.9	119.7	0.0	0.65	2.9	0.39	41390	1729270	8470	353890	84490	1788610	5.1	51	633680	121	63	1	71430	159

Fig. 4 Screenshot of a more detailed TBM performance prediction and cutterhead model for a ~6 m diam. TBM with 42 cutters

charts, the maximum penetration rate of an open mode TBM can be estimated for rocks of different strength. This chart clearly reflects the capabilities of the machine and what limits the performance of the machine in each rock strength range. The performance of the TBMs is often limited by the machine propel rate or the maximum rate that the thrust pumps can push the machine forward. Similarly, if the muck transport of the TBM or in the tunnel (tunnel conveyor) has a set limit, such restrictions would be the ceiling of machine penetration rate and ROP in low strength rocks. This is assuming that the machine has sufficient thrust and power as is customary on the TBMs of its size since most manufacturers install sufficient thrust on the machines to offer full cutter load (in other words, cutterhead thrust = Number of cutters × nominal cutter load capacity). In such situations, machines are not going to run thrust limited in soft rock. As for the machine power, again, manufacturers typically install sufficient power/torque on the machines to prevent cutterhead stalling at near full cutter load in the anticipated rock types. This provides sufficient torque capacity to rotate the cutterhead in soft to medium rock types where deep penetration can be achieved at low cutter load or normal forces.

Passing through the weaker rocks where pump limit is the barrier to machine performance, the TBM enters a range of rock strength where the torque is the limit. This typically occurs in the range of ~70–120 MPa rocks where higher cutter load and rolling forces are required to break the rock. In higher strength rocks say around UCS >120–150 MPa, the cutter load or thrust limit is often the limiting factor for

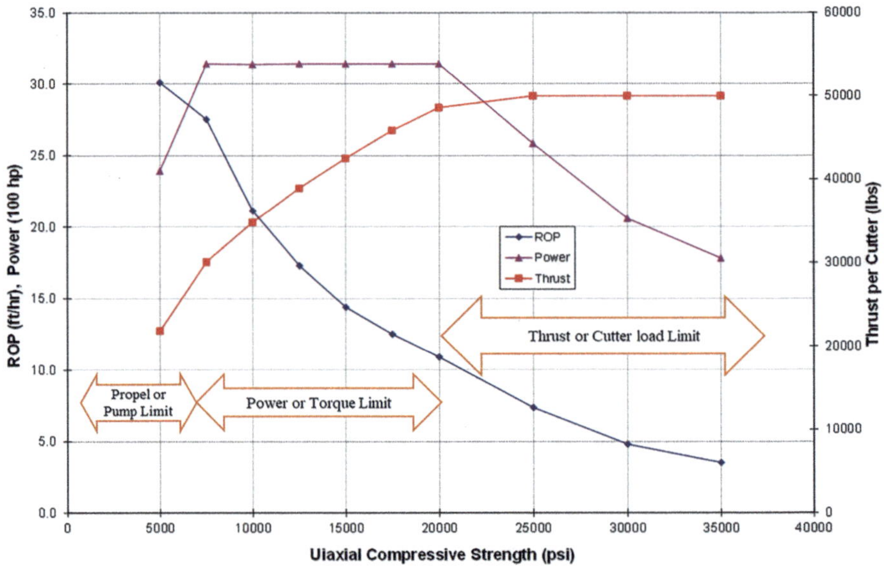

Fig. 5 A typical TBM performance chart that can be generated by these models for a given TBM design to show potential machine performance across various rock strength ranges. *Note* Conversion values for the units 145 psi = 1 MPa for UCS, 1 ft/h = 0.305 m/h for ROP, 2000 lbs = 1 ton for cutter load or thrust per cutter

machine performance. This is the range where higher normal load is required to penetrate the rock and hence the limited thrust and cutter load limits penetration and with lower penetration, the torque capacity of the machine is not fully utilized.

One should be aware that there are two different parameters to observe in this range, one is the overall machine thrust, the other is the cutter load. As noted before, typically manufacturers install sufficient thrust capacity to offer full cutter load to all cutters. This means that more often than not the TBM has sufficient thrust capacity, which is rather inexpensive feature for the machine to have. However, if the full thrust is applied on the cutterhead, the cutters in the gage area and those with smaller spacing, will take lower cutter load, meaning that the cutters at the face and center with higher spacing will take higher load and can be easily overloaded. Overloading the cutter leads to premature failures of the cutter bearing and cutter rings. Hence in the program there are two limitations in the higher strength rock that has to be monitored and checked, thrust and cutter load on individual cutters. The computer models developed for this purpose often have indicators and notifications of such limits (see Fig. 4 in the upper right corner of the spreadsheet).

These models are suitable for performance prediction of the open mode TBMs including Open TBM, Double Shield, Single shield where there is no face pressure applied. They can also be used on single shield pressurized face machines (EPB or Slurry Shield) since their performance in a given formation is determined by the cutting forces. The difference is the adjustments for torque in EPB machines due to movement of the muck in cutting chamber, and similar adjustments in slurry machines, plus proper adjustments for shield friction when calculating the available machine thrust or cutterhead thrust.

This class of models allows for playing a lot of what if scenarios with the rock types and tradeoff between the machine capacities in different ranges of rock strength. They can monitor the load on individual cutters and offer the flexibility to use a mixture of cutter sizes and configurations on the head (i.e. using a wider cutter tip at the gage). It also allows for monitoring the cutter wear per individual cutter and hence the estimation of the cutter life (hours) per cutter using pertinent cutter life formulas (using rock abrasivity). To conduct a full performance prediction using these models, one can break down the tunnel alignment into different formations and run the model to estimate ROP and other parameters per formation, play tradeoffs between machine specifications, and estimate tunnel completion time. The level of details also enables the user to see what machine limits are hit at given formation. This can be helpful in evaluating machine specifications offered by manufacturers for a given project and suggesting possible changes in specifications to fit the project. Adding capacities that are unused on the machine for a given alignment is an unnecessary cost to the project. Alternatively, increasing machine capabilities could prove to be very helpful in improving the overall performance of the TBM in a given project and a cost that can be well justified. Using these models can offer the quantitative reasoning for such discussions.

4 Detailed Cutterhead Design, Optimization and Performance Prediction Models for Rock TBMs

The models introduced in previous section have been around since 1990's as the first version of the CSM model for cutterhead design was discussed in Rostami and Ozdemir (1993). During the same year, computer models for full cutterhead design have been introduced where the cutters were programmed individually with all the details noted above, as well as the third dimension of the cutterhead and angular distribution to show the location of the cutters in a 3D space. The new model that was introduced for cutterhead design had the ability to allow for rotation of the head relative to the face, hence allowing to see the variations in forces and moments during the cutter rotation. One of the first uses of the new detailed models were in balancing of the cutterhead, as this was not necessarily a strong suit of some manufacturers due to complexity of allocating cutters around the head and limited space between the cutter locations and buckets, access to the head etc.

Figure 6 shows a screenshot of an example program with cutterhead details, which includes all the features of the previous models, in addition to the ability to locate the cutters in 3D space, allowing to monitor the forces in 3D space (not limited to thrust) and hence ability to calculate the lateral forces as well as thrust. The same programs allowed the detailed calculation of moments caused by cutting forces acting on individual cutters in 3D, where one component would constitute torque, but the other two moments would show the eccentric moments imposed on the main bearing. Prior to using this concept and cutterhead balancing programs, there were incidents of high lateral force and moments leading to main bearing damage or failure, which could be easily addressed in this program. Cutterhead optimization on such platform could allow for optimizing the cutterhead profile, as well as cutter distributions around the head to avoid eccentric forces. This was due to the fact the all features of cutter placement on the cutterhead could be individually controlled to yield the best results relative to cutterhead balance. With minimized eccentric forces and moments, cutter life and wear could be improved, and unnecessary lateral loading of the main bearing could be minimized. More details on the cutterhead modelling program and the process of using such models in cutterhead design can be found in Rostami and Chang (2017). Figure 7 is a screenshot of two different cutterheads with identical profile but different distributions of cutters around to head exaggerated to show the possibility of unbalanced forces on the head.

These programs can also be used for TBM performance prediction. As noted before, the location and characteristics of the cutters can be programmed individually. Program allows for using various cutters at various positions and implements the calculation of the cutting forces on each individual cutter based on rock properties and depth of penetration. The normal, rolling and side forces are estimated from the formulas and models that were discussed before. In this model, cutting forces are then decomposed into Fx, Fy, and Fz, along the cartesian reference coordinate system. This allows the model to sum up the forces in X, Y, and Z direction, where sum of Fz components is the machine thrust, and sum of forces in X and Y directions are

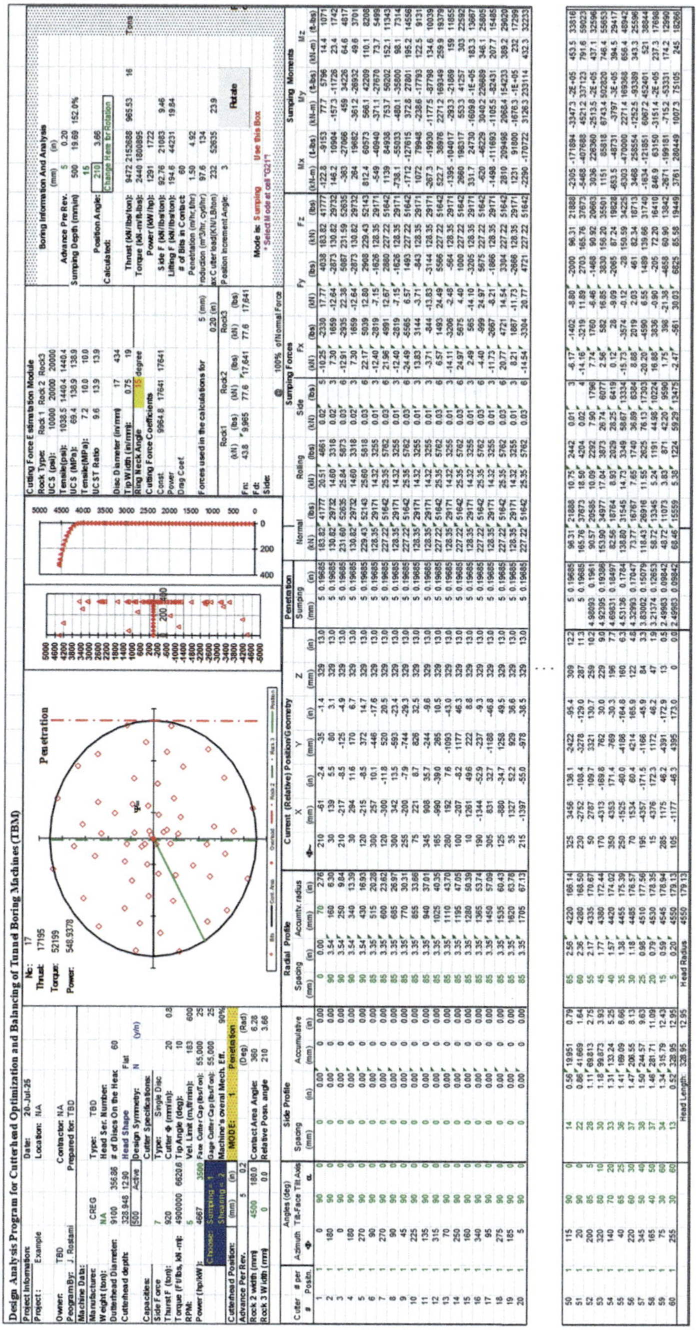

Fig. 6 Screenshot of the 3D cutterhead design and optimization model

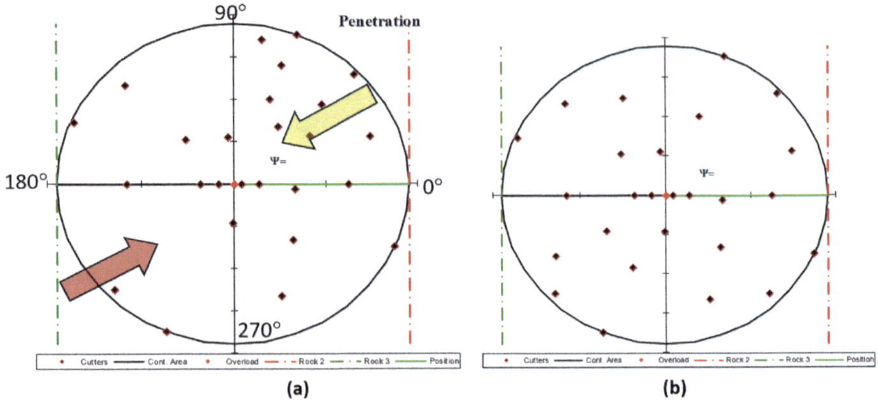

Fig. 7 Cutter distribution on a cutterhead with **a** cluster in 0°–90° quarter, and resultant force (exaggerated), **b** a more reasonable and uniform distribution (Rostami and Chang 2017)

the lateral forces. Similarly, the force components and spatial location of the cutters are used to estimate the moments Mx, My and Mz, for each cutter, and sum them up, where sum of Mz components is machine torque, and other two components represent the excentric moments. The eccentric moments cause the extra loading on the main bearing and also tends to push the machine off the alignment and impose additional side loads on the cutter bearings as well. These are the features that allows an end user to compare the design of various cutterhead arrangements and patterns to get to a reasonable setting where eccentric forces and moments are minimal.

To use these models for performance prediction comes handy where the penetration can be changed to hit the penetration rate where the machine limits are hit. The model allows for more detailed accounting of the forces and moments on the head compared to the previous models and more accurate predictions can be expected. Similar to cutterhead profile modeling, one can develop machine performance characteristics across various rock strength values. The added benefit here is that the user can keep track of the lateral forces and their ranges in different rock type.

One of the features of this group of cutterhead modeling programs is the fact that the cutter positions on the cutterhead are based on a cylindrical coordinate system which can be projected into a reference cartesian coordinate system. This feature allows the user to rotate the head and track the location and position of each cutter in 3D space. Therefore, as the head rotates the direction of the lateral forces change and can be tracked. In other words, as the cutterhead rotates, the magnitude of Fx, Fy and Mx and My for the cutterhead changes and it can be linked to the cutterhead rotational position (0°–360°). Along with this capability, there is a built-in feature in the program that allows the modeling of mixed face conditions, in that the face could be a composition of one rock type (uniform face), two rock types (mixed face with split), or three rock types (a triple mixed face condition). The internal program keeps track of location of the cutters on the face and uses proper rock properties for calculation of the cutting forces. This allows the user to see the impact of different combination of rock types on the cutterhead and the entire machine, including total cutterhead thrust, torque, power, and lateral forces and moments.

The ability of the model to simulate mixed face conditions is of great importance since other models are unable to offer an insight into cutterhead/disc cutter conditions in mixed face situations. Figure 8 shows an example of simulation of the cutterhead in mixed face conditions for a small cutterhead. No other model offers the ability to understand what happens to the total thrust that is delivered to the cutterhead, as it can only track cutterhead thrust and not the actual load taken by any individual cutters. In other words, if a thrust amount of 1000 ton is delivered to the cutterhead with 50 cutters, it is not possible to say which cutters are taking the load, since surely cutters are not equally taking 20 ton each. Surely the cutters in stronger formation take higher load, as the cutters in the softer rock will take lower load, simply indicating the possibility of overloading the individual cutters. However, in this type of model the cutting forces are traced and the cutters taking extra load for a given penetration rate are marked. This prevents overloading of the cutters in the mixed face conditions. In practice, it is important to limit the penetration rate of the TBM to a level where individual cutters are not overloaded. This simply means that the harder formation at the face will determine and limit the penetration rate. This is despite the fact that overall machine thrust and torque limits are not hit and machine has plenty of torque and thrust, but penetration rate is limited to avoid premature cutter failures.

As noted, not only can these models be used in mixed face conditions, but they can also be used for cutterhead optimization and while there is no way to balance the eccentric forces in a mixed face condition, it is possible to calculate the magnitude of lateral/eccentric forces and moments for design and selection of the main bearing.

Figure 9 shows the simulation of a cutterhead in a mixed ground condition where the split between the two rock types is at the middle or center. This is practically the worst-case scenario where the maximum disparity of forces acting on the cutterhead can be experienced. Modelling allows for moving the cutters around to see what arrangement offers the best balance of the cutterhead forces and where the changes in forces do occur due to mixed face conditions and variation in rock strength, it offers the estimated magnitude of forces and moments. In this case, the design of the cutterhead is symmetrical, with optimum use of angular spacing, which provides for the best distribution of the cutters around the head and hence the reduced loading shocks and lower variability of lateral forces.

5 Summary and Future Outlook of the Cutterhead Modeling for Rock TBMs

As discussed in this chapter, cutterhead modeling has had a major role in catapulting the use of rock TBMs from art to engineering science. These models have allowed for better design of the cutterhead as well as ability to perform forensic engineering to see the conditions prevailing on the cutters and cutterhead design in a given

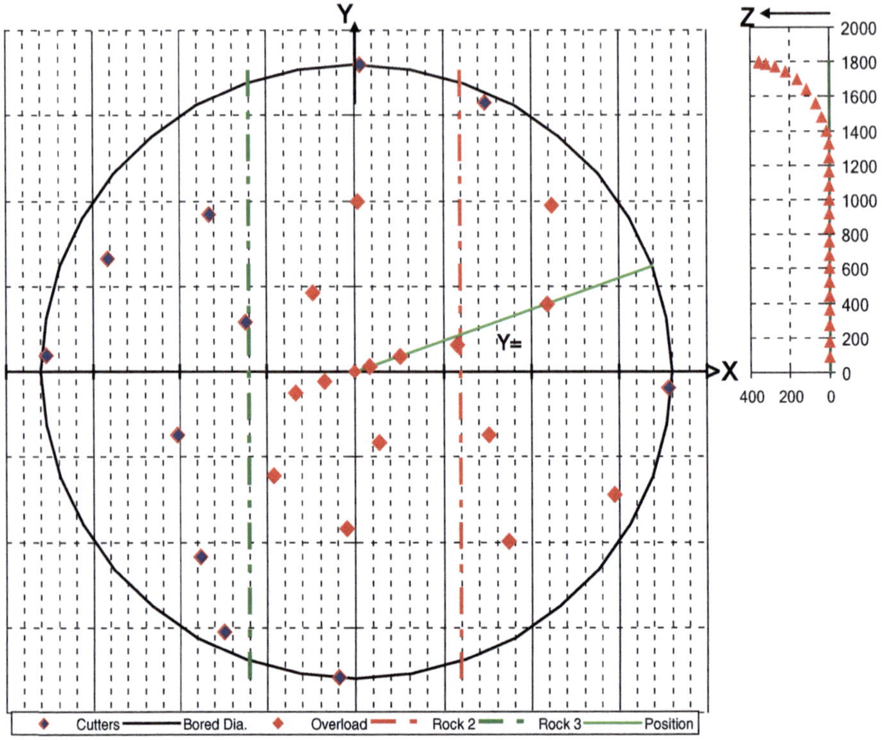

Fig. 8 Graphical representation of the cutterhead in computer modeling where a mixed face condition in modeled with 3 rock types at the face

project. They allow us to analyze the recorded performance of the TBMs in various projects while facilitating the ability to implement the new capabilities and features of modern machines in the predictions and estimates of machine performance and cutter life. As such, they have provided a solid platform for improving the cutterhead design and performance prediction of rock TBMs in various projects in recent years. Yet, a quick review of available literature shows that the power of these models is underutilized and limited number of people, mainly the manufacturers of the machines have partially used these models.

One of the features of these models, especially 2nd and 3rd groups with cutterhead design included in the calculation is that they can be used in reverse calculations of the ground conditions. In other words, if the performance and operating parameters of a TBM are recorded and is available, it is possible to back calculate rock properties such as UCS or RQD etc. This feature has been the focus of many of the recent studies (Xie et al. 2023; Dong et al. 2025; Samadi et al. 2024a, b; Samadi and Hassanpour 2025; Pourhashemi et al. 2022).

The future of cutterhead modeling lies in the implementation of AI and ML algorithm for this application. Many of the recently developed TBM performance

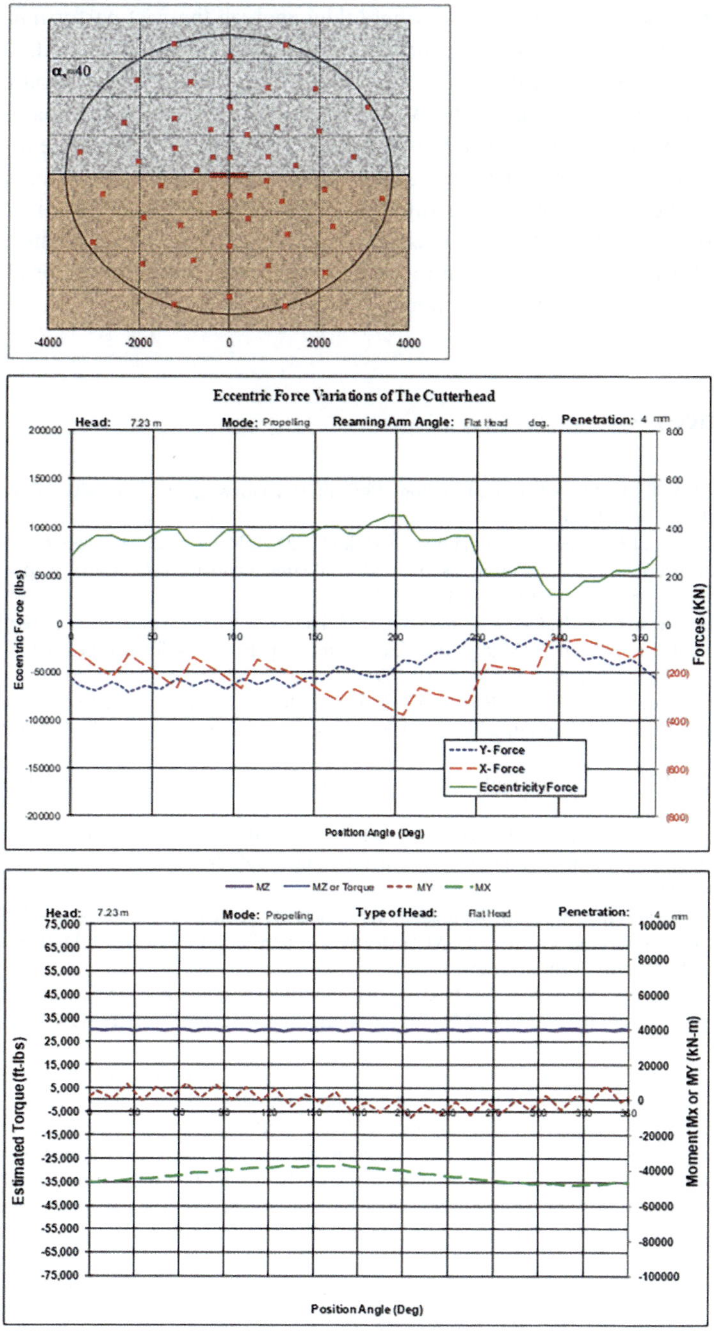

Fig. 9 An example of variation of the estimated eccentric forces and moments for lacing pattern using evenly distributed cutters on the head in mixed face with two rock types split at the middle of the face, for a 7.2 m diameter TBM

models are based on these systems where they analyze the past performance of the machines to project the rate of penetration and advance of open mode TBMs in future projects. However, it is imperative to develop the next generation of cutterhead modeling programs using the AI/ML algorithm to expedite the analysis of the cutterhead layout and cutter installation pattern, as well as performance predictions in various tunneling project with variable ground and alignment features. This is especially true for use of AI in estimating machine performance in mixed ground conditions, where AI systems can generate a model and use it to estimate the anticipated performance of a TBM with a given cutterhead configuration, or alternative designs that could improve the potential performance of a machine.

References

Bruland A (1998) Hard rock tunnel boring. PhD thesis, Norwegian University of Science and Technology, Trondheim

Copur H, Shaterpour-Mamaghani A, Tumac D, Balci C (2024) Relationships between performances of a button cutter and a disc cutter based on full-scale linear cutting tests. Acta Geotech 19:2731–2752. https://doi.org/10.1007/s11440-023-02146-2

Dong Z, Yu H, Tian G, Li X, Cao R, Zhang Y (2025) Data visualization strategy for predicting rock mass classification with TBM key rock-fragmentation parameters. Transp Geotech 51:101489. https://doi.org/10.1016/j.trgeo.2025.101489

Entacher M, Rostami J (2019) TBM performance prediction model with a linear base function and adjustment factors obtained from rock cutting and indentation tests. Tunn Undergr Space Technol 93:1–12

Gehring K (1995) Leistungs-und Verschleißprognosen im maschinellen Tunnelbau. Felsbau 13:439–448

Hassanpour J, Rostami J, Zhao J (2011) A new hard rock TBM performance prediction model for project planning. Tunnel Undergr Space Technol 26(5):595–603. https://doi.org/10.1016/j.tust.2011.04.004

Hassanpour J (2009) Investigating the impact of engineering geological parameters on TBM performance and modifying existing TBM performance prediction models. Ph.D. Thesis, Tarbiat Modares University, Tehran, Iran

Macias FJ, Jakobsen PD, Seo Y, Bruland A (2014) Influence of rock mass fracturing on the net penetration rates of hard rock TBMs. Tunn Undergr Space Technol 44:108–120. https://doi.org/10.1016/j.tust.2014.07.009

Macias FJ (2016) Hard rock tunnel boring: performance predictions and cutter life assessments. PhD thesis, Norwegian University of Science and Technology, Trondheim

Pourhashemi SM, Ahangari K, Hassanpour J, Eftekhari SM (2022) TBM performance analysis in very strong and massive rocks; case study: Kerman water conveyance tunnel project, Iran. Geomech Geoeng 17(4):1110–1122. https://doi.org/10.1080/17486025.2021.1912410

Rostami J (2013) Study of pressure distribution within the crushed zone in the contact area between rock and disc cutters. Int J Rock Mech Min Sci 57:172–186. https://doi.org/10.1016/j.ijrmms.2012.07.031

Rostami J, Chang SH (2017) A closer look at the design of cutterheads for hard rock tunnel-boring machines (TBMs). Engineering 3(6):892–904. https://doi.org/10.1016/j.eng.2017.12.009

Rostami J, Ozdemir L (1993) Computer modeling for cutterhead design and layout of mechanical excavators. In: Proceedings of 12th annual technical meeting of the institute of shaft drilling technology (ISDT), Las Vegas, May 3–5, 1993

Rostami J (1997) Development of a force estimation model for rock fragmentation with disc cutters through theoretical modeling and physical measurement of crushed zone pressure, vol 38. Colorado School of Mines, Golden, pp 56–64

Samadi H, Hassanpour J (2025) Application of the stacking-ensemble machine learning algorithm to investigate the rock mass cuttability index in different lithological units using tunnel boring machine performance data. Rock Mech Rock 58(7):8445–8475. https://doi.org/10.1007/s00603-025-04501-z

Samadi H, Karimi H, Khamehchiyan M, Hassanpour J (2024a) Prediction of engineering characteristics of rock masses using actual TBM performance data with supervised and unsupervised learning algorithms (a case study in strong to very strong igneous and pyroclastic rocks). Rock Mech Rock Eng 57(9):7223–7252. https://doi.org/10.1007/s00603-024-03903-9

Samadi H, Hassanpour J, Rostami J (2024b) Assessment of the relationship between rock mass classification systems using hybrid XGBoost algorithms: insights from the mechanized tunnelling project database. Rock Mech Rock Eng 603:4227. https://doi.org/10.1007/s00603-024-04227-4

Sanio HP (1985) Prediction of the performance of disc cutters in anisotropic rock. Int J Rock Mech Min Sci 22:153–161. https://doi.org/10.1016/0148-9062(85)93229-2

Sato K, Gong F, Itakura K (1991) Prediction of disc cutter performance using a circular rock cutting ring. In: Proceedings of the international symposium on mine mechanization and automation, Golden (Colorado), Colorado School of Mines

Xia Y, Guo B, Tan Q, Zhang X, Lan H, Ji Z (2018) Comparisons between experimental and semi-theoretical cutting forces of CCS disc cutters. Rock Mech Rock Eng 51:1583–1597. https://doi.org/10.1007/s00603-018-1400-x

Xie WQ, Zhang XP, Tang SH, Liu XL, Li XF, Zhang Q, Yan FY, Xu C, Liu QS (2023) Fast perception of rock mass strength and integrity in TBM tunnelling using in-situ penetration test. Tunnel Undergr Space Technol 141:105358. https://doi.org/10.1016/j.tust.2023.105358

The NTNU Prediction Model for Hard Rock Tunnel Boring: Performance Predictions and Cutter Life Assessments

Javier Macias

1 Introduction

This chapter presents the most recent edition of the NTNU prediction model for hard rock TBMs. The model has been reviewed and updated by means of several ongoing and recently completed TBM projects, combined with revisions of the model in response to developments in TBM technology. The review has led to an extension of the range of application of the model in the light of the influence of recently defined parameters (Macias 2016).

The chapter includes the complete methodology and necessary data required for the prediction of performance, time consumption and cutter wear for hard rock tunnel boring projects. A brief presentation is given of the relevant rock mass properties and TBM parameters to determine the main factors influencing predictions of penetration rate and cutter wear.

The chapter presents the following data and models:

- The rock mass and machine parameters required as input to the estimation models
- Net penetration rate model
- Cutter life model
- Estimations of machine utilisation and weekly advance rate

Accurate prediction of TBM performance is essential for reliable scheduling, cost estimation, and risk management. Special care should thus be taken when predicting TBM performance and excavation costs in projects involving extensive sections of hard-to-very hard and abrasive rock masses exhibiting low levels of fracturing, where penetration rate and cutter consumption become the dominant issues. An appropriate geological pre-investigation of the rock mass conditions is essential for successful

J. Macias (✉)
JMConsulting-Rock Engineering AS, Oslo, Norway
e-mail: javier.macias@jmc-rockeng.com

J. Hassanpour and J. Rostami (eds.), *Performance of Tunnel Boring Machines (TBM) in Rock*, Springer Series in Geomechanics and Geoengineering,
https://doi.org/10.1007/978-3-032-20424-0_3

accomplishment of tunnelling projects (Macias et al. 2025). The consequences of inaccurate predictions extend beyond schedule deviations to financial risks that can jeopardize the viability of entire projects.

The NTNU prediction model for hard rock tunnel boring has a long history, with many professionals from different areas of the hard rock TBM world involved. It is materially impossible to acknowledge every individual who has contributed, to a greater or lesser extent, to the development of hard rock tunnel boring knowledge and the NTNU model over the past 50 years. Numerous people have played important roles, including doctoral and master's students, researchers, professors, consultants, laboratory technicians, engineers from project owners and contractors, TBM manufacturers, cutter tool suppliers, industrial partners, tunnel workers, and other personnel involved in the tunnelling industry. Their combined expertise and dedication, especially at the Norwegian University of Science and Technology (NTNU), form the foundation of this work.

I would like to express my special gratitude to Professor Amund Bruland, my mentor, who generously shared his extensive knowledge of hard rock tunnel boring and guided me through the latest developments of the NTNU model. His insight and guidance have been invaluable.

This acknowledgment is offered with appreciation to all who have contributed—past and present—to the advancement of hard rock tunnel boring, and, in particular, to the NTNU prediction model.

2 The NTNU Prediction Model

2.1 General Remarks

The philosophy of the NTNU model is to achieve reliable predictions by combining relevant rock properties and machine parameters. The model involves several steps in the NTNU prediction model for hard rock TBMs to estimate time and costs involved in tunnel excavation using factors such as net penetration rate, cutter life as well as advance rate and excavation costs. The model has been the subject of progressive development since the first version was used in 1976 by the NTNU (formerly NTH). Table 1 shows the successive editions of the NTNU prediction model to date.

Estimation models for hard rock tunnel boring should be applied with care. In combination with other models published in the Project Report Series issued by the NTNU Department of Civil and Transport Engineering, the Hard Rock Tunnel Boring reports provide reliable and practical tools for the following processes:

- Estimating net penetration rate and cutter wear
- Estimating time consumption and excavation costs, including risk
- Assessing risk linked to variation in rock mass boreability and machine parameters
- Establishing and managing contract price regulation
- Verifying machine performance

Table 1 History of the NTNU prediction model for hard rock TBMs

Edition	Year	References
1st edition	1976	NTH (1976)
2nd edition	1979 (published in 1981)	NTH (1981)
3rd edition	1983	NTH (1983)
4th edition	1988	NTH (1988)
5th edition	1994	NTH (1994)
6th edition	2000	Bruland (2000)
7th edition	2016	Macias (2016)

- Verifying variation in geological conditions

The estimation models are designed for use during several stages of a given project:

- Preliminary and feasibility studies
- Evaluation of the excavation method
- Project design and optimisation
- During tendering and contract processes
- Documentation during construction
- Possible disputes and claims

The models were developed based on on-site project studies, combined with analysis and statistics from tunnelling projects carried out in both Norway and abroad, involving more than 40 sites and more than 300 km of tunnel. The data have been systematised and normalised, and the results are regarded as representative for well-organised tunnelling projects (Macias 2016).

2.2 *Input Parameters*

The values of net penetration rate and cutter life depend on rock properties and machine parameters. Rock properties consist of intact rock and rock mass parameters. These are combined to generate a single rock mass boreability parameter, called the equivalent fracturing factor (k_{ekv}), while the machine parameters are combined into a single parameter, the equivalent thrust (M_{ekv}).

Cutter life, measured in hours, is equivalent to the cutter life in rolled distance at a given cutterhead velocity, measured in rpm. Cutter life in hours is combined with the penetration rate (m/h) and the TBM diameter to calculate cutter life in terms of tunnel metres excavated per cutter (m/cutter) and solid cubic metres excavated per cutter (sm^3/cutter). Table 2 lists the machine and rock parameters that influence net penetration rate.

Cutter wear depends on the rock properties and machine parameters listed in Table 3.

Table 2 Machine and rock parameters that influence net penetration rate

Rock properties		Machine parameters
Intact rock	Rock mass	
Drilling rate index, DRI	Rock mass fracturing factor (k_s)	TBM diameter
		Cutter diameter
Porosity		Number of cutters
		Gross average cutter thrust
		Average cutter spacing
		Cutterhead rpm

Table 3 Machine and rock parameters that influence cutter wear

Rock properties	Machine parameters
Cutter life index, CLI	TBM diameter
	Cutter diameter
Content of abrasive minerals	Number of cutters
	Cutterhead rpm
	Gross average cutter thrust

The gross advance rate (e.g. meter per week) is estimated based on three input parameters:

- Net penetration rate
- Machine utilisation
- Number of working hours in each period (e.g. a week).

The machine utilisation is in turn based on time consumption for the various operations involved in the tunnel excavation process.

2.2.1 Rock Parameters

This section provides a brief description of the geological parameters required as input to the prediction model. It also provides geological information which may be of help during a project's early planning phase. Estimates of time consumption and costs require an engineering geological pre-investigation, adapted to hard rock TBM tunnelling.

Rock Drillability

The rock drillability is evaluated based on the Drilling Rate Index (DRI) and the Cutter Life Index (CLI). The DRI is assessed based on two laboratory tests, the Brittleness Value S_{20}' and 'Sievers' J-value'. The DRI is an indirect measure of the work required to break the rock and is thus an effective measure of the rock-breaking process under a cutter.

The CLI evaluates the ability of the rock to produce cutter wear and is assessed on the basic of the 'Sievers' J-value' and the 'Abrasion Value Steel' (AVS). The CLI is an expression of cutter disc ring life in terms of boring hours. Recorded variations in the DRI and CLI for some rock types are shown in Figs. 1 and 2. The influence of rock porosity is discussed in Sect. 3.2.

The following Table 4 shows the classification for drillability indices (DRI and CLI).

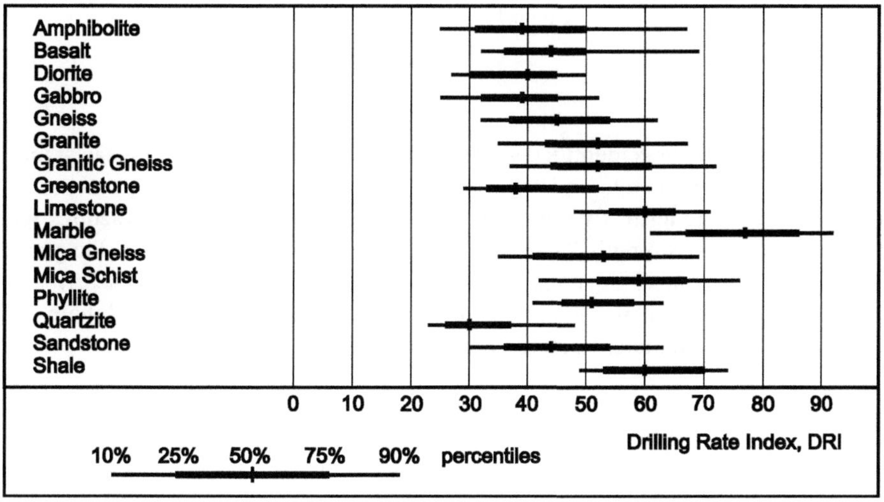

Fig. 1 Recorded Drilling Rate Index values for some rock types (Bruland 2000)

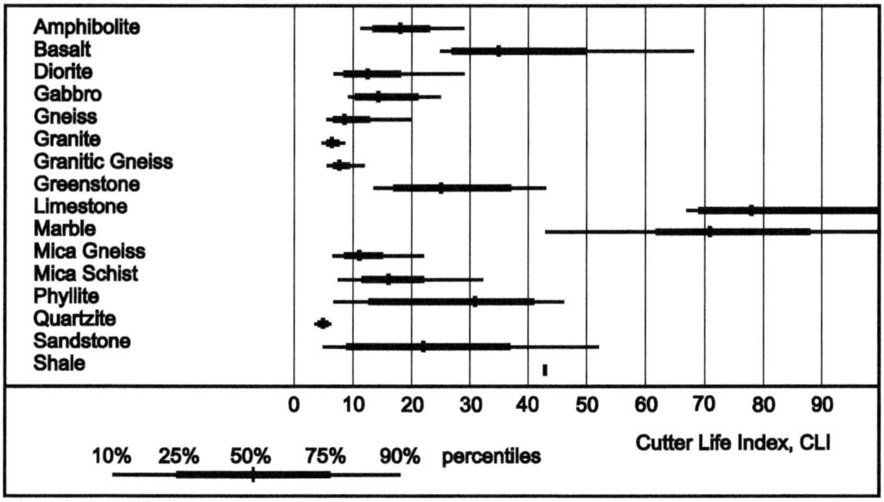

Fig. 2 Recorded Cutter Life Index values for some rock types (Bruland 2000)

Table 4 Category intervals for drillability indices (Macias 2016)

Category	DRI	CLI
Extremely low	<26	<4.8
Very low	26–32	4.8–5.8
Low	33–41	5.9–8.2
Medium	42–52	8.3–15.6
High	53–62	15.7–35
Very high	63–73	36–72
Extremely high	>73	>72

Degree of Fracturing

Rock mass fracturing is identified as the geological factor that has the greatest influence on net penetration rate, and thus also on project costs during hard rock tunnelling. In this context, the term 'fracturing' refers to planes of weakness or fractures with limited or no shear strength. The less the distance between the fractures, the greater the influence on penetration rate. The rock mass fracturing parameter is described in terms of degree of fracturing and the angle between the tunnel axis and the planes of weakness in systematic fractured rock mass.

Fractures

The term 'fractures' incorporates all continuous and non-continuous fractures that are capable of representing relatively weak areas of rock under a cutter disc, and which can thus influence performances. Fractures may be open, such as joints in granite, or filled with clay or low shear strength minerals, such as calcite, chlorite or similar minerals. In metamorphic rocks, such as schists and gneisses, schistose fabrics can also represent planes of weakness exhibiting low shear strength.

Non-fractured Rock Mass (Class 0): This class includes massive rock without fractures, such as may be the case for intrusive dykes, sills or batholiths, etc. Rock masses with joints that are filled or healed with minerals exhibiting high shear strength properties, such as quartz or epidote, may also approach Class 0.

Marked Single Joints (MSJ): MSJ includes highly distinctive discontinuities. These may be entirely open, may conduct water or may be filled with clay. Minor faults filled with gouge may be classified as MSJ.

Single joints are recorded individually. They are marked discontinuities in the rock mass. As the name indicates, they will appear as a singular phenomenon at the tunnel level, even if they belong to a systematically occurring joint set on a larger scale. Marked Single Joints may result in very high net penetration rates locally. Due to the risk of stability problems and damage to the cutters and the cutterhead, one should evaluate the effect carefully and consider possible delays if the Marked Single Joints may result in mixed face conditions, heavy rock support, water inflows, etc. For detailed descriptions, the reader is referred to Bruland (2000) and Macias (2016). Figure 3 presents a proposed length-based classification of the main types of

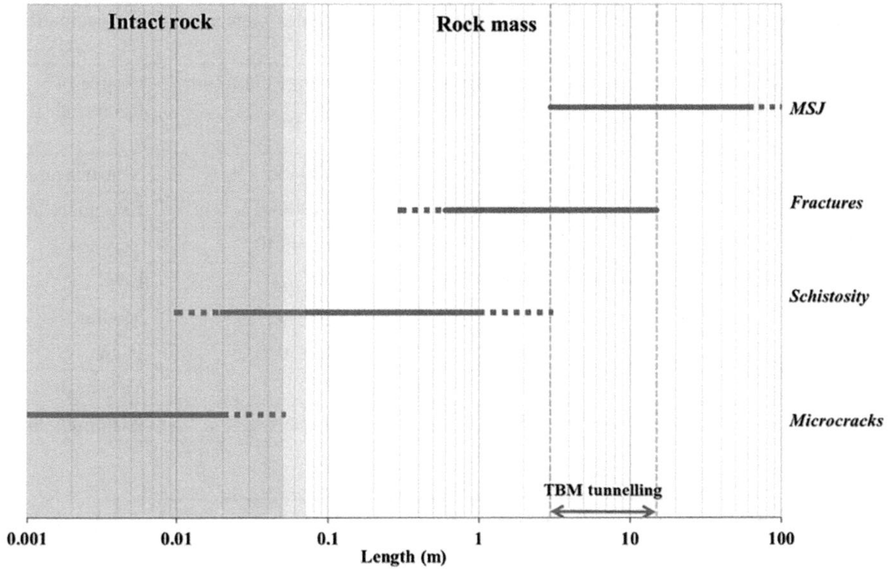

Fig. 3 The main types of intact rock and rock mass discontinuities encountered during back-mapping in hard rock TBM tunnelling. Types are categorised according to length based on the NTNU methodology. The shaded area indicates where data have been derived from laboratory tests carried out on intact rock

discontinuities from an engineering geological perspective according to the NTNU back-mapping methodology.

For practical mapping purposes in connection with TBM tunnelling projects, the degree of fracturing in a systematically fractured rock mass is divided into classes, shown in Table 5. The definition of classes also includes the average distance between the plane of weakness. Figure 4 shows recorded degrees of fracturing for some rock types. Geological mapping in connection with tunnelling projects is limited by tunnel size. Fractures also vary in length.

2.2.2 Machine Parameters

Several machine parameters have been selected to estimate penetration rate and cutter life. Machine parameters are subdivided into two groups; 'machine specifications', such as machine and cutter diameters, the number of cutters or cutterhead power, and 'machine operation' parameters such as gross cutter thrust and applied rpm.

Machine specifications

During the early stages of project planning, some parameters must be assumed based on general machine specifications.

Table 5 Fracture class as defined by the spacing between planes of weakness in systematically fractured rock mass

Fissure class (St) (Bruland 2000)	Fracture class (Sf)	Average spacing between fractures a_f (cm)	Range class (cm)	Degree of fracturing
0	0	∞	480–∞	Non-fractured
0	1	320	240–480	Extremely low
0–I	2	160	120–240	Very low
I	3	80	60–120	Low
I	4	40	30–60	Medium
II	5	20	15–30	High
III	6	10	7.5–15	Very high
IV	7	5	4–7.5	Extremely high

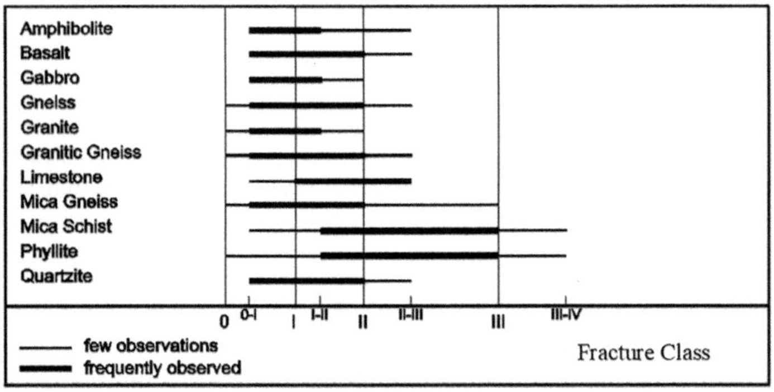

Fig. 4 Recorded degrees of fracturing for some rock types (Bruland 2000)

TBM Diameter

TBMs developed for tunnelling in hard rock vary in diameter from 1.2 to 15 m. Our database indicates that the estimation models currently in use are applicable for TBM diameters ranging from 3 to 12 m.

Cutter Diameter

Currently, cutter diameters of 483 mm (19 in) are commonly frequently used in hard rock tunnel projects, although diameters of 432 mm (17 in) are standard for smaller TBMs. Cutter discs with diameters of 508 mm (20 in) have recently been introduced, for use on larger diameter TBMs, involving the use of 483 mm cutters with enlarged cutter rings.

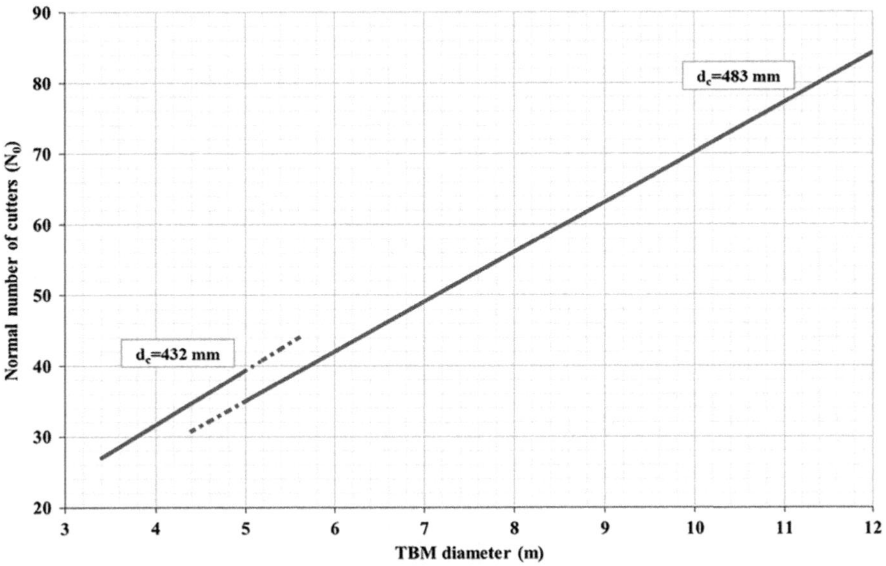

Fig. 5 Standard number of cutters on a cutterhead (N_0) as a function of TBM cutterhead diameter

Standard Number of Disc Cutters on the Cutterhead

Figure 5 shows the proposed number of cutters on a cutterhead as a function of TBM and cutter diameter. For smaller TBM cutterhead diameters, the space available on the cutterhead is limited and the number of cutters indicated is the higher end.

Installed Cutterhead Power

Installed cutterhead power is not considered to be a limiting factor for modern TBMs, assuming that they are adequately designed. However, since insufficient cutterhead power may result in limitations, Fig. 6 can be used as a guideline on the recommended installed power for cutterhead rotation as a function of TBM and cutter diameter. The installed power indicated should provide sufficient torque to produce a penetration rate of approximately 14 mm/rev at full cutter load in many cases.

Machine Operation Parameters

Cutter Thrust

Figure 7 is a plot showing recommended applied gross average thrust per cutter disc as a function of cutter diameter and TBM diameter. The term 'gross average thrust' refers to the thrust that can be applied on the cutterhead that supports cutter loads over an extended period of time. It does not incorporate peak loads that occur over short time intervals. At present, the quality of the steel in the cutter ring and bearing capacity of the hub remains the limiting factor on cutter thrust. Estimations of penetration rates and related factors require that the thrust capacity of the cutter rings

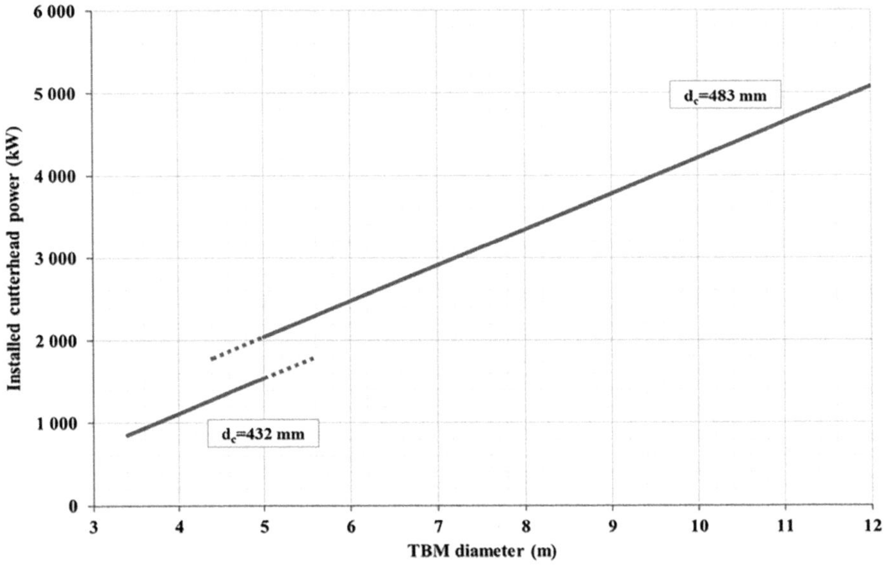

Fig. 6 Recommended installed cutterhead power (P_{tbm}) as a function of TBM cutterhead diameter

be considered, together with the capacity of the cutter bearings and the cutterhead main bearing.

The upper limit applies to boring operations in a non-fractured, hard rock mass, while the lower limit applies to moderately to very fracture rock masses. Tunnelling operations along curved tunnels or on downward slopes may require lower thrust values (e.g. to avoid overloading gage cutters).

Cutterhead Velocity

Cutterhead velocity (rpm) is inversely proportional to cutterhead diameter. This is because the rolling velocity of the peripheral cutter must be limited. Figure 8 shows maximum and recommended cutterhead rpm as a function of TBM diameter and cutter size. The figure shows a range of available applied rpm. A low optimal velocity in indicated in situations involving low degrees of rock mass fracturing and/or low thrust.

3 The Net Penetration Rate Model

Net penetration rate is defined as the length of tunnel bored (in metres per hour) during cutterhead rotation against the rock face, or the actual boring time. Figure 9 is a flowchart showing the components of performance prediction in the latest version of the NTNU model.

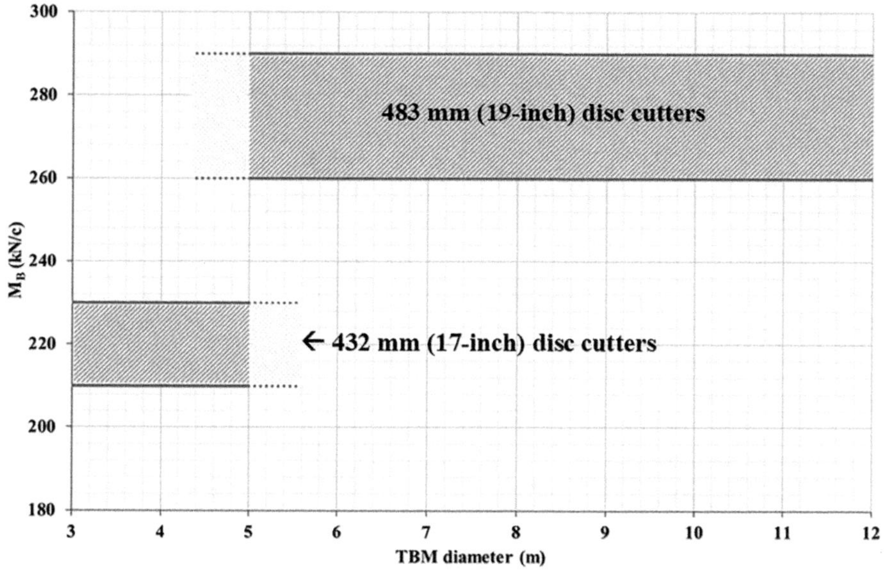

Fig. 7 Recommended gross average thrust per cutter disc. The upper limit applies to boring operations in a non-fractured rock mass, while the lower limit applies to moderately to very fracture rock masses

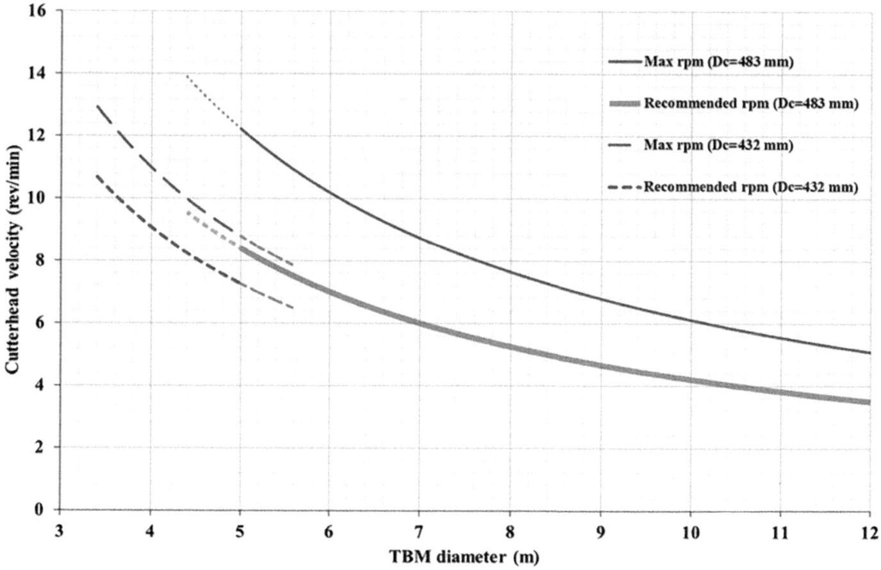

Fig. 8 Maximum and recommended cutterhead velocities (rpm) as a function of TBM and cutter diameter

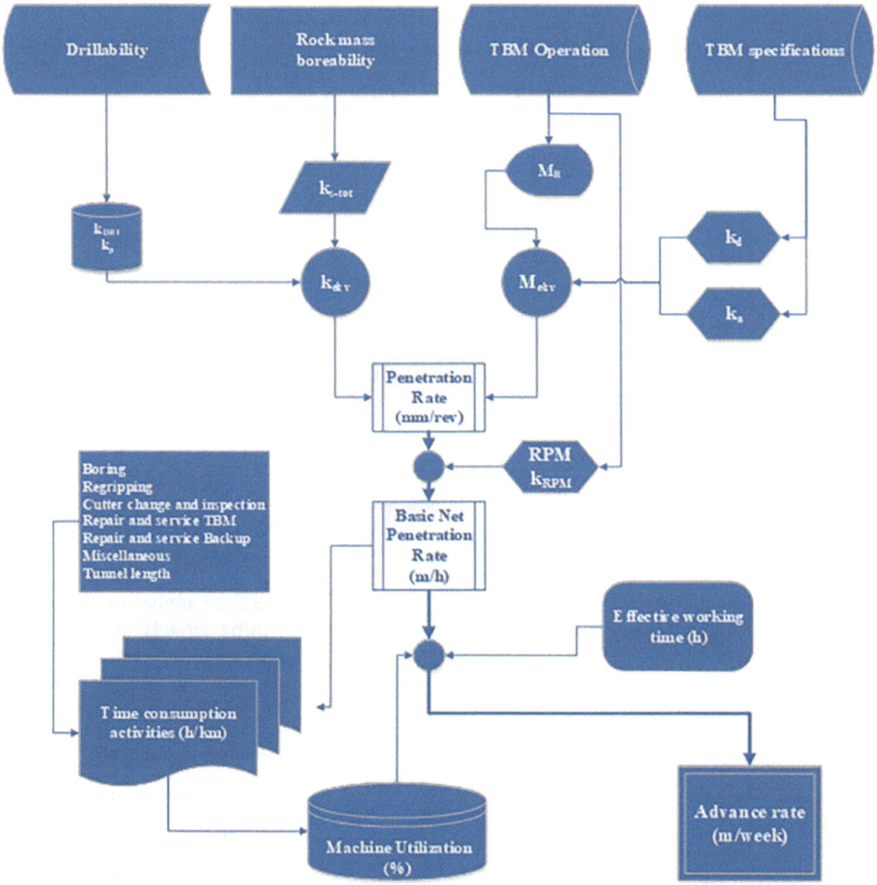

Fig. 9 Flowchart for performance prediction by the latest version of the NTNU model

The penetration rate estimation model is based on normalised penetration curves obtained from penetration tests under different rock mass conditions. The best fit is found to be a power function. The Eq. 1 is the formula for expressing basic penetration rate:

$$i_0 = \left(\frac{M_{ekv}}{M_1}\right)^b \quad (\text{mm/rev}) \tag{1}$$

where:

i_0 = basic penetration rate (mm/rev).

M_{ekv} = equivalent cutter thrust (kN/cutter).

M_1 = critical cutter thrust (kN/cutter) required to achieve 1 mm/rev.

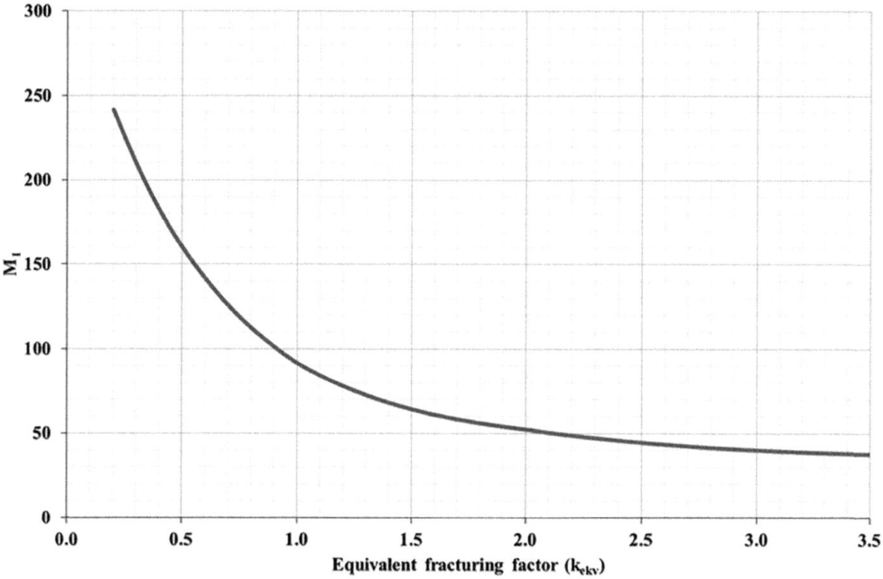

Fig. 10 Critical thrust (M_1) as a function of the equivalent fracturing factor

b = penetration coefficient.

Figures 10 and 11 show the relationships between the equivalent fracturing factor (k_{ekv}) and critical thrust (M_1), and the penetration coefficient (b), respectively.

3.1 *Rock Mass Fracturing*

The rock mass fracturing parameter is expressed by the fracturing factor (k_s), which is dependent on the degree of fracturing and the angle between the tunnel axis and the planes of weakness (α). For practical mapping purposes in connection with TBM tunnelling projects, the degree of fracturing in a systematically fractured rock mass is divided into classes, Table 6.

The orientation of the fractures is determined from measurements of strike and dip and expressed in the following equation:

$$\alpha = \arcsin \cdot \left(\sin \alpha_f \cdot \sin(\alpha_t - \alpha_s) \right) \quad (^\circ) \tag{2}$$

where:

α_s = strike angle of the planes of weakness.

α_f = dip angle of the planes of weakness.

α_t = azimuth of the tunnel axis.

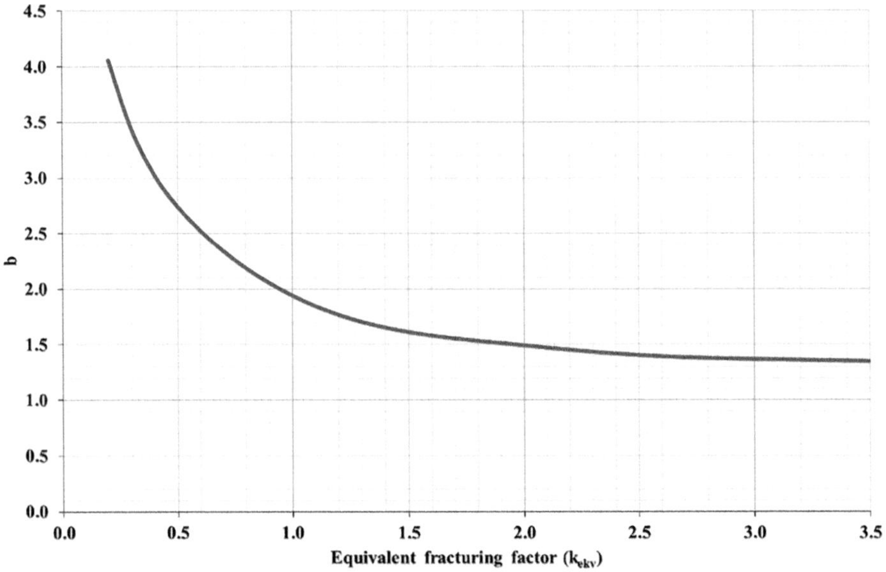

Fig. 11 Penetration coefficient (b) as a function of the equivalent fracturing factor

Table 6 Fracture class as defined by the spacing between planes of weakness

Fracture class (Sf)	Average spacing between fractures a_f (cm)	Range class (cm)	Degree of fracturing
0	∞	480–∞	Non-fractured
1	320	240–480	Extremely low
2	160	120–240	Very low
3	80	60–120	Low
4	40	30–60	Medium
5	20	15–30	High
6	10	7.5–15	Very high
7	5	4–7.5	Extremely high

The fracturing factor k_s is shown in Fig. 12 as a function of Fracture Class and the angle between the tunnel axis and the fractures. Figure 13 shows the fracturing factor for detailed calculations of rock masses with low degrees of fracturing. For more than one set of fractures, the total fracturing factor is as follows:

$$k_{s-tot} = \sum_{i=1}^{n} k_{si} - (n-1) \cdot 0.36 \tag{3}$$

where:

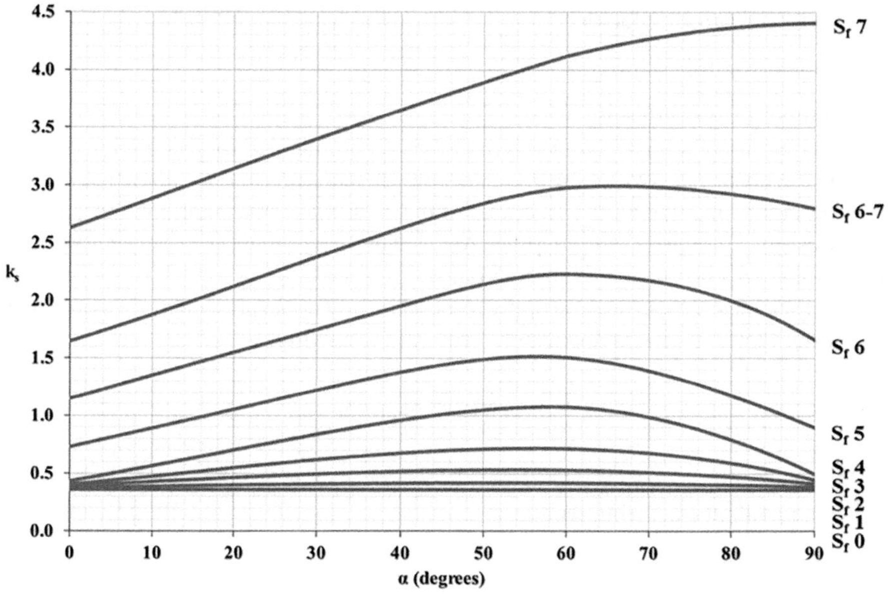

Fig. 12 Rock mass fracturing factor (k_S) as a function of the angle between the tunnel axis and the fractures

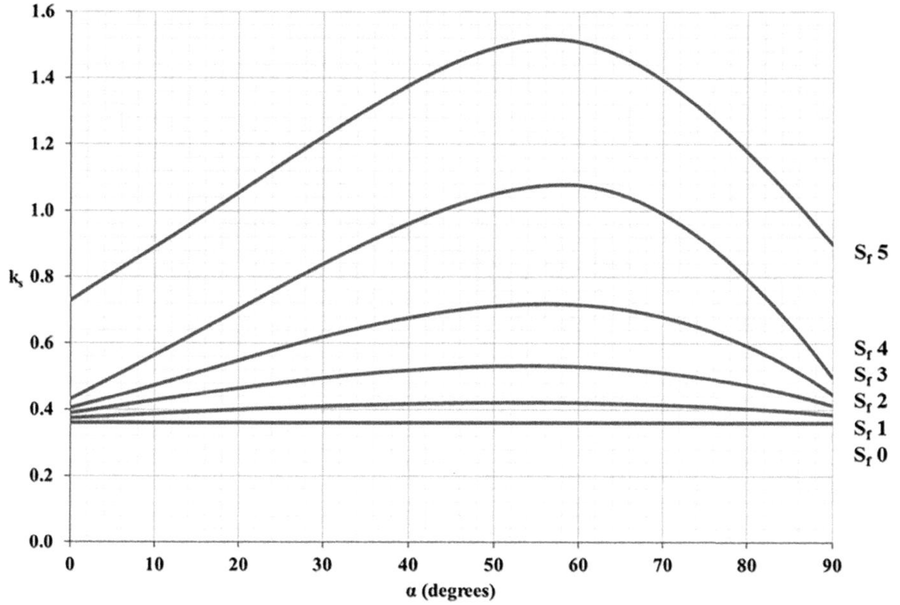

Fig. 13 Rock mass fracturing factor (k_S) as a function of the angle between the tunnel axis and the fractures (for detailed calculations of rock masses with low degrees of fracturing)

$k_{s\text{-}tot}$ = total fracturing factor.

k_{si} = fracturing factor for set no. i.

n = number of fracture sets.

A maximum of three fracture sets (n = 3) is recommended for calculation of $k_{s\text{-}tot}$, possibly with one or two main sets and one set including the random fractures. It is important to include all the fractures in the rock mass interpretation, i.e. random fractures (except marked single joints, MSJ). The $k_{s\text{-}tot}$ should not exceed a value of 4 since this will be outside the model range.

The rock mass properties that apply to TBM boring are expressed by the equivalent fracturing factor, which is determined from the following:

$$k_{ekv} = k_{s-tot} \cdot k_{DRI} \cdot k_{por} \tag{4}$$

where:

k_{ekv} = equivalent fracturing factor.

$k_{s\text{-}tot}$ = total fracturing factor.

k_{DRI} = correction factor for the DRI of the rock as shown in Fig. 14.

k_{por} = correction factor for the porosity of the rock.

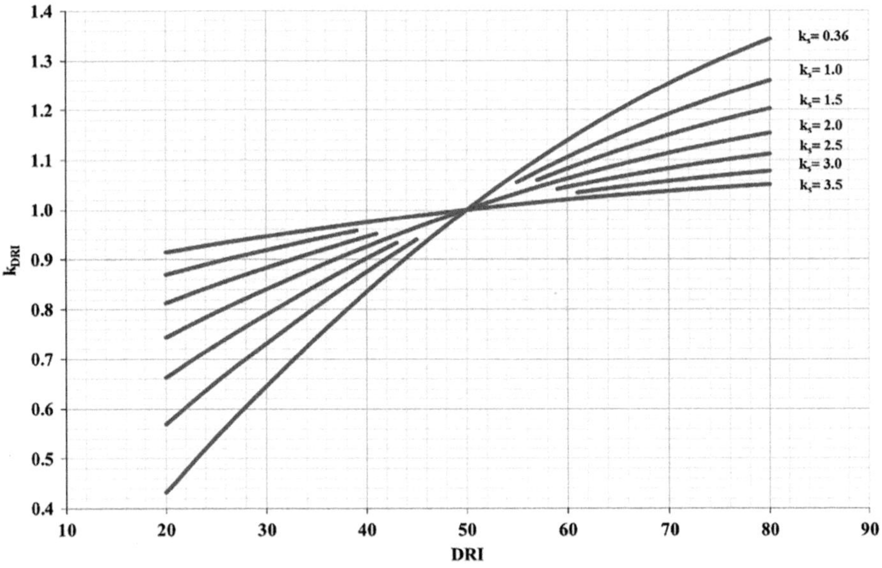

Fig. 14 Correction factor for DRI values ≠ 50

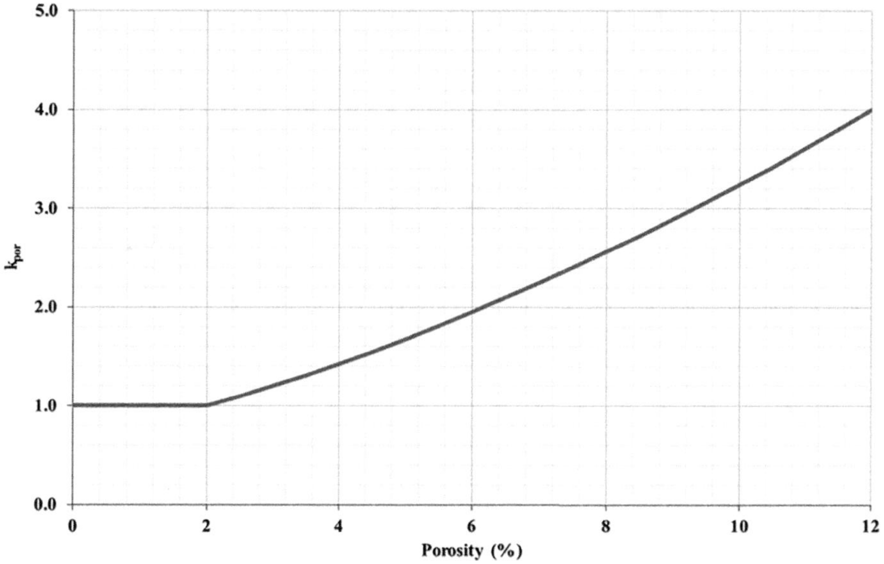

Fig. 15 Correction factor for porosity of the rock (k_{por})

3.2 *Rock Porosity*

The model requires that porosity be measured for rock types with porosities higher than approximately 2%. The influence of porosity on the DRI is negligible for porosities less than about 10–12%. For this reason, porosity must be incorporated in the penetration rate model as an independent parameter. TBM projects in North Atlantic basalts in the Faeroe Islands have shown that porosity has a significant influence on net penetration rate (Bruland 2000). The experience data are limited to approximately 15% porosity. A rough estimate of the influence of porosity on penetration rate (Correction factor for porosity of the rock, k_{por}) is shown in Fig. 15.

3.3 *Basic Penetration*

Basic penetration (i_0) is a function of the equivalent thrust and equivalent fracturing factor parameters. Equivalent thrust is given by Eq. 5.

$$M_{ekv} = M_B \cdot k_d \cdot k_a \quad \text{(kN/c)} \tag{5}$$

where:

M_B = applied cutter thrust (kN/cutter).

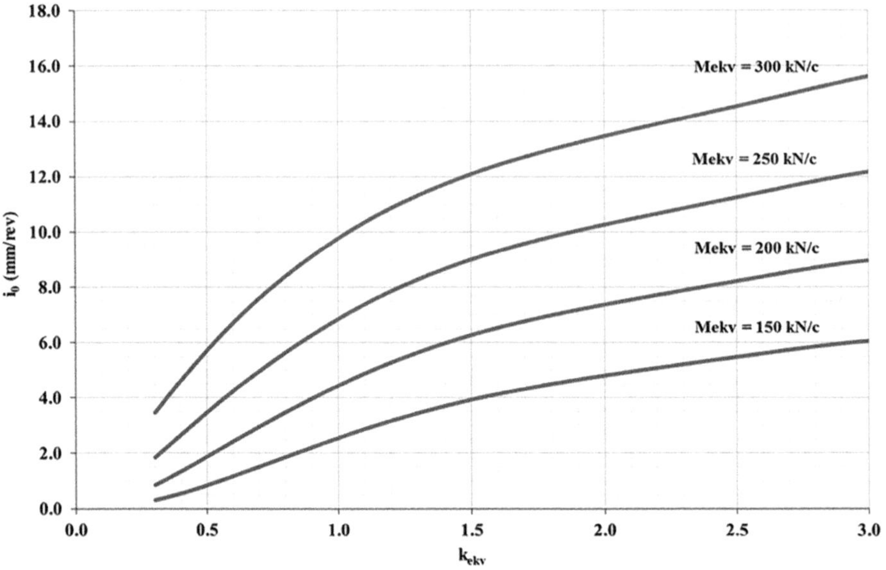

Fig. 16 Basic net penetration rate $d_c = 483$ mm, $a_c = 70$ mm and recommended cutterhead rpm

k_d = correction factor for cutter diameter (Fig. 18).

k_a = correction factor for average cutter spacing (Fig. 19).

Figure 16 shows the basic penetration rate (i_0) as a function of the equivalent thrust and equivalent fracturing factor parameters when cutter diameter (d_c) is 483 mm, average cutter spacing (a_c) is 70 mm and applied recommended rpm.

Figure 17 illustrates the expected basic penetration rate (mm/rev) as a function of the rock mass equivalent fracturing (k_{ekv}) values for standard machine parameters (i.e. 19-in cutters, recommended number of cutters, cutter average spacing). The expected penetration rate is given according to recommended gross cutter thrust (i.e., increased when there is low rock boreability while it is substantially decreased when boreability increases).

3.4 Basic Net Penetration Rate

The basic net penetration rate (I_0) is a function of the basic penetration rate and the cutterhead rpm (Eq. 6).

$$I_0 = i_0 \cdot rpm \cdot \left(\frac{60}{1000}\right) \cdot k_{rpm} \quad (\text{m/h}) \tag{6}$$

where:

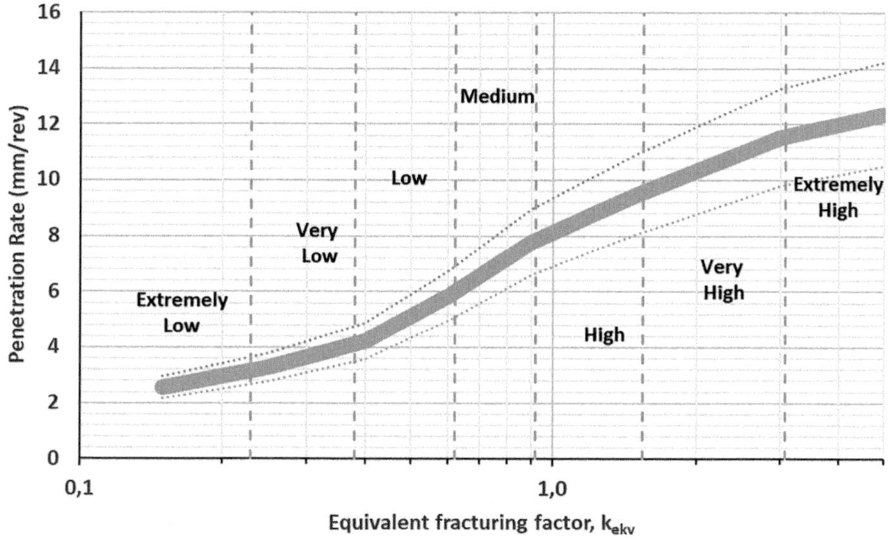

Fig. 17 Expected penetration rate (mm/rev) as a function of the rock mass equivalent fracturing (k_{ekv}) values for 19-in cutters and standard machine parameters. The categorization refers to rock boreability

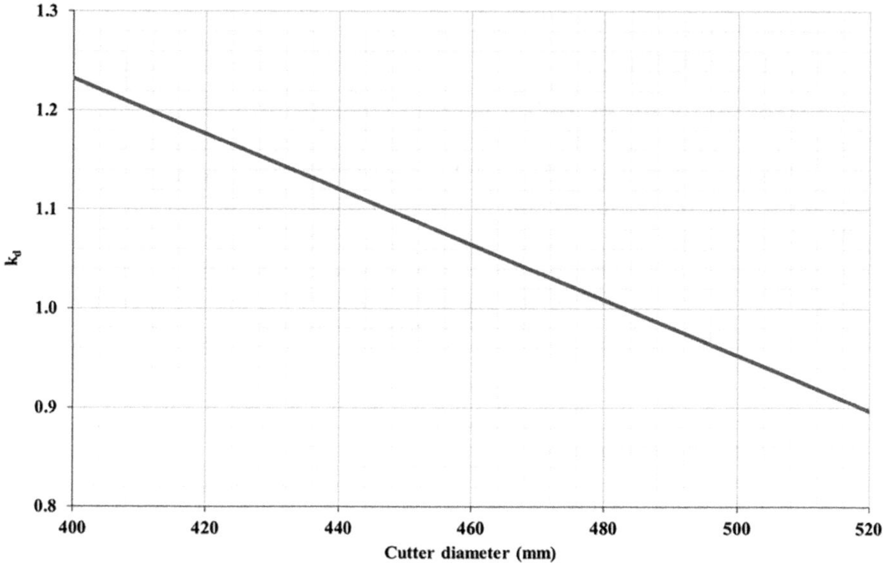

Fig. 18 Correction factor for cutter diameter where $d_c \neq 483$ mm. (k_d)

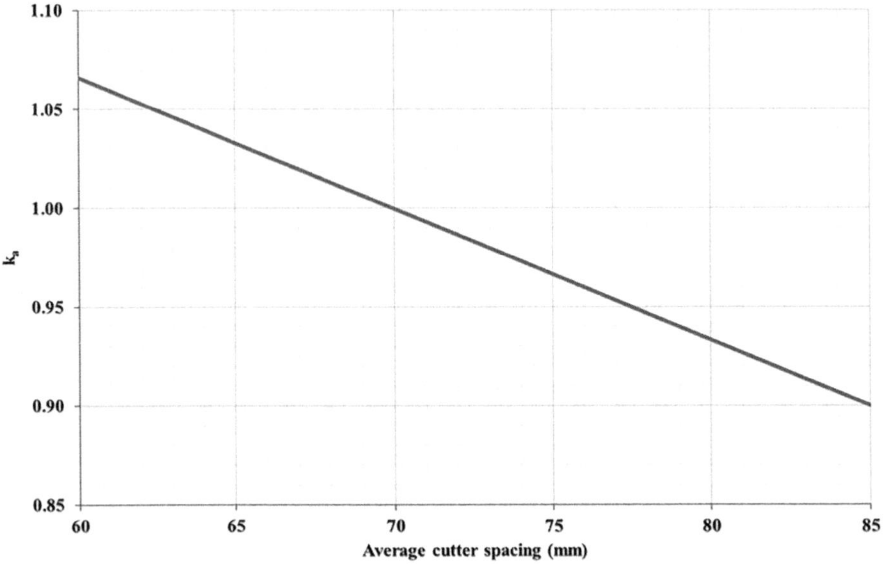

Fig. 19 Correction factor for average cutter spacing where $a_c \neq 70$ mm. (k_a)

i_0 = basic net penetration rate (mm/rev).

rpm = applied cutterhead rpm.

k_{rpm} = correction factor for applied cutterhead rpm (Fig. 20).

The basic net penetration rate is applicable for systematically fractured rock mass without Marked Single Joints (MSJ). Influence of MSJ on the basic penetration rate is included in the following section.

The cutterhead velocity correction factor is based on limited information and must be applied with caution. The optimal cutterhead velocity appears to be influenced by rock drillability, rock mass fracturing and/or thrust level. Low values of and applied cutter thrust indicate a lower optimal cutterhead velocity.

3.5 *Marked Single Joints*

An interpretation of the influence of Marked Single Joints (MSJ) is based on a relatively small number of observations. MSJ are marked discontinuities in the rock mass which may be completely open, lead water or be filled with clay (Bruland 2000).

The theoretical penetration addition is shown in Fig. 21, and the normalised correction factor in Fig. 22.

In Eq. 7, the length (l_{esp}) of tunnel that is influenced by Marked Single Joints is estimated from the tunnel diameter (d_{tbm}), the angle between the Marked Single Joints

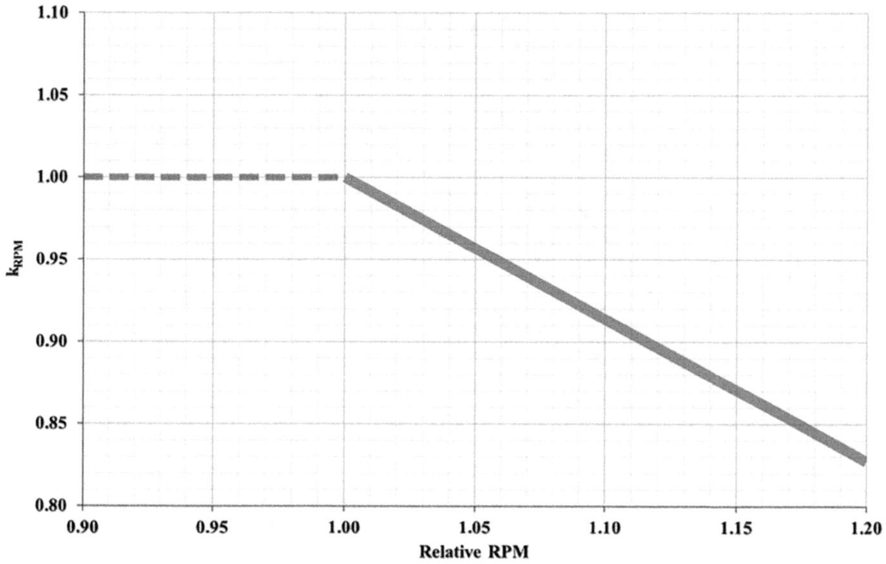

Fig. 20 Correction factor for cutterhead velocity (rpm) illustrating where it differs from the recommended value

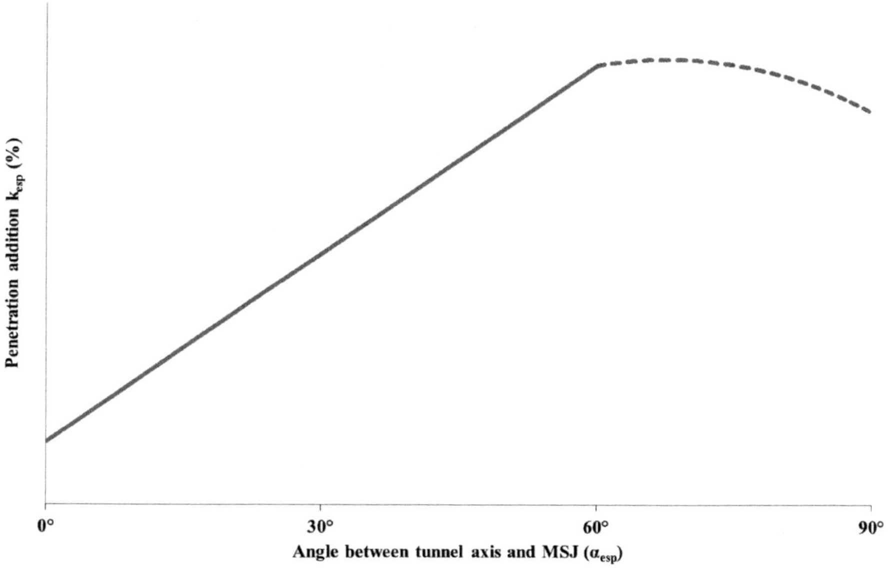

Fig. 21 Theoretical averaged penetration addition for marked single joints

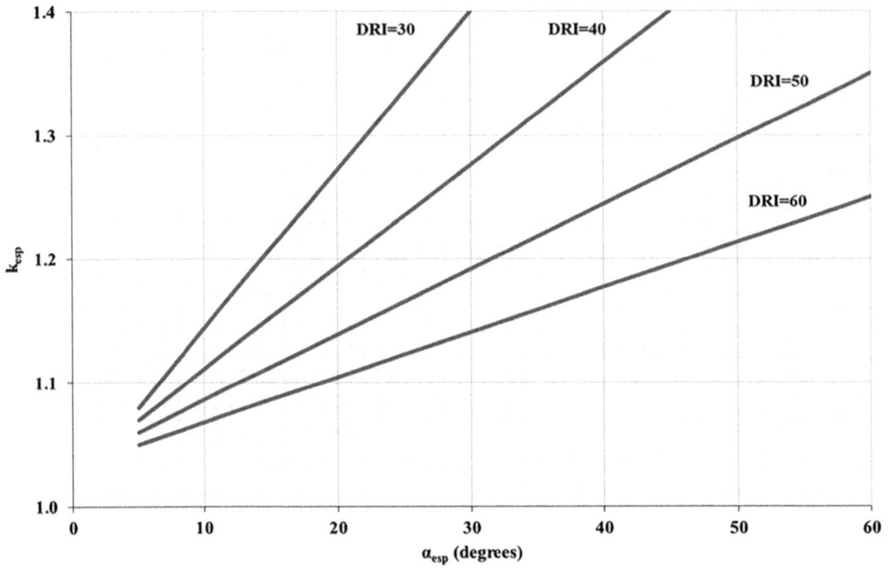

Fig. 22 Correction factor for marked single joints (k_{esp})

and the tunnel axis (α_{esp}), and the number (n) of similar Marked Single Joints occurring in the geological zone in question. The angle (α_{esp}) can be found by manipulating Eq. 7.

$$l_{esp} = n \cdot d_{tbm} / \tan \alpha_{esp} \quad \text{(m)} \tag{7}$$

The net penetration rate achieved when boring through zones characterised by Marked Single Joints (I_{esp}) is found by:

$$I_{esp} = I_0 \cdot k_{esp} \quad \text{(m/h)} \tag{8}$$

where:

k_{esp} = correction factor for Marked Single Joints (Fig. 22).

The correction factor (k_{esp}) is dependent of the angle between the Marked Single Joints and the tunnel axis (α_{esp}) and the DRI of the rock. k_{esp} should not be assigned a value >1.4. At such high penetration increases, the vibration level of the cutterhead and the peak forces on the cutters are very high, requiring a reduction in the level of thrust.

The average net penetration rate (I_n) within a rock mass zone of a total length (l_j) is given by:

$$I_n = \frac{l_j}{\frac{l_j - l_{esp}}{I_0} + \frac{l_{esp}}{I_{esp}}} \quad \text{(m/h)} \tag{9}$$

In rock mass with low boreability (e.g. DRI = 30 and a low degree of fracturing) the practical limit of penetration addition will be reached at an angle as low as 20°–30°. When the angle between the tunnel axis and the plane of the Marked Single Joint is 80°–90°, the influence of the joint will most likely be negative due to heavy vibrations and damage of the cutter rings. In any case, a Marked Single Joint at an angle of 80°–90° will have influence over a very short distance and should therefore not be included in the penetration rate estimations.

3.6 Torque Demand

At high net penetration rates, or when boring in fractured rock, it is important to check for sufficient cutterhead power being installed to utilise the estimated thrust. In situations where installed power is too low, the TBM torque is limited and will be unable to rotate the cutterhead to provide the required penetration. In such cases, the thrust must be reduced, and also the torque so that it becomes less than the torque capacity of the cutterhead drive. The required torque is given by:

$$T_n = r_{mc} \cdot d_{tbm} \cdot N_{tbm} \cdot M_B \cdot k_c/2 \quad \text{(kNm)} \tag{10}$$

where:

r_{mc} = the relative position of the average cutter on the cutterhead.

$$r_{mc} = \frac{\frac{\sum_{i=1}^{N_{tbm}} r_i}{N_{tbm}}}{0.5 \cdot d_{tbm}} \tag{11}$$

where:

r_i = radius to position of cutter no. i (m). This factor can be calculated once the cutterhead design is known. Normally, the value of r_{mc} is approximately 0.59.

d_{tbm} = cutterhead diameter (m).

N_{tbm} = number of cutters on the cutterhead.

M_B = gross average cutter thrust (kN/c).

k_c = cutter coefficient (rolling resistance, see Eq. 12).

$$k_c = c_c \cdot \sqrt{i_0} \tag{12}$$

Figure 23 shows the cutter constant (C_c) as a function of cutter diameter. The required torque determines the required level of installed power (P_n) which is given by Eq. 13.

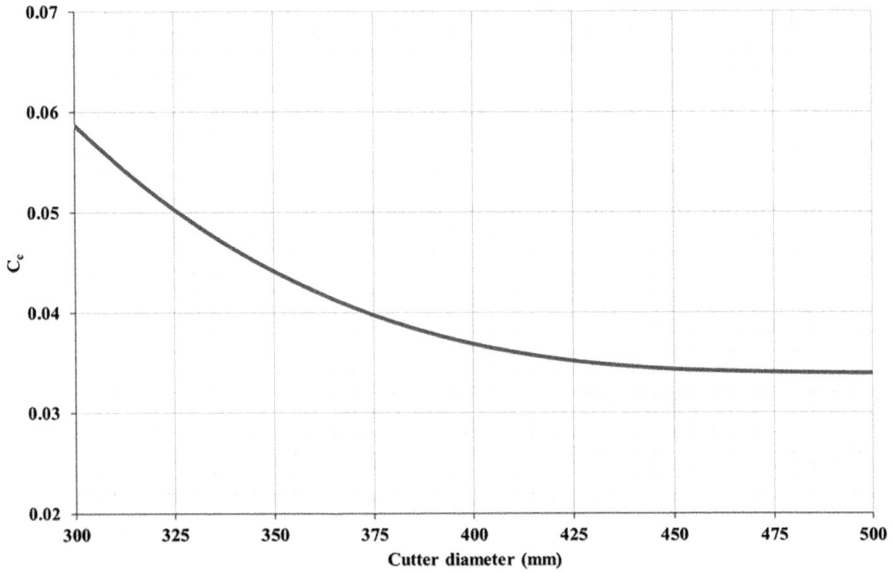

Fig. 23 Plot showing the variation in the cutter constant (C_c) with cutter diameter

$$P_n = \frac{T_n \cdot 2 \cdot \pi \cdot rpm}{60} \quad (\mathrm{kW}) \tag{13}$$

If the installed cutterhead power is less than P_n, the estimation of penetration rate must be recalculated using a lower cutter load value (M_B).

3.7 *Limitations on Net Penetration Rate*

At high penetration rates, or when boring in fractured rock, installed cutterhead power and torque may represent a limitation on net penetration. Such limitations should be considered during the machine design phase.

The capacity of the system (the TBM, back-up for muck removal and the tunnel muck transport systems) may limit the net penetration rate. This is especially true in the case of muck removal systems linked to large diameter machines.

When boring through Marked Single Joints or heavily fractured rocks, it may be necessary to reduce thrust due to too high machine vibration level and very high instantaneous cutter loads are experienced.

Tunnels exhibiting sharp curves and/or steep gradients may also place limits on applied thrust levels and cutterhead velocity, especially in hard rock conditions.

4 The Cutter Life Model

The NTNU prediction model presupposes that a given TBM is operated at a thrust level that results predominantly in abrasive wear of the cutter rings. The number of cutters replaced due to normal abrasive cutter ring wear typically represent more than 70% of total replacements. Cutter rings worn by ring chipping and/or mushrooming should account for <5–10% of the total number of replacements. Cutter replacements due to bearing damage (blockage, leakage, damaged hubs, etc.) should also account for <20–30% of the total number of replaced cutters. The number of abnormal cutter wears due to blocked bearing increases as the required cutter load increases in very hard rock with low DRI/high UCS exhibiting low levels of fracturing.

4.1 *Cutter Ring Life*

Basic cutter ring life, in boring hours, is proportional to the Cutter Life Index (CLI). Figure 24 shows basic cutter ring life (H_o) as a function of the CLI and cutter diameter.

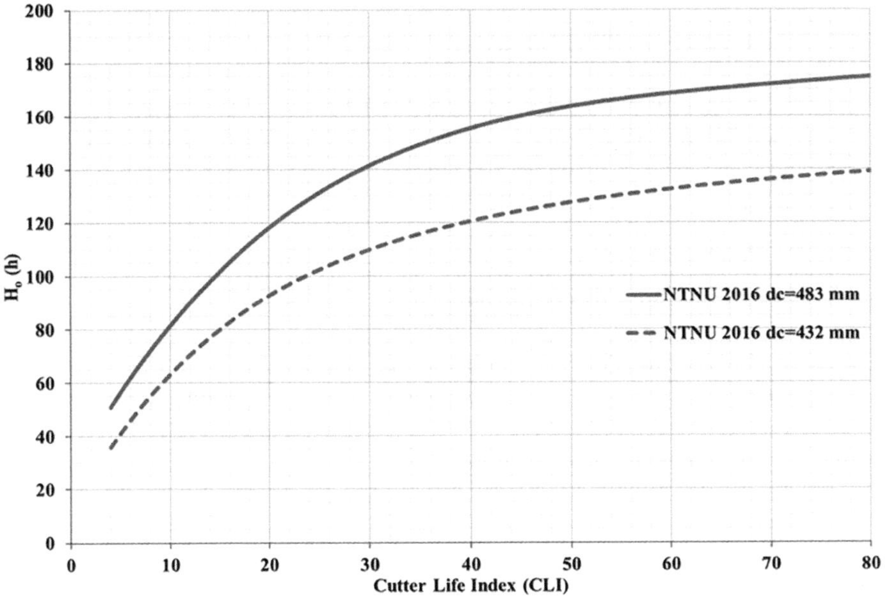

Fig. 24 Basic cutter ring life (H_0) as a function of the CLI and cutter diameter

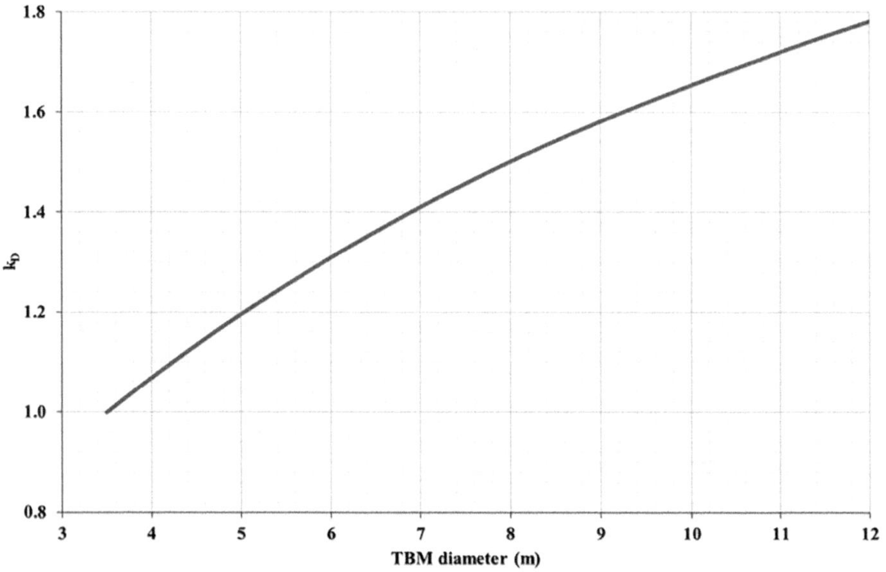

Fig. 25 Correction factor for TBM diameter

4.1.1 Correction for TBM Diameter

Figure 25 illustrates the correction factor for TBM diameter. Centre and gage cutters have shorter lifetimes than face cutters. With increasing TBM diameter, the ratio of centre and gage cutters to face cutters decreases, so the average cutter will have a longer lifetime.

4.1.2 Correction for Cutterhead Velocity

Cutter ring life is inversely proportional to cutterhead velocity (rpm) based on the assumption that cutter ring life, for a given set of rock mass conditions, is proportional to rolling distance but independent of rolling velocity. The correction factor applied under varying cutterhead RPM is shown in Eq. 14.

$$k_{rpm} = \frac{42/d_{tbm}}{rpm} \tag{14}$$

where:

d_{tbm} = TBM cutterhead diameter (m).

rpm = cutterhead velocity.

4.1.3 Correction for the Standard Number of Cutters

In cases where the actual number of cutters on a cutterhead differs from the model, the average cutter life parameter will change. The correction factor applied for the actual number of cutters used is:

$$k_N = \frac{N_{tbm}}{N_0} \tag{15}$$

where:

N_{tbm} = actual number of cutters.

N_0 = standard number of cutters (Fig. 5).

4.1.4 Correction for Abrasive Minerals

Cutter ring life varies with the quartz content and other hard and abrasive minerals (i.e. garnet, epidote, olivine, pyrite and others.) in the rock. Figure 26 shows a correction factor that is applied as a function of abrasive minerals content, including quartz, garnet, epidote, pyrite and olivine; others similar abrasive minerals may be included in the abrasive mineral content parameter when estimating cutter ring life using the plot in. The correction factor shown in Fig. 26 is based on normalised field and laboratory data.

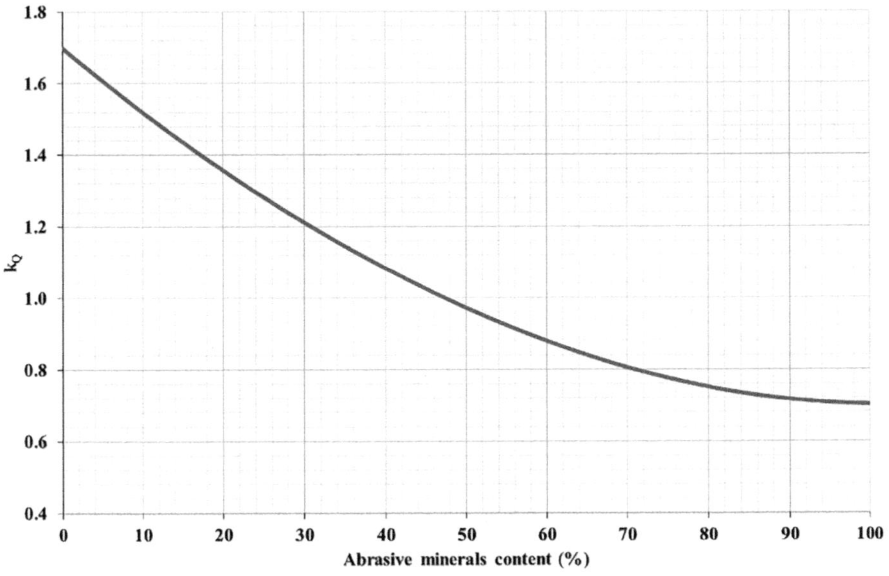

Fig. 26 Correction factor (k_Q) for abrasive minerals

4.1.5 Correction for Gross Cutter Thrust

The influence of cutter thrust on cutter life has been recently recognised. It is also anticipated that cutter life will be lower when boring in non-fractured rock masses, where high applied cutter thrust is needed, with extremely and very abrasive properties (4.5 < CLI < 6.0).

A correction factor (Fig. 27) has been introduced based on observations of the relationship between cutter life and gross cutter thrust data in extremely and very abrasive rock masses (Macias 2016). Cutter life, expressed in cubic metres per cutter, may vary in response to variations in penetration rate.

The average life of a cutter ring is given by the following Eqs. (16, 17, and 18).

$$H_h = \left(H_0 \cdot k_D \cdot k_Q \cdot k_{rpm} \cdot k_N \cdot k_T\right)/N_{tbm} \quad (h/c) \tag{16}$$

$$H_m = H_h \cdot I_n \qquad \text{(m/c)} \tag{17}$$

$$H_f = H_h \cdot I_n \cdot \pi \cdot d_{tbm}^2/4 \quad \left(\text{m}^3/\text{c}\right) \tag{18}$$

where:

H_0 = basic average cutter ring life (hours) from Fig. 24.

H_h = average cutter ring life (hours).

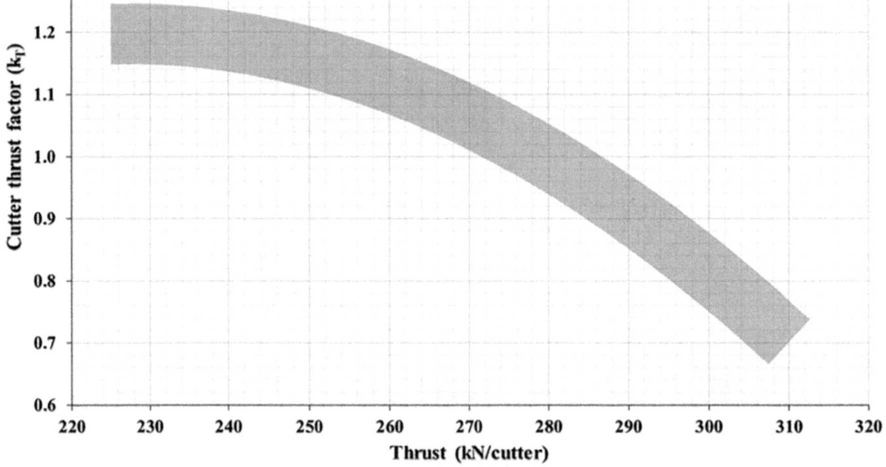

Fig. 27 Correction factor (k_T) for cutter thrust level in extremely and very abrasive rock masses

H_m = average cutter ring life (metres).

H_f = average cutter ring life (solid cubic metres).

I_n = net penetration rate (m/h).

d_{tbm} = TBM diameter (m).

The H_0 is an expression of the lifetime, in machine hours, of one individual cutter ring located in the average cutter position ($\approx 0.59 \cdot r_{tbm}$). For example, for a CLI value of 8 and a quartz content of 35%, a single 483 mm diameter cutter ring in position 29, on a 7-m diameter TBM with standard machine parameters, will have a lifetime of approximately 116 h.

The H_h, H_m and H_f parameters are expressions of averaged cutter life for a given machine or tunnel. For example, if H_m = 10 m/c, this means that for each 10 m of tunnel bored, the total wear on all cutters on the cutterhead corresponds to one complete cutter ring. In other words, this means that, on average, a cutter must be replaced for every 10 m of tunnel bored.

4.2 Considerations for Cutter Life Assessments

Rock surface hardness or resistance to indentation is the rock property that exerts the greatest influence on cutter ring life at low SJ values (equivalent to extremely high, very high and high surface hardness values, 2.0 < SJ < 6.9).

For rock types where CLI > 30 (very high and extremely high), a high proportion (60–90%) of cutter replacements are due to bearing-related issues. Cutter life is thus influenced most by factors other than the prolonged life of the cutters installed on the cutterhead. The high percentage of cutters change due to set bearing or others is a reason of the almost flat estimated basic cutter ring life in rock masses with high CLI values. Therefore, the bearing life plays a significant role for rock masses with exhibit low abrasivity and/or high degree of fracturing.

In fractured rock masses, or in situations where extremely good rock chipping occurs, the cutters will be exposed to large instantaneous loads. Under such conditions, a cutter ring exhibits a tendency to chip along its edge. Extensive ring chipping and high cutter thrust may result in bevel edge wear, loosened rings and blocked bearings. Additional loads will result in higher abrasion on protruding cutters when the difference in diameter between adjacent cutters is too large as a result of deficient wear height control. Heavy vibration of the cutterhead results in high lateral forces on the cutters, which in turn causes additional wear and failures.

5 Gross Advance Rate

The gross advance rate is expressed in metres per week, representing an average calculated over a longer period. Gross advance rate depends on net penetration rate, machine utilisation and the number of working hours during the period in question. The NTNU model is primarily based on data derived from a shift system that totals approximately between 100 and 160 working hours per week.

The machine utilisation parameter includes only a limited amount of rock support work. For this reason, the model cannot be applied directly to tunnels that require substantial amounts of rock support, and the additional extra time must be evaluated separately. The possibilities of changing cutters, performing maintenance and carrying out repairs on the TBM and the back-up systems, etc. while rock support is being installed, should be evaluated. In the case of double-shield TBMs, when boring in rock masses with very high boreability and therefore high net penetration rate, the extra time for ring building must be evaluated separately. The same applies to pre-excavation grouting and other additional works.

5.1 Machine Utilization

The machine utilisation parameter is expressed in terms of net boring time as a percentage of total tunnelling time. Total tunnelling time includes time consumed performing the following activities:

- Boring—T_b
- Re-gripping—T_r
- Cutter replacement and inspection—T_c
- Repair and service of the TBM—T_{tbm}
- Repair and service of the back-up systems—T_{back}
- Miscellaneous—$\mathrm{T_m}$
- Tunnel length factor—$\mathrm{T_l}$

The time consumed in performing these activities is expressed in hours per kilometre. The time consumption used here is representative of the methodologies practised in most modern tunnelling projects. The machine utilisation parameter is expressed as follows:

$$u = \frac{100 \cdot T_b}{T_b + T_r + T_c + T_{tbm} + T_{back} + T_m + T_l} \quad (\%) \tag{19}$$

5.1.1 Boring

Boring time depends on the average net penetration rate (In), and is calculated using the following equation:

$$T_b = \frac{1000}{I_n} \quad \text{(h/km)} \tag{20}$$

5.1.2 Re-gripping

The time taken for re-gripping depends on the stroke length of the thrust cylinders and unit time per re-grip. It is calculated using the following equation:

$$T_r = \frac{1000 \cdot t_r}{60 \cdot l_s} \quad \text{(h/km)} \tag{21}$$

where:

l_s = stroke length, typically 1.5–2.0 m.

t_r = time per re-grip.

On average, time per re-grip t_r is 4.5 min. Time consumption varies with gripper hold, the stroke length, the TBM diameter, boring along curved sections and declines, and the capacity of the hydraulic system. Under favourable conditions, time consumption will be somewhat lower but may increase substantially under difficult conditions.

5.1.3 Cutter Replacement

The time taken to carry out cutter replacements and inspections (T_c) depends on the cutter ring life (H_h), net penetration rate (I_n) and the time used to replace each individual cutter (t_c). It is calculated using the following equation:

$$T_c = \frac{1000 \cdot t_c}{60 \cdot H_h \cdot I_n} \quad \text{(h/km)} \tag{22}$$

The time taken to replace a cutter (t_c) will vary with cutter size. Typically, time consumption per replaced cutter is listed in Table 7.

t_c is based on data under favourable working conditions. Water ingress, unstable rock conditions, high rock temperature and other factors may contribute towards major changes of the time consumed.

The inspection time per changed cutter increases when boring in rock types with low abrasivity. This results in an increase in the total time per changed cutter. This

Table 7 Time per changed cutter

Cutter diameter	Front loaded (min)	Back loaded (min)
≤432 mm (17 in)	45	40
≥483 mm (19 in)	–	50

parameter also depends on the number of cutters replaced at any given time. If only a few cutters are changed during a break in operations, the unit time (t_c) increases. This may also result in shorter cutter life.

5.1.4 Other Activities

The time consumption for repair, maintenance and service on the TBM and the back-up systems, as well as other miscellaneous activities, is shown in Fig. 28. These data are regarded as being representative for well-organised tunnelling operations. The downtime involved in the event of possible main bearing failure and other extended breaks in operations is not included. Such risks must be evaluated separately.

Miscellaneous activities may include the following:

- Normal rock support work in good rock conditions, such as the installation of support while boring is in progress without having to increase the number of tunnelling personnel
- Transport system downtime

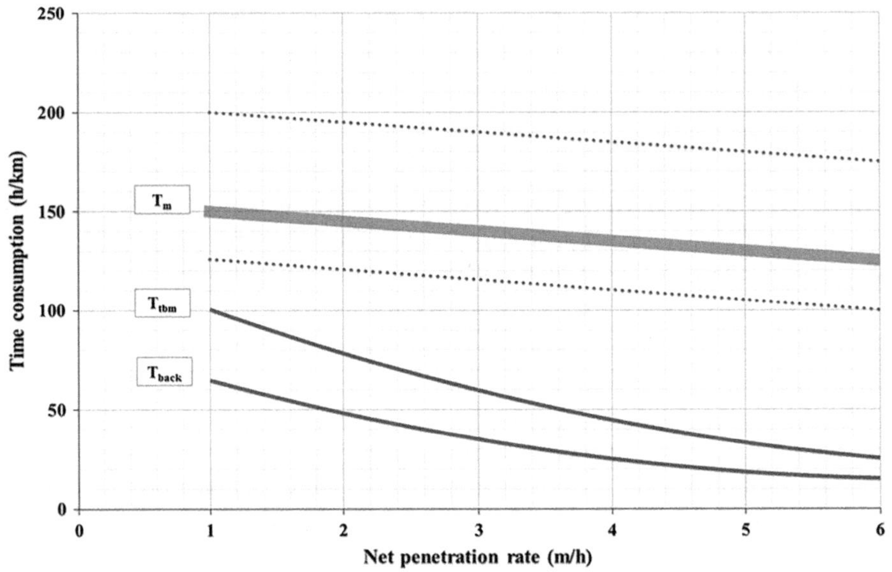

Fig. 28 Time consumption for various activities

- Installation and maintenance linked to the transport system
- Surveying, moving of laser
- Installation, or extension and maintenance linked to water supply, water drainage, ventilation and electrical cables
- Washing and cleaning of the TBM and back-up system components
- Other (crew changes, incidental lost time, etc.)

In addition to the activities listed above, time will be consumed in connection with application of the tunnelling method and project organisation.

5.1.5 Length Factor

Tunnel length exerts a relevant influence on time consumption during tunnelling operations. During the initial "learning curve" period of a tunnelling project, more time is spent on any given activity as the crew's skills levels are built up, and the quality assurance system adapts to the project in question. The longer the tunnel (>8 km), the more likely it will be that problems will be encountered with muck transport, the ventilation, electricity and water systems, and delays linked to supplies. If the capacity of the transport system is inadequate, waiting times for transport services will increase substantially.

Problems with the ventilation, electricity and water systems will continue to increase with tunnel length. The transport system remains dependent on TBM penetration rate and factors such as the volume of material that requires removal. The higher the penetration rate, the greater these problems linked to downtime and delays in connection with the transport system and supply logistics will be. The data in the NTNU model database is derived from tunnels of up to 10 kms in length. Expressions of time consumption in previous versions of the model represent the average of the total tunnel lengths. Low skills levels among the crew, and an inadequate tunnel quality assurance system, will result in additional losses in time. The term 'skills levels' used here refers not only the crew, but also to system manufacturing competence.

The updated version of the model proposes the introduction of an additional time consumption (h/km) related to the tunnel length. The 'additional time consumption' is illustrated in Fig. 29, and includes boundary limits for low and high skills levels, and the quality assurance system. The values for every kilometre correspond to the extra time during the last km.

Figure 30 illustrates the cumulative additional time involved for the total tunnel length under consideration. This graph might be used for overall rough estimations.

5.1.6 Weekly Working Hours

Most of the data used as a basis for the model are derived from tunnelling operations involving approximately 100 working hours per week. Thus, it is assumed that there

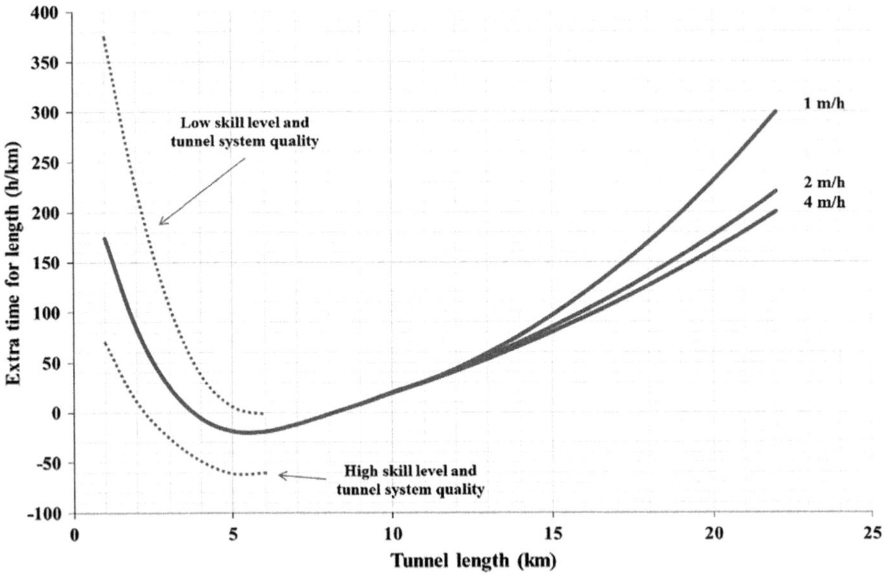

Fig. 29 The 'additional time consumption' factor, based on tunnel length T_l

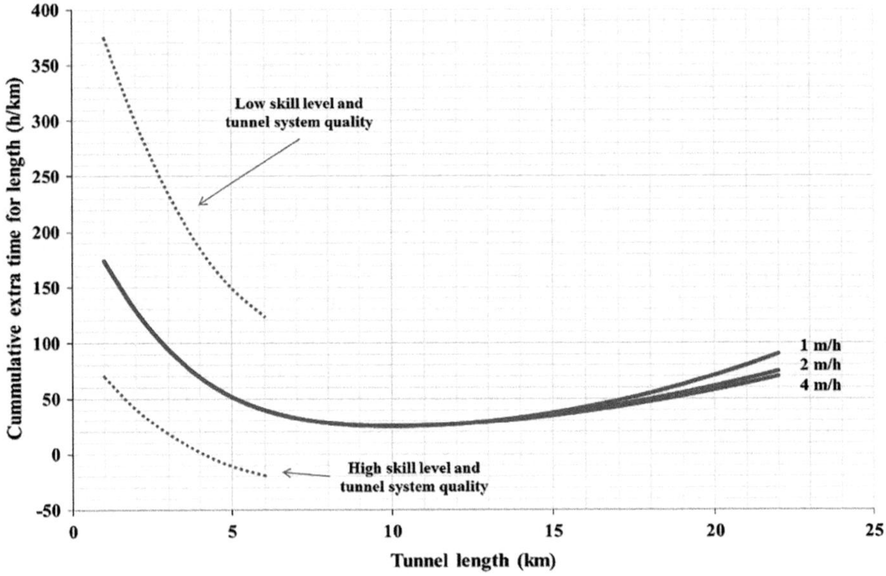

Fig. 30 Cumulative additional time involved for the total tunnel length under consideration

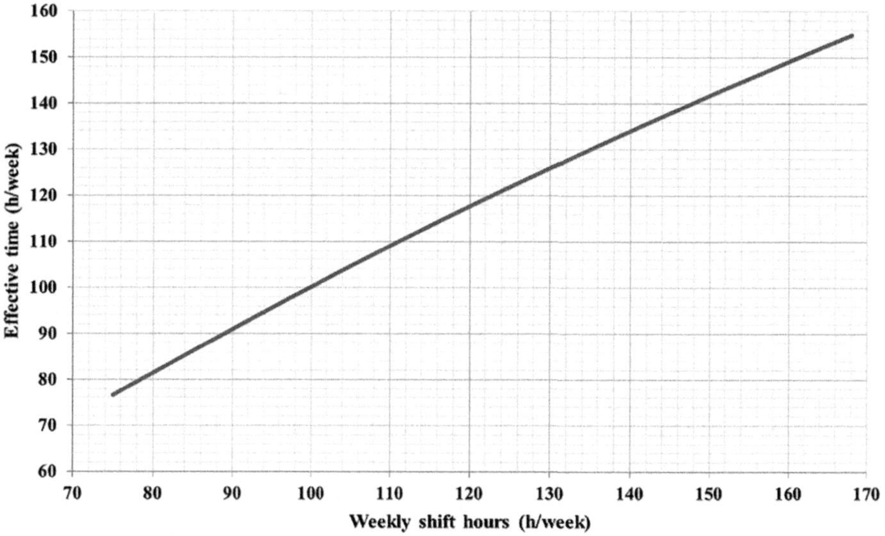

Fig. 31 Plot showing effective working hours per week

exists some time available outside standard working hours to address unforeseen and critical incidents such as major repair work. Some aspects of downtime of this type are not recorded in shift logs and are therefore not included in Fig. 28.

The plot in Fig. 31 indicates that opportunities to address unforeseen and critical incidents in a flexible manner become fewer as the number of hours worked per week (T_u) increases towards 168 h. The parameter 'effective time' (T_e) provides an expression of working hours that are effectively available when the number of hours deviates from 100, which forms the basis for the model. The curve in the figure is believed to represent a conservative estimate of losses in effective working hours.

The weekly advance rate (I_u) will be:

$$I_u = u \cdot T_e \cdot I_m / 100 \ \ (\text{m/week}) \tag{23}$$

where:

u = machine utilisation.

T_e = effective working hours per week.

I_m = average net penetration rate over the tunnel.

5.2 Additional Time Consumption

Estimates of time consumption for a given tunnel are derived from the weekly advance rate parameter, which is in turn based on net penetration rate and machine utilisation. Furthermore, additional time must be allowed the following factors:

- The excavation of the underground assembly and start-up area, muck transfer station, etc., if necessary
- The assembly and disassembly in the tunnel of the TBM and the back-up systems. This normally takes from 4 to 8 weeks, depending mainly on the TBM diameter
- The excavation of cross passages, niches, branching, etc.
- Pre-excavation grouting works: drilling and grouting
- Boring through and stabilising zones of poor rock quality
- Encounters with unexpected rock mass conditions
- Permanent rock support and lining work
- TBM downtime (due to possible major breakdowns)
- The dismantling of tracks, the ventilation system, invert clean-up, etc.

Furthermore, the time spent to manufacture and/or refurbish the TBM, and the back-up systems may exert an influence on overall project schedules, depending on the duration of necessary activities that precede actual excavation operations. The refurbishment of a previously used TBM may take from 3 to 6 months. The manufacture of a completely new TBM may take from 6 months to 1 year, depending mainly on the TBM diameter. Transport of the TBM and the back-up systems to the site may also influence the overall project schedule, for the same reasons as above. Transport of a TBM is a demanding operation because the largest component may weigh as much as 20 tons or more, even for a 3.5-m diameter machine, hence limiting the roads and bridges that can be used for transporting the machine to the jobsite. Additionally, there might be extra time needed for assembly of the machine in a launch shaft as part of the accessibility of cranes to move the parts to the assembly chamber.

6 Limitations of the NTNU Model

A model is a simplified description of a given phenomenon, which can be used to make predictions. A good model is both as accurate as possible and as simple as possible, which makes it not only powerful but also easy to understand. However, no matter how good a model is, it will usually have limitations. Since models must be simple enough to enable them to be used to make predictions, they often lack the ability to simulate all the details of the phenomenon in question.

In the case of the NTNU prediction model for hard rock TBMs, the following limitations should be considered.

The NTNU model does not cover the entire range of rock masses that occur in nature:

- Rock drillability expressed by the Drilling Rate Index (DRI) parameter should be within the range 20–80, which roughly corresponds to the UCS parameter values 25–350 MPa. The strength parameter applied is in accordance with the ISRM classification 'medium to extreme'.
- The rock type in question should exhibit medium to low porosity—less than approximately 10% by volume.
- The rock mass degree of fracturing, expressed by the average spacing between planes of weakness, should be greater than approximately 50 mm.
- The rock-breaking process must be mainly by brittle failure (chipping) between the disc cutters.
- Rock mass quality should be such that the excavated tunnel will generally only require light-duty support mechanisms (rock bolts and/or shotcrete). Exceptions will occur in zones of weakness or because of other specific phenomena.
- Rock mass qualities roughly related to Q and RMi values should be somewhere >0.1, and >20 for the RMR.
- Boreability assessments do not include the possible influence of groundwater and/or rock mass stress.

The NTNU/SINTEF drillability test method has gained international recognition. There are few laboratory facilities in the world that can reproduce these tests, unlike other test methodologies such as UCS and BTS that are common rock mechanics tests performed by practically all rock mechanics labs in the world. There are other laboratories worldwide with the necessary equipment, such as in Australia, South Korea, Spain, South Africa, Turkey and the USA that can perform NTNU/SINTEF drillability test method. The trademark acronyms and terms relating to the indices DRI™ (Drilling Rate Index™) and CLI™ (Cutter Life Index™) are unique to these tests performed directly by the NTNU/SINTEF laboratory in Trondheim in Norway.

A cutter diameter of 483 (19 in) and 508 mm (20 in) are currently commonly used in hard rock tunnelling, especially in hard-to-very hard and more abrasive rock masses. For smaller TBMs, and in general most applications in medium to strong rocks the disc diameters of 432 mm (17 in) are used. The NTNU model makes no distinction between cutter ring design or quality and assumes in all cases that cutter ring quality is tailored to the rock conditions.

The machine type assumed by the model is a hard rock TBM (open or shield). However, the model is based on data derived from many open TBM types tunnelling in rock mass conditions summarised in the foregoing. No distinction is made between machine types. The gross cutter thrust parameter may interfere with the influence of TBM type, diameter, and/or cutter type. TBM cutterhead design is not included in the model as an independent parameter.

Geological assessments from blasted and/or exposed rock areas (e.g., D&B tunnels, surface cuts) must be considered with caution when extrapolating and using the values for rock boreability evaluations (Macias and Barton 2022).

The machine utilisation parameter (expressed as a percentage of tunnelling time) used in the model does not include additional time spent on operations such as TBM and back-up assembly and disassembly, excavation work (niches, branching, portal, shaft etc.), pre-excavation grouting works, tunnelling through zones of poor quality, additional time for unexpected rock mass conditions, permanent rock support and lining operations, and additional TBM downtime. These activities need to be quantified separately.

References

Bruland A (2000) Hard rock tunnel boring, vol 1–10. PhD thesis. Norwegian University of Science and Technology (NTNU), Trondheim

Macias FJ, Barton N (2022) The NTNU and Q_{TBM} methods of prognosis with emphasis on hard-rock mixed-face problems. Fjellsprengningsteknikk conference, Bergmekanikk, Oslo

Macias FJ, Hansen AM, Skjeggedal T (2025) On the efficiency when boring in hard-to-very hard rock mass conditions. In: Proceedings of the ITA-AITES world tunnel congress 2025 (WTC 2025), 9–15 May 2025, Stockholm

Macias FJ (2016) Hard rock tunnel boring: performance predictions and cutter life assessments. PhD Thesis, Norwegian University of Science and Technology (NTNU), Trondheim

NTH (1976) Hard rock tunnel boring, project report 1–76. Norwegian Institute Technology, Div. of Construction Engineering, Trondheim

NTH (1981) Hard rock tunnel boring, project report 1–81. Norwegian Institute Technology, Div. of Construction Engineering, Trondheim

NTH (1983) Hard rock tunnel boring, project report 1–83. Norwegian Institute Technology, Div. of Construction Engineering, Trondheim

NTH (1988) Hard rock tunnel boring, project report 1–88. Norwegian Institute Technology, Div. of Construction Engineering, Trondheim

NTH (1994) Hard rock tunnel boring, project report 1–94. Norwegian Institute Technology, Div. of Construction Engineering, Trondheim

AR, Time, Tunnel Length and Geology Using Q_{TBM}

Nick Barton

1 Introduction

In this chapter we will focus more on advance rate (AR) than penetration rate (PR), and focus on the effect of tunnel length and geological variations, and therefore the fundamental question of time (T). Case records from 1000 km of mostly open-gripper TBM will be supplemented with a synthesis of world records for different sized TBM and their variation with time. Two fundamental questions will finally be addressed: is it correct to use TBM 'because the tunnel is so long'? And is it correct to use TBM: 'because conditions will be bad'? Logic suggests that in general the rock mass quality statistic, for instance described by the Q-value, will contain more outliers such as more serious fault zones, or more cases of exceptionally hard massive rock, the longer the planned tunnel. It is the outliers that will have the strongest influence on AR and therefore total project time. The Q_{TBM} prognosis method will be described and illustrated with examples, including Follobanen where four Herrenknecht double-shield TBM drove approximately 9 km each. The use of seismic refraction results and correlation of P wave velocity (V_P) with Q and therefore Q_{TBM} prognosis will be demonstrated.

The following symbols will be used: PR (m/h), means the penetration rate with continuous boring, while AR is the actual advance rate e.g. in m/week, m/month, m/year. U is utilization which depends on machine-type, diameter, geological conditions and on the time period considered, while time T is total time: such as 24 h, 168 h, 730 h, 8730 h, for a day, week, month and year. If each AR time period is given in m/hr then the influence of time and tunnel length can be better understood. U is very much a time-dependent variable, and 24 h tells little.

N. Barton (✉)
NB&A, Oslo, Norway
e-mail: nickrbarton@hotmail.com

J. Hassanpour and J. Rostami (eds.), *Performance of Tunnel Boring Machines (TBM) in Rock*, Springer Series in Geomechanics and Geoengineering,
https://doi.org/10.1007/978-3-032-20424-0_4

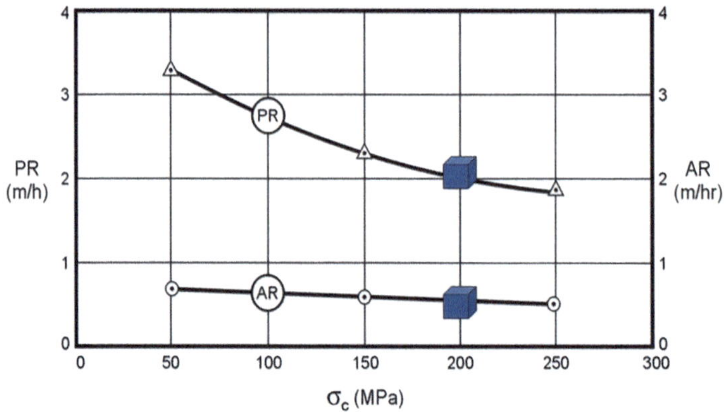

Fig. 1 These two trend curves for PR and AR are from tabulated data given by Fawcett (1993). The resulting figure was published in Barton (2000). The data points refer to average PR and AR (m/week) in shale, tillite and hard sandstone with respective UCS = 50, 150 and 250 MPa. They are symptomatic of TBM behavior trends. More delays in AR in weaker rock because stability may be an issue, and possibly severe cutter change delays in AR in very hard rock

The two blue cubes in Fig. 1 correspond almost exactly to the overall mean values of PR, AR, and UCS (200 MPa) for the four double-shield TBM driving approximately 9 km each at the Follobanen rail tunnels south of Oslo. They also apply very closely to four TBM at the Guadarrama rail tunnels which each tunnelled 14 km, though here the UCS was more variable and the overall AR was 0.55 m/h. The above (Fig. 1) means that AR (in m/week) has a more rapid decline in these long tunnels.

In Fig. 2a we see PR trends for 2.8 km of TBM tunnelling in variably jointed granites (UCS from 130 to 250 MPa), correlated with Q-values. Logging was performed on a continuous basis in the open-gripper TBM back-up area.

It should be noted that Fig. 2b has, at present, inappropriate adjectives for TBM. When the Q_{TBM} equation is presented later, they will be made appropriate. For example, when Q > 100, 'tough boring' will be seen, not 'extremely good'. One further temporary 'error' is shown: there should be multiple AR curves for e.g. 1 week, 1 month, 3 months, 1 year etc. These will be shown later in relation to Q_{TBM}.

2 Trends from Thousands of Kilometers of TBM

In Fig. 3a, b we see a synthesis and then the raw data from the Barton (2000) analysis of some 1000 km of data from 145 well-documented lengths of mostly open-gripper case records. PR (m/h) applies only to the left axis in each of the figures. The remaining area of the diagrams and all the data apply to AR and time T, each on a log-scale. The deceleration gradients (−m) range from about − 0.15 to − 0.25 when Q-values are above 1.0, but as seen by the dotted curves they decelerate

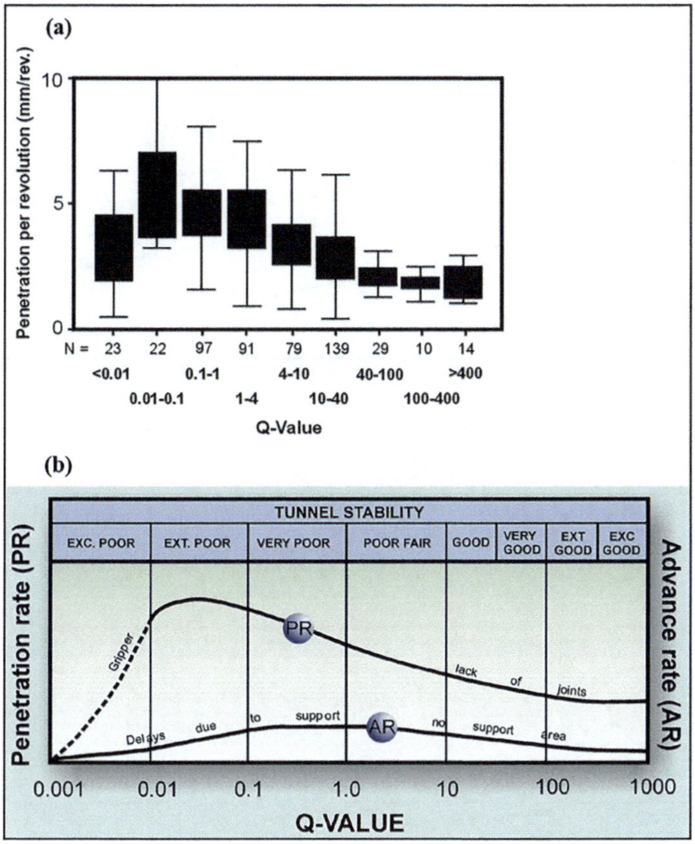

Fig. 2 **a** Comparison of penetration per revolution with the regular Q-value in granites, except that J_r and J_a were averaged for all joint sets. (Sundaram and Rafek 1998). **b** PR and AR trends graphed on a regular Q-value 'tunnel stability' chart (Barton 2000)

more rapidly (more negative gradient m) as Q-values decline to 0.1, 0.01 etc. Crosses: real TBM.

Perhaps of surprise is that even world record TBM performances (mostly from Robbins) when gathered in common size ranges (Barton 2013) show the same trends of deceleration, which mostly match the open-gripper case record trends. An obvious scale effect is also seen in Fig. 4 and clearly needs to be incorporated in PR and AR prognoses, as suggested in Macias and Barton (2022).

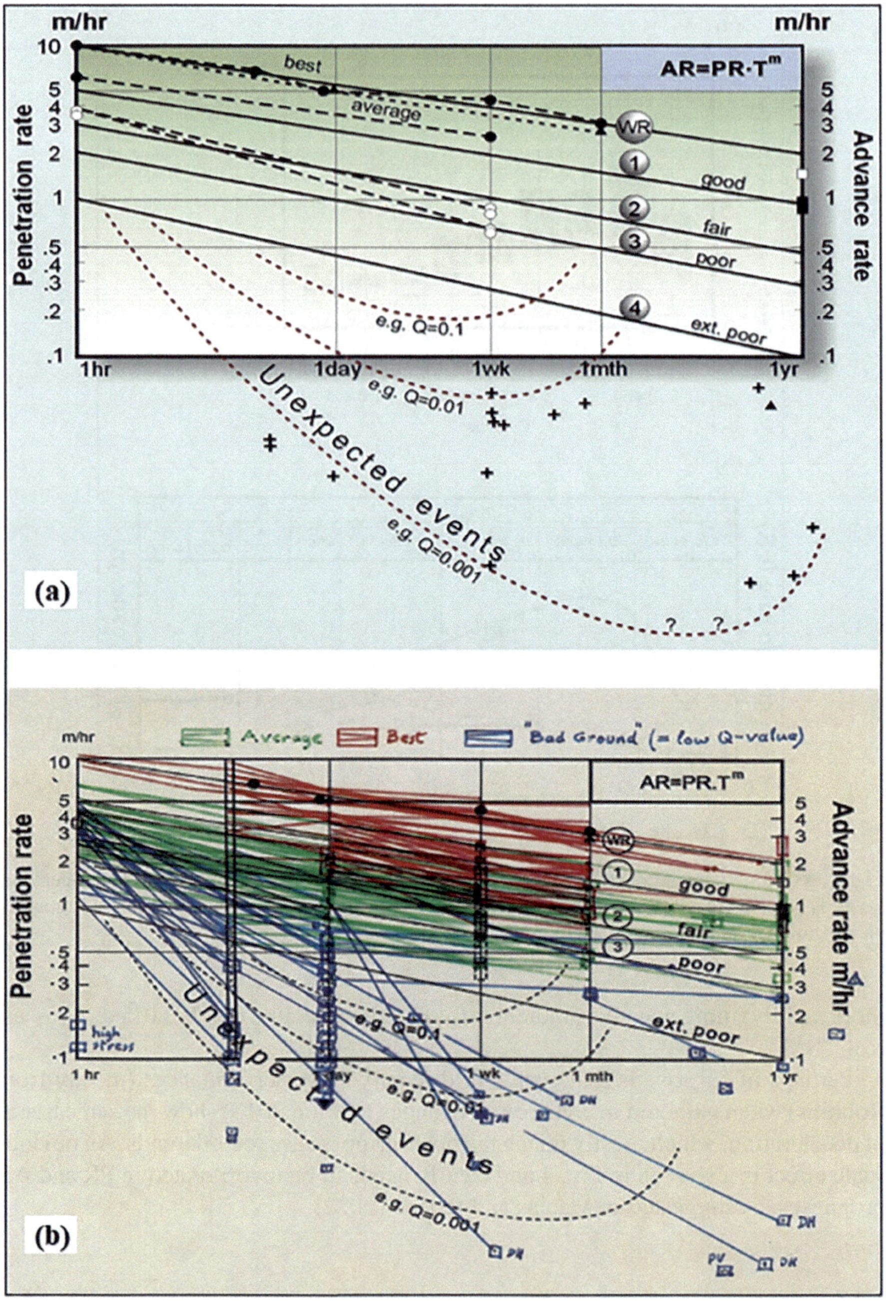

Fig. 3 Open gripper data from Barton (2000). **a** synthesis of trends with suggested nomenclature WR (best), good, fair, poor. **b** Raw data: best, average, worst. The black crosses (+ far right) are permanently buried TBM

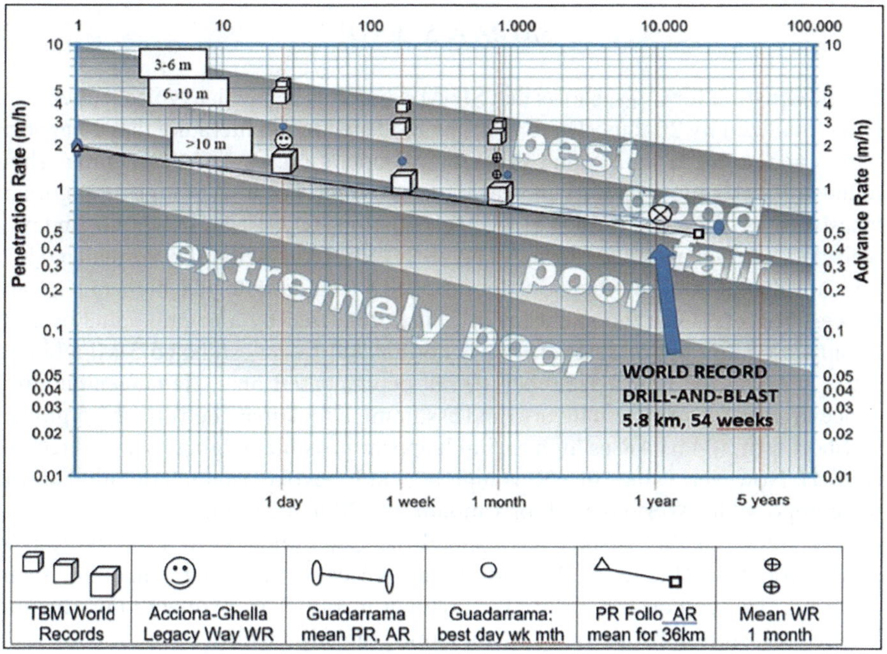

Fig. 4 In addition to the world record data for three diameter ranges (see cubes, derived from Fig. 5) a solid black and nearly parallel thin blue line are shown. These stretch from mean PR ≈ 2.0 m/h out to final AR of 0.5 and 0.55 m/h and apply to the 4 × 9 km Follobanen TBM and the 4 × 14 km Guadarrama TBM rail tunnels in Norway and Spain. Each of these were driven most of the time in double-shield mode and show shallower deceleration gradients, at best approximately half those of open-gripper TBM. The world records suggest a strong scale effect, with smaller diameter TBM clearly the fastest

3 The Q_{TBM} Prognosis Method

When developing the Q_{TBM} method in 1999–2000 (Fig. 6) it appeared that rock mass quality was sufficiently described with the Q-system. There were however two important additional factors to be addressed. Firstly, the RQD term needed to be specifically estimated in the tunnelling direction, so an oriented RQD_0 was required. Also, although it was obvious that UCS was an important parameter, the action of UCS within a jointed rock mass would be less direct so a rough estimate: SIGMA was suggested, based on $Q_c = Q \times UCS/100$. The equation for SIGMA gives a 'rock mass strength' estimate ranging from about 1 to 100 MPa depending also on how massive the rock is. This may be realistic. However, nobody knows *UCS ROCK MASS*.

$$Q_{TBM} = \frac{RQD_o}{J_n} \cdot \frac{J_r}{J_a} \cdot \frac{J_w}{SRF} \cdot \frac{SIGMA}{\left(F^{10}/20^9\right)} \cdot \frac{20}{CLI} \cdot \frac{q}{20} \cdot \frac{\sigma_\theta}{5} \quad (1)$$

$$SIGMA \approx 5 \cdot \gamma \cdot Q_c^{1/3} \tag{2}$$

The equation for Q_{TBM} shows that SIGMA is compared with a typical cutter force (F) normalized by 20 tons. There is strong evidence in the literature that comparing 'stress and strength' is a sound geotechnical operation, as TBM sometimes show reduced PR despite operators increasing cutter force because of high strength and massive rock. Moving further along the equation, CLI is the cutter life index, a well-documented parameter from NTNU/SINTEF, Trondheim. It is normalized by a typical value of 20. Low values, as for instance 5 for quartzite give higher Q_{TBM} and slower progress. Quartz content (q) in % is normalized by a neutral value of 20%. Finally, a confining pressure term σ_θ is normalized by an estimated 5 MPa biaxial stress state on the tunnel face, per 100 m depth, based on a 2D tangential stress concentration factor (Kirsh theory) and assuming a rock mass density of 2.5 gr/cm^3.

Short-term (24 h) AR as high as 5 m/h appear possible judging by world records (Fig. 4). However, values of AR as high as 3 m/h beyond 1 month duration appear to be impossible (World record for 1 month, $\approx$ 720 h = 2163 m).

The author of this chapter has applied the Q_{TBM} method to TBM projects in several countries (Brazil, Chile, Peru, South Africa, Hong Kong, Australia, Spain, Norway). The demonstration of Q_{TBM} shown in Fig. 7 is a suitable introduction. Figure 8 illustrates the use of Q_{TBM} for a mine tunnel in Chile.

4 Lessons Learned via Q_{TBM} at Follobanen 4 × 9 km

The Follobanen TBM tunnels, which were driven from a central assembly hall, saw two 9.9 m diameter Herrenknecht double-shield TBM boring to the north towards Oslo, and two boring south towards Ski. The tunnels were mostly driven during 2016 and 2017. In 2009, some two years into the BaneNor investigations for the future tunnels, the author of this chapter was invited to perform Q-logging of relevant exposures mostly along road-and-motorway cuttings, at major building sites and in quarries, both in the neighborhood of the planned northern tunnels, and as near as possible to the more laterally distant southern tunnels (Figs. 9 and 10). The planned tunnels would be mostly 70–100 m deep, but with a range of 30–160 m.

Clearly, surface exposure logging will not pick up reliable data for typical 70–100 m tunnel depths. A partial improvement could be the use of seismic refraction data and drill core. Unfortunately, both (as available in 2009) were focused on weakness zones. These were assumed by the owner to be the source of potential delay. In fact, with the benefit of subsequent hind-sight, it was the more dominant rather massive, high V_P and high UCS, 'hard mixed-face' nature of the rock mass that was the source of many cutter ring and bearing problems (Macias and Barton, 2022). Water inflow problems and delaying pre-injection due to unusually high horizontal permeability were also a factor in the southern tunnels. Such problems were not yet identified at the time of e.g. Barton and Gammelsæter (2010) prognoses.

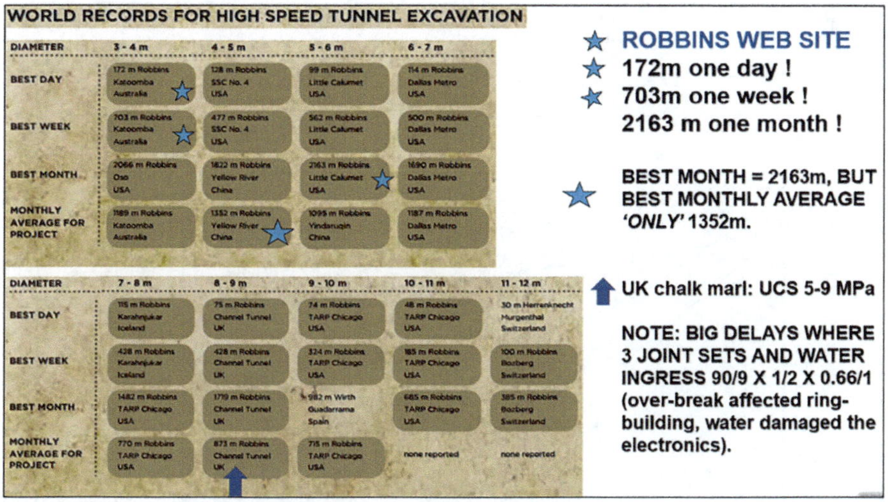

Fig. 5 Robbins web-site data: TBM world records. Mostly Robbins, occasional Wirth, and Herrenknecht. Bottom-right: some details of problems at the Channel Tunnel (UK-France) outside the majority of the less-jointed chalk marl where world records were broken

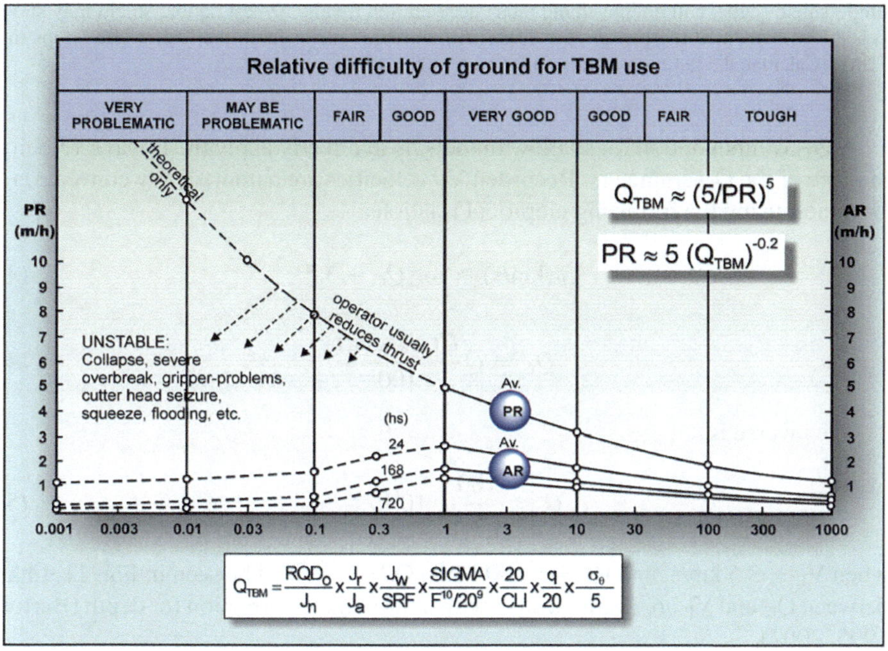

Fig. 6 The Q_{TBM} method of estimating PR and a time-dependent AR, here drawn for 1 day, 1 week and 1 month. Higher short-term AR can be obtained with higher PR, but conveyor belt capacity may be compromised beyond the 6 to 8 m/hr range of PR

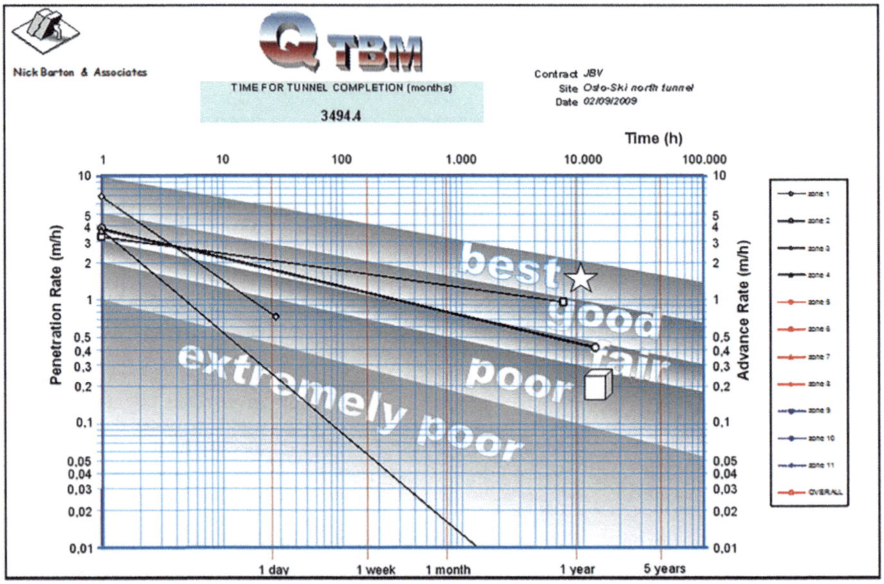

Fig. 7 Left axis: demonstration of PR prognosis (with F = 28 or 26 tnf), and the differentiation of single-shield (cube) and double-shield (star) deceleration trends. At best double-shield may show ½ × (−)m compared to open gripper TBM. An untreated major fault: the lowest line 'stops the TBM' (Calculated 3494 months in simulation ≈ 300 years)

A Q_{TBM} input option for shallow tunnels, is to directly apply the P-wave velocity in place of the Q-parameters. Recorded V_P velocities are automatically converted to Q-values, using the following empirical relations:

$$V_P(\text{km/s}) \approx \log Q_C + 3.5 \tag{3}$$

$$Q_C = Q\frac{UCS(\text{MPa})}{100} \tag{4}$$

The inverse is:

$$Q \approx \frac{100}{UCS} \cdot 10^{(V_P - 3.5)} \tag{5}$$

when $V_P = 3.5$ km/s, and UCS ≈ 100 MPa, Q ≈ 1. As will be seen in Fig. 11, links between Q_C and V_P are available also with approximate correction for depth (Barton 1995, 2007).

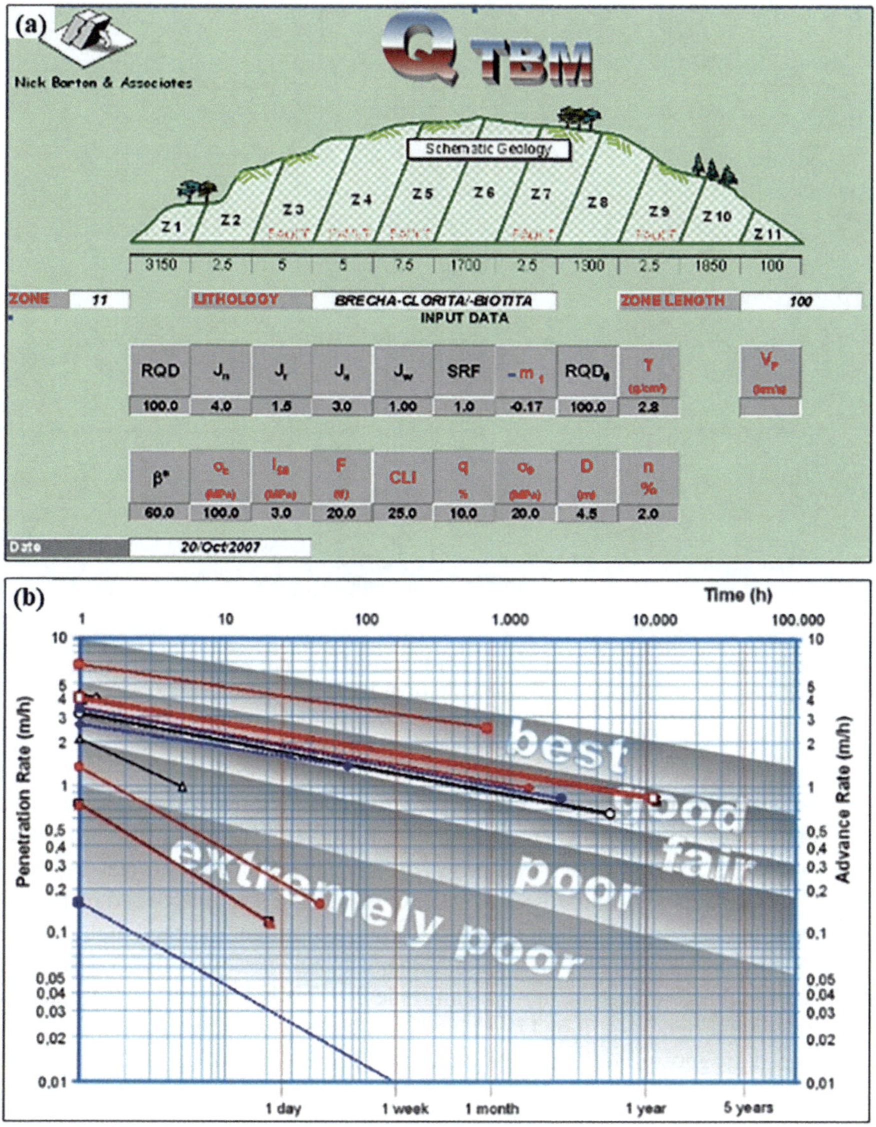

Fig. 8 Application of Q_{TBM} for a mine axis tunnel in Chile. In this unusual case there were several very deep drill-core available, drilled mainly to analyse the orebody. An early faulted zone with adverse J_n/J_r of 15/1.5 and $J_a = 2$ (weathered) caused a 1 month delay due to excessive overbreak so 'over-boring' in terms of volume of cuttings and blocks (Imagine as lowest blue line). The prognosis of some 60 weeks (longest red line, final AR = 0.8 m/h) for 5 km proved over-optimistic due to severe labour laws with workers being transported several kilometers to a miner's hostel each lunch 'hour'! If the reader expands the scale, the left-hand input data screen shows four narrow faults. These give the steepest deceleration. The last 'zone 11' of 100 m length has the assumed Q-parameters shown

Fig. 9 Four of more than 300 examples of surface exposures/cuttings/quarries with locations generally within 1 km of the future Follobanen rail tunnels. Hard, pre-Cambrian tonalitic gneisses, granites, pegmatites and amphibolites, with UCS from 150 to 300 MPa

Weakness zones which had been identified by low velocities in seismic refraction profiles along the planned rail tunnel corridor (as of 2009) were modelled with Q_{TBM}. An example of an input data screen is shown in Fig. 12. The appearance of one of these weakness zones is shown in Fig. 13. Blue pencils mark the soft clay core.

Preliminary seismic data provided by the Owner in 2009 is reproduced in Table 1. There was then no further update from the Owner. The author returned to the project in 2018, now assisting the contractor in a major claim. Seismic evidence of major lengths of high velocities was now seen (Table 2). This has a significant impact on more relevant prognosis if it had been made available in a timely manner.

Updated prognoses could now be made which were confirmed by the results of tunnelling the 4 × 9 km (Tables 3 and 4). Figure 14 is one example. There was now a focus on longer sections of high velocity, high Q, and higher Qc due to the challenging hard rock UCS.

Q - VALUES:	(RQD	/	Jn)	*	(Jr	/	Ja)	*	(Jw	/	SRF)	=	Q
Q (typical min)=	75	/	15.0	*	1.0	/	4.0	*	0.50	/	1.0	=	0.625
Q (typical max)=	100	/	3.0	*	3.0	/	1.0	*	1.00	/	1.0	=	100.0
Q (mean value)=	96	/	7.1	*	1.8	/	1.3	*	0.83	/	1.0	=	15.36
Q (most frequent)=	100	/	9.0	*	1.5	/	1.0	*	1.00	/	1.0	=	16.67

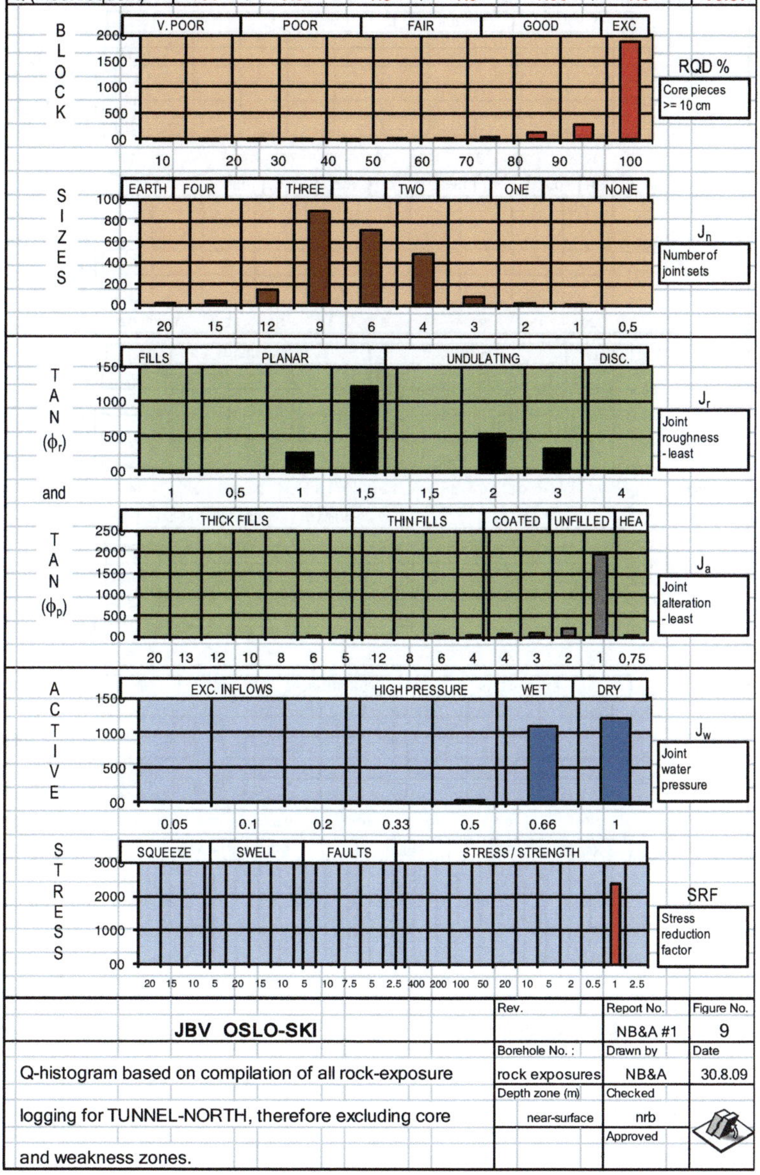

Fig. 10 Q-histogram logging of more than 300 rock-cuttings and rock exposures in the general area of the North and South tunnels, but often as much as 1 km distance in the case of the southern tunnels due to a lack of rock exposures caused by a diverging motorway

Q - VALUES:	(RQD	/	Jn)	*	(Jr	/	Ja)	*	(Jw	/	SRF)	=	Q
Q (typical min)=	75	/	15.0	*	1.0	/	5.0	*	0.50	/	1.0	=	0.500
Q (typical max)=	100	/	4.0	*	4.0	/	1.0	*	1.00	/	1.0	=	100.0
Q (mean value)=	98	/	8.4	*	1.7	/	1.3	*	0.75	/	1.0	=	11.07
Q (most frequent)=	100	/	9.0	*	1.5	/	1.0	*	0.66	/	1.0	=	11.00

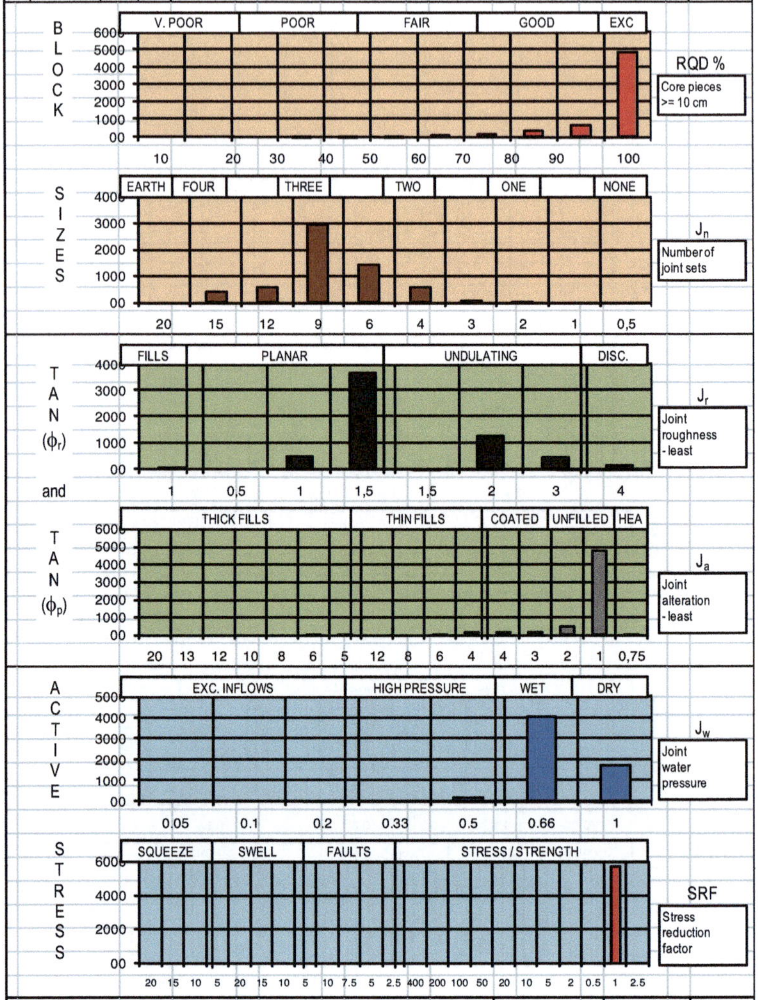

	Rev.	Report No.	Figure No.
JBV OSLO-SKI		NB&A #1	10
Q-histogram based on compilation of all rock-exposure logging for TUNNEL-SOUTH, therefore excluding core and weakness zones.	Borehole No. : Rock slopes	Drawn by NB&A	Date 31.8.09
	Depth zone (m) near-surface	Checked nrb	
		Approved	

Fig. 10 (continued)

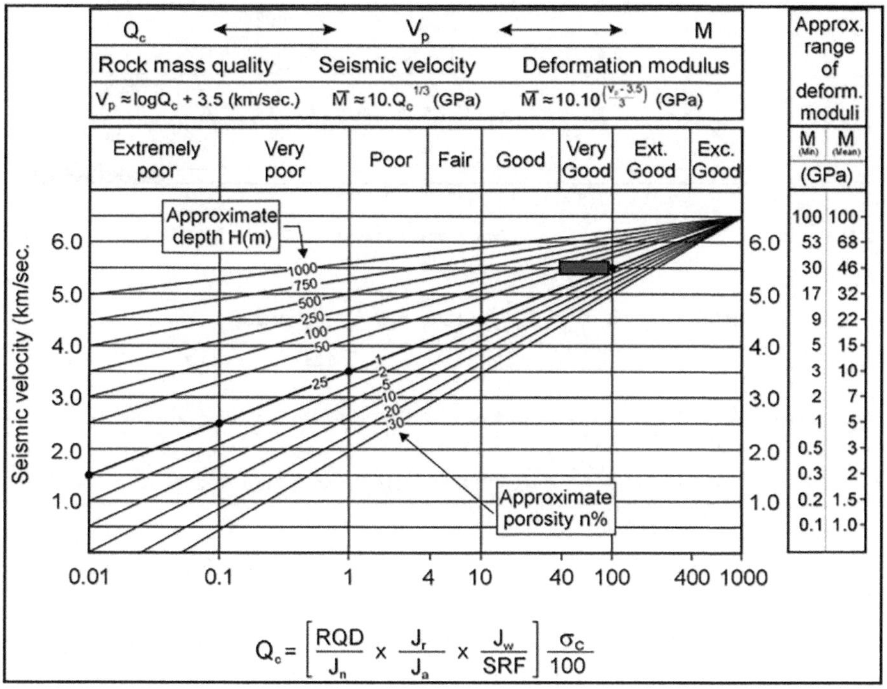

Fig. 11 Empirical links between P-wave velocity and Q or Q_c with approximate correction for depth or stress level (Barton 1995, 2007). Blue rectangle: $V_P = 5.5$ km/s, depth 25–100 m, Q_c = 40–100, Q = 20–50 if UCS = 200 MPa

5 Fault Zones and Their Challenge to TBM

Fault zones are of course a challenge to all methods of tunnelling. For some years TBM manufacturers have been advertising that TBM, in particular double-shield, are faster than drill-and-blast, both for bad rock and for long tunnels. In connection with such claims the gradually improved ability to perform pre-injection which has been noted during several decades, clearly brings TBM closer to tackling the worst of fault zones. Three examples of challenging fault and fractured zones follow here, with an assessment of how they impacted the TBM tunnelling. See Figs. 15, 16, 17, 18, 19 and 20.

The TBM manufacturers claim are tested frequently in some tunnels, and sometimes TBM have to be replaced by drill-and-blast when the complexities of some faults even overcome the machines' and the operators' and the designers' ingenuity. On other occasions pre-injection, if it is planned and therefore possible, can save the project. There are many examples of this.

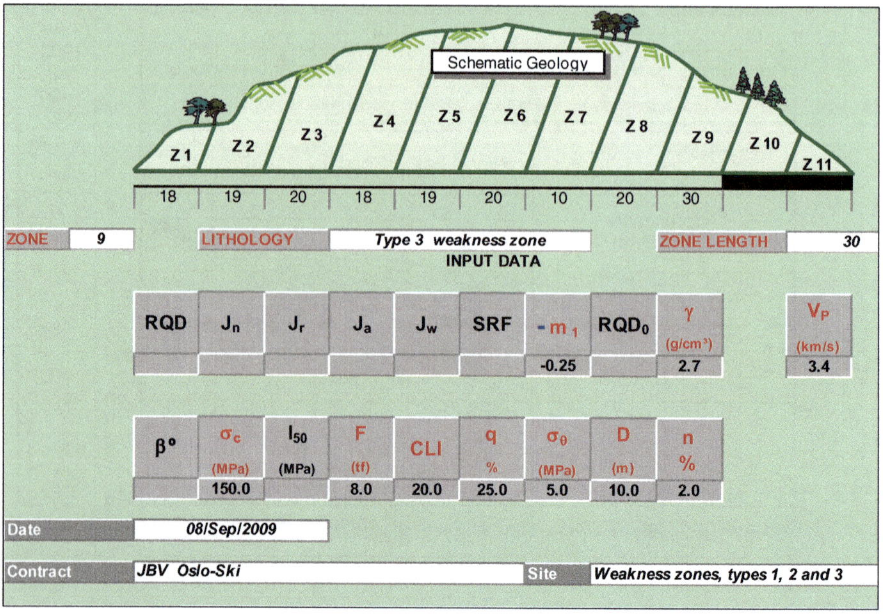

Fig. 12 Direct use of shallow refraction VP velocities in place of Q parameters. Zone 9 input data assumptions for machine-rock interaction are shown along the bottom keys

The contractor's solution for void formation was filling with foamed resin, an after-the-event measure that is actually too late, as a few meters per day was the limits of advance and the zone of over-excavation continued for several weeks in Chile. Note that a tunnel fire can be caused if too large volumes of chemical grout mix are needed for void filling. 'Waiting for the smoke' (to clear), was unexpectedly different from 'waiting for the train' on a pie-diagram. The problem with this TBM was that ring-mounting equipment had to be dismantled before pre-injection drilling could be performed, clearly not the ideal choice of priority if faulted rock is expected. How to do both (build rings and pre-inject) when both were needed was not solved.

A particularly challenging series of fault 'swarms' was experienced at the Pont Ventoux hydroelectric project in Northern Italy (see Figs. 18 and 19). These were not expected but were gradually encountered at the down-stream end of the tunnel as the TBM gradually curved towards the upstream and became sub-parallel to the swarm of faults.

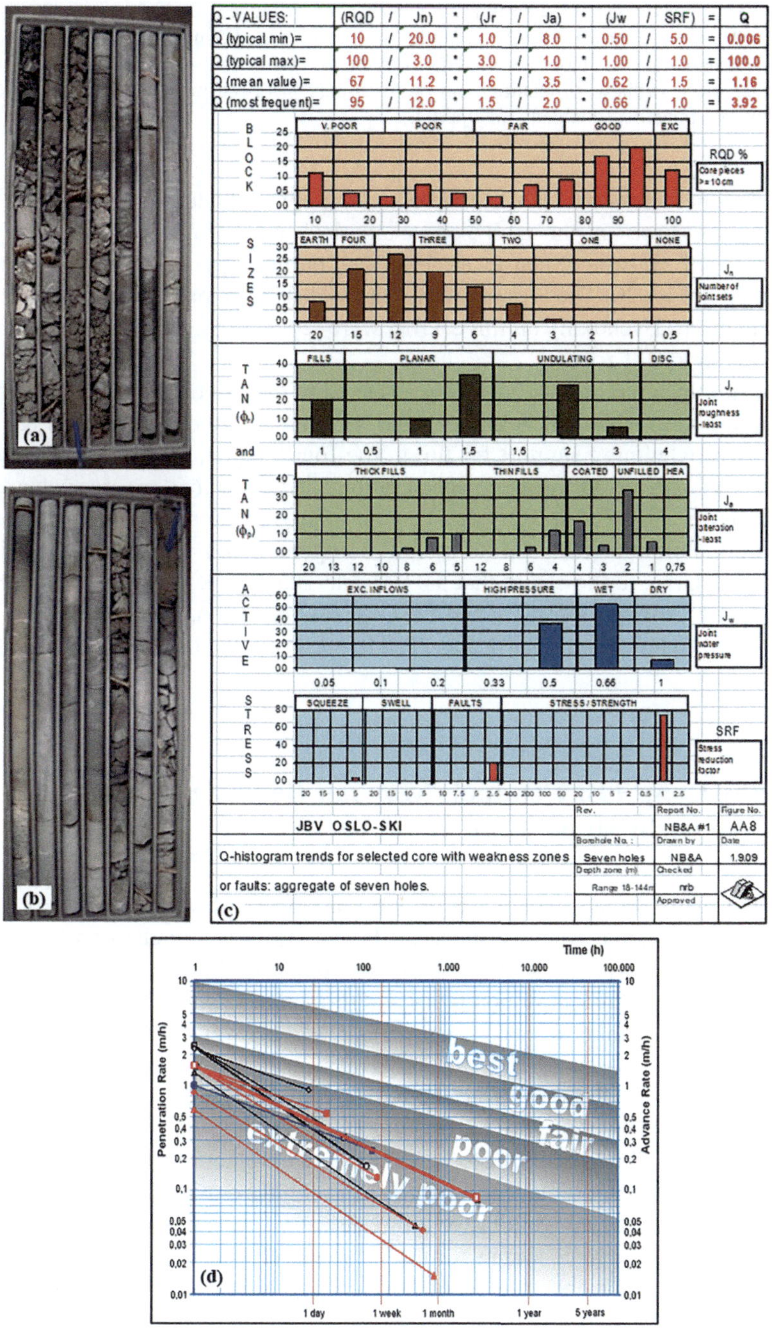
(a)
(b)
Q - VALUES: (RQD / Jn) * (Jr / Ja) * (Jw / SRF) = Q
Q (typical min)= 10 / 20.0 * 1.0 / 8.0 * 0.50 / 5.0 = 0.006
Q (typical max)= 100 / 3.0 * 3.0 / 1.0 * 1.00 / 1.0 = 100.0
Q (mean value)= 67 / 11.2 * 1.6 / 3.5 * 0.62 / 1.5 = 1.16
Q (most frequent)= 95 / 12.0 * 1.5 / 2.0 * 0.66 / 1.0 = 3.92
BLOCK
V. POOR
POOR
FAIR
GOOD
EXC
RQD %
Core pieces >=10 cm
SIZES
EARTH
FOUR
THREE
TWO
ONE
NONE
Number of joint sets
TAN (ϕ_r)
and
FILLS
PLANAR
UNDULATING
DISC.
Joint roughness - least
TAN (ϕ_a)
THICK FILLS
THIN FILLS
COATED
UNFILLED
HEA
Joint alteration - least
ACTIVE
EXC. INFLOWS
HIGH PRESSURE
WET
DRY
Joint water pressure
STRESS
SQUEEZE
SWELL
FAULTS
STRESS / STRENGTH
SRF
Stress reduction factor
JBV OSLO-SKI
Q-histogram trends for selected core with weakness zones
or faults: aggregate of seven holes.
Rev.
Borehole No.: Seven holes
Depth zone (m) Range 18-144m
Report No. NB&A #1
Drawn by NB&A
Checked nrb
Approved
Figure No. AA8
Date 1.9.09
(c)
Time (h)
Penetration Rate (m/h)
Advance Rate (m/h)
best
good
fair
poor
extremely poor
1 day
1 week
1 month
1 year
5 years
(d)

◂**Fig. 13** The photo is one of the weakness zones that was Q-logged. The Q-histogram statistics for seven such zones are shown. The Q_{TBM} model shows a predicted 3 months delay (due to steep -m gradients). Input data was based on V_P data from focused refraction seismic above nine simulated weakness zones. In these simulations, no pre-treatment has been modelled, meaning the delay caused by pre-injection (doubled – m, approx.)

Fault zone delays on TBM can be convincingly explained by a 'theo-empirical' equation, as shown in the fault zone example from the ultra-challenging Pinglin tunnels in NE Taiwan (Shen et al. 1999). A particularly challenging situation is illustrated in Fig. 20. The writer was able to visit Pinglin project during several by-pass situations. Eventually all three TBM were replaced by drill-and-blast. One was destroyed in a fault zone.

The simple equations shown in Fig. 20 need to be understood. The first three are obvious assuming that U is acknowledged to be time-period dependent (Figs. 3 and 4 show overwhelming support for this). The fourth equation explains why serious fault zones (with low Q and steep deceleration due to negative -m) are such a serious threat, unless the effective Q-value can be improved by successful pre-injection. Without this improvement the 'too large' component $1/(1 + m)$ in Eq. 4 (Fig. 20) may be too large, to the extent that a TBM may need to be left in the mountain and a drill-and-blast bypass constructed for D + B completion. A comparison of TBM and D + B is shown in Figs. 21 and 22. The mean Follobanen experience is drawn in Fig. 22.

6 Why Long TBM Tunnels May not Be Faster Than D + B – Hybrid Suggestion

Each and every tunnel, whether drill-and-blast or TBM will have a geologically determining rock mass quality statistics. In Fig. 23, it is modelled on hypothetical Q-system rock classes. A shorter, shallower tunnel is shown and a deeper, longer tunnel. The former may have boreholes and refraction seismic, the latter usually has a lot of relatively unknown hydro-geological details, even if the expected geology seems to be well understood. For the above reasons the two tunnels are given slightly different Q-class predictions. Note the 'tails' (low Q, high Q) that are different.

Although only hypothetical, it is quite likely that the major (FF) faults shown in Fig. 23 have allowed valleys to be carved out by glacio-fluvial processes. If so, then multiple 'attack points' (and ventilation) could be available for an early start by drill-and-blast, while waiting for TBM delivery. Reserve the better investigated and perhaps lower-cover sections for the subsequently delivered TBM. They will be tunneled much faster once they start boring. This deliberate hybrid approach was recommended previously in a Chinese Academy of Engineering workshop (Barton 2012).

Table 1 Preliminary seismic refraction measurements focused on low Q-value-and-low-velocity weakness zones where potential delays were assumed

Approx Q-value in 25 m depth	0.01–0.03	0.03–0.1	0.1–0.3	0.3–1.0	1.0–3.2	3.2–10	10–32	32–100	100–320
Vp (km/s)	1.5–2.0	2.0–2.5	2.5–3.0	3.0–3.5	3.5–4.0	4.0–4.5	4.5–5.0	5.0–5.5	5.5–6.0
No. of measurements in range	I (1)	III (3)	IIIII IIII (9)	IIIII (5)	IIIII IIIII IIII (14)	IIIII IIIII IIIII II (17)	IIIII IIIII IIII (14)	IIIII IIIII IIIII IIIII IIIII (25)	IIIII (5)
Thickness of weakness zones ≈ (m)	15 22	13 20 31 30 8	8 35 6 10 11 9 17 40	16 40 15 8	8 15				
Sum of thicknesses ≈ (m)	37	102	136	79	23	Σ = 377 m recorded with various seismic profile directions, typically NNE and WSW in 'focused' lows			

Data available in 2009

Table 2 A more complete description of the rock mass to be tunneled through with seismic refraction measurements also along sections of the future tunnels with high velocities and high Q-values (Made available to the author in 2018, since now working for the contractor)

Approx Q-value in 25m depth	0.01–0.03	0.03–0.1	0.1–0.3	0.3–1.0	1.0–3.2	3.2–10	10–32	32–100	100–320
V_P (km/s)	1.5–2.0	2.0–2.5	2.5–3.0	3.0–3.5	3.5–4.0	4.0–4.5	4.5–5.0	5.0–5.5	5.5–6.0
Number of measurements in given V_P range	II	III II I II I II	IIIII II I III IIII IIII	I II II IIIII IIIII IIII	I I I IIII IIIII	IIIII IIII III III I I	IIIII II IIIII IIII IIIII IIIII IIIII IIIII I IIIII IIII	IIIII IIIII IIIII IIIII II IIIII IIIII II IIIII II IIIII IIIII II	IIII IIIII I IIIII
Total number	2	11	19	19	12	17	49	53	15
						Well jointed	Sparse jointed	Quite massive	Massive rock
Sum of thicknesses ≈ (m)	138	221	252	242	311	975	2769	3229	771

Note the dominance of $V_P = 5.0–5.5$ km/s, and many assumed Q-values of approx. 30–100. This helps to explain the greater difficulties with the four Herrenknecht TBM, since many more kilometers of hard, high-V_P, high-Q, high-Q_c and in particular high Q_{TBM} rock masses

Table 3 The overall results for the four Follobanen Herrenknecht TBM are tabulated here

TBM	m	Start date	Finish date	Weeks/hrs (appr)	PR (mean) m/h	AR (mean) m/h
TBM 1 N	8.898	5 Sept 2016	11 Sept 2018	105/17,640	1.9	0.50
TBM 2 N	8.908	6 Oct 2016	11 Sept 2018	101/16,970	1.9	0.52
TBM 3 S	9.142	30 Nov 2016	26 Feb 2019	116/19,490	2.0	0.47
TBM 4 S	9.129	7 Nov 2016	26 Feb 2019	119/19,990	1.9	0.46

Records show that all the four TBM were struggling to reach up to 0.4–0.6 m/h AR and the final mean result after approximately 2 years was 0.49 m/h (Macias and Barton 2022)

With their huge high-speed rail network and record number of TBM tunnels, the experience within China is exceptional. Nevertheless, to give 'geology' the influence it deserves, it is a fact that a Millenium Tunnel project in China required acceleration by drill-and-blast to achieve completion within the deadline. This was a case of an unplanned hybrid solution.

7 The Ultimate Challenge of TBM Tunnels that Are Too Deep

The author has had the opportunity to review and suggest solutions for some of the deepest TBM tunnels, meaning 2 km or more, including the Olmos Tunnel through the Andes in Northern Peru, and the massive Jinping II headrace tunnels in China with four parallel tunnels: two TBM and two drill-and-blast, and the tragically ending pilot TBM of smaller diameter, in which there were many fatalities due to a massive rock burst. This had reached the high overburden region first.

The problem is initially two-fold: too high cover (several kilometers at > 2 km depth) and in the case of Jinping II insufficiently strong marble. A simple Kirsch-based elastic-medium calculation showed that the most simplistic $2 \times 2000 \times 2.5/100$ MPa = 100 MPa estimate of tangential stress, even higher with stress anisotropy, was too close to the approximately 90–120 MPa UCS of the marble.

The standard (but actually misinterpreted) rule-of-thumb is that 'stress-induced' fracturing begins when σ_θ maximum reaches $0.4 \times$ UCS ($\pm$ 0.1) from Martin et al. (1999) and shown in Fig. 24. The observed onset of 'stress' induced fracturing at a similar stress/strength ratio of '0.4' has been part of the Q-system SRF factor since earlier (Grimstad and Barton 1993).

Table 4 Some of the input data and calculations for the 6 zones modelled, showing Q values ranging from 37.5 to 150, and Qc values ranging from 70 to 315.

Zone	Lithology	Stability	Oriented	Rock mass strength					Q_{TBM}	Gradient
		Q	Q_0	Q_c	Q_T	$SIGMA_{CM}$	$SIGMA_{TM}$	SIGMA		m
1	Tonalitic gneiss	150.00	150.00	315.00	0.00	95.29	0.00	95.29	**345.49**	– 0.17
2	Granitic gneiss	71.25	71.25	171.00	0.00	77.71	0.00	77.71	**130.04**	– 0.18
3	Biotite gneiss	45.00	45.00	69.75	0.00	59.69	0.00	59.69	**67.36**	– 0.15
4	Amphibolite	37.50	37.50	103.13	0.00	70.34	0.00	70.34	**17.75**	– 0.11
5	Granitic gneiss	45.00	45.00	108.00	0.00	66.67	0.00	66.67	**64.43**	– 0.15
6	Tonalitic gneiss	47.50	47.50	97.38	0.00	64.41	0.00	64.41	**65.70**	– 0.15
7									**0.00**	0.00
8									**0.00**	0.00
9									**0.00**	0.00
10									**0.00**	0.00

Zone	Lithology	Penetration		Time to advance		Overall performance		
		PR	AR	T	check	ΣL	ΣT	
1	Tonalitic gneiss	1.55	0.54	458.85	250.00	9000.00 m 9.00 km	16382.39 h 22.75 months	PRL (av)
2	Granitic gneiss	1.89	0.67	373.19	250.00			21609.00
3	Biotite gneiss	2.22	0.94	264.89	250.00			
4	Amphibolite	2.81	1.61	155.23	250.00			
5	Granitic gneiss	2.17	0.75	1326.33	1000.00			
6	Tonalitic gneiss	2.16	0.51	13803.89	7000.00			ART (av)
7		0.00	0.00	0.00	0.00			0.5494
8		0.00	0.00	0.00	0.00			

(continued)

Table 4 (continued)

Zone	Lithology	Stability	Oriented	Rock mass strength					Q_{TBM}	Gradient
		Q	Q_0	Q_c	Q_T	$SIGMA_{CM}$	$SIGMA_{TM}$	SIGMA		m
9		0.00	0.00	0.00	0.00					
10		0.00	0.00	0.00	0.00					

The range of calculated Q_{TBM} values, with input for zone 6 shown in Fig. 14a, is 18–345 (see bold). Calculated PR for the six modelled zones ranges from 1.55 to 2.81 m/h, and AR from 0.51 m/h (the dominant 6 km) up to 1.61 m/h (for a modelled 250 m long zone)

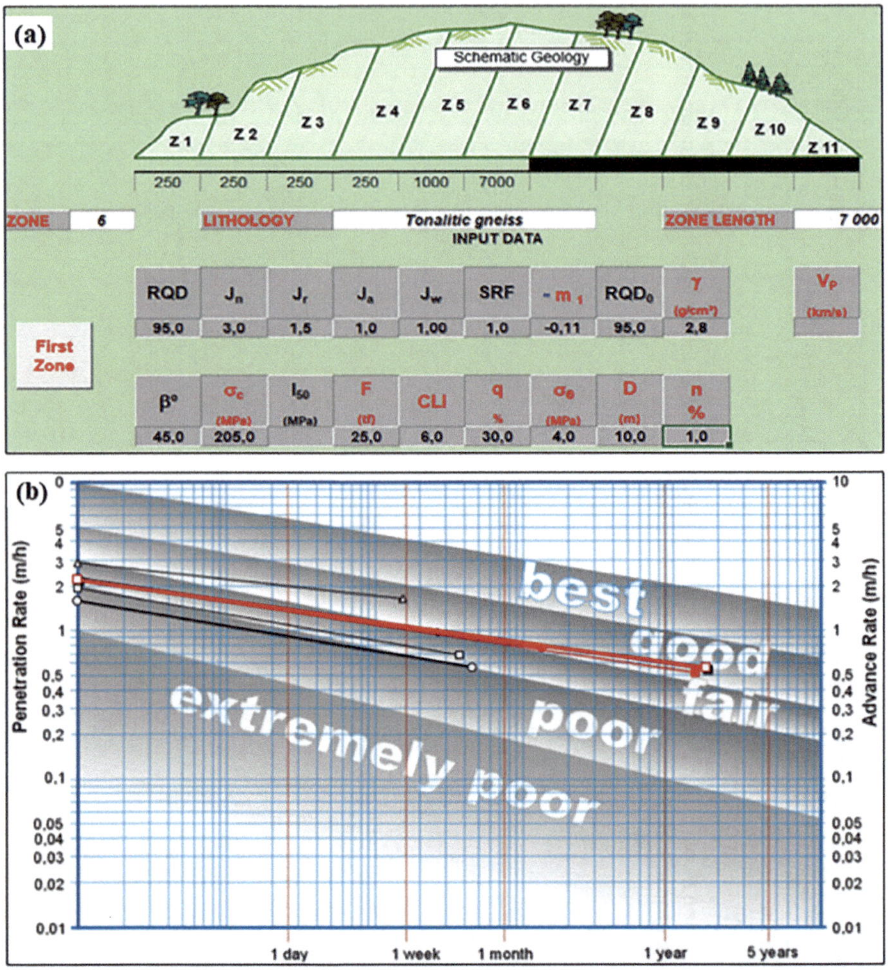

Fig. 14 The six modelled lengths demonstrate mean PR = 2.3 m/h (slightly high). The final AR = 0.5 m/h, is identical to that experienced on average in the 36 km of Follo Line TBM tunnelling, with four Herrenknecht TBM. Note that the 'zone 6' dominant 7 km, as shown on the input data screen, has an assumed Q = 95/3 × 1.5/1 × 1/1 = 47. In fact, the most frequent J_n observation in TBM face logging was 2.0 (one set), and $J_n = 4$ (two sets)

This is shown in Table 5 by the rapidly 'accelerating' values of SRF as the ratio 0.4 was exceeded. In fact, the fracture initiation explanation is due to something else. The uniaxial compressive strength of rock is not its lower-bound strength, nor is cohesive strength, while tensile strength is, when combined with Poisson's ratio. The explanation is shown in Fig. 25, from Barton and Shen (2017).

Fig. 15 Mine access tunnel in Chile driven with double-shield SELI machine. In an unexpected first one month, the TBM operators struggled with 'over-excavation' in a fractured zone with 'too many' joint orientations. When $J_n/J_r \geq 6$, over-break or over-boring is likely, and in this particular zone $Q \approx 40/15 \times 1.5/4 \times 1.0/2.5 = 0.4$ 'very poor' (i.e. $J_n/J_r \geq 6$) (see Fig. 16 for 'overbreak' rule from Q)

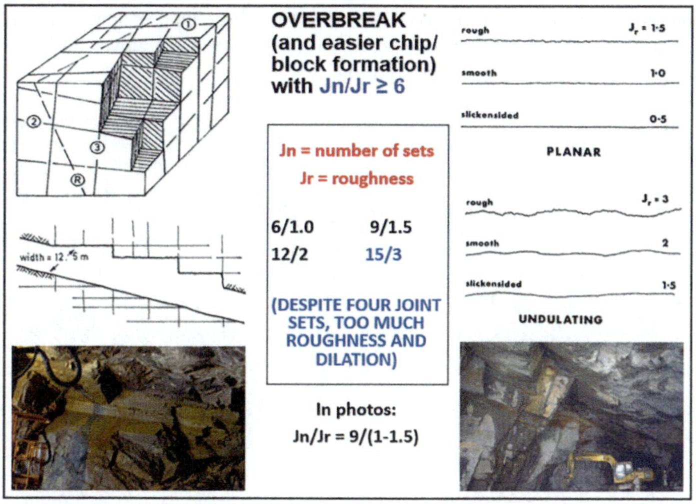

Fig. 16 In drill-and-blast tunnels the rule-of-thumb for almost inevitable overbreak is when the Q-system based ratio $J_n/J_r \geq 6$. This will tend to affect TBM also, especially if assisted with adverse J_a (weathering, clay coatings, clay fillings). These important factors J_n, J_r and J_a are not adequately represented when just adding joint sets as in the NTNU prognosis method. It is hard to compare the NTNU and Q_{TBM} methods of prognosis because the former is missing the reasons for over-break. This is not just the sum of joint frequency or spacing

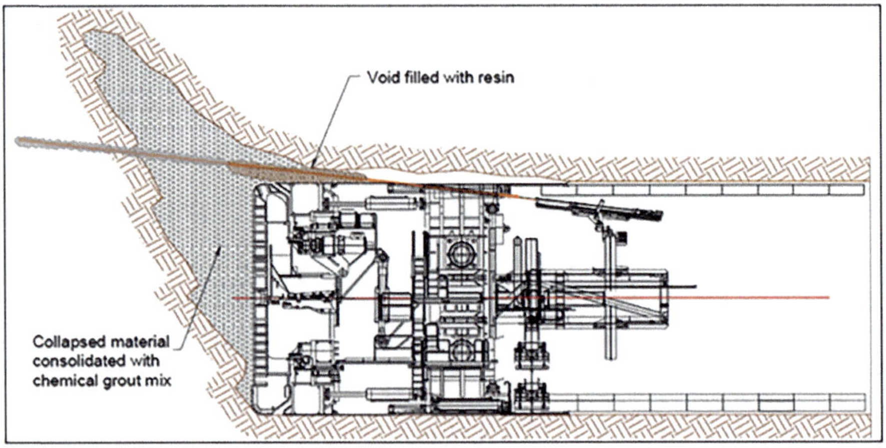

Fig. 17 Over-boring due to over-breaking fractured rock without significant clay is a problem that was encountered in the Chile mine access TBM, Fig. 15. Filling the frequent voids in the arch and between fallen blocks is not solving the reason for, or preventing, the over-boring. Pre-injection with cement grouts should have been the solution

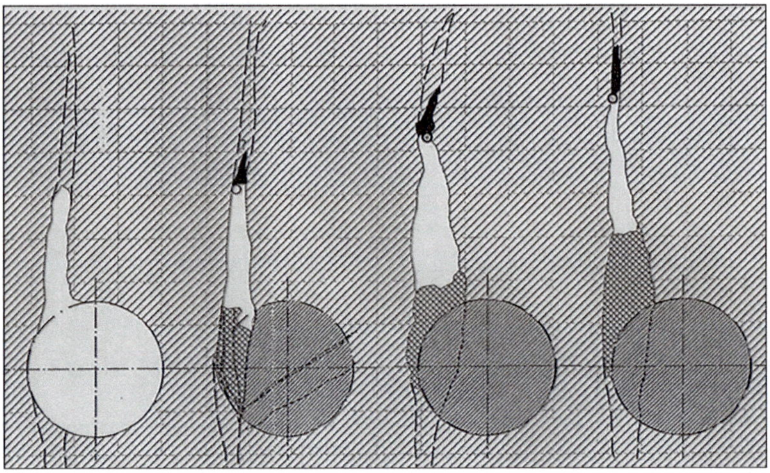

Fig. 18 An example of problems encountered with one of the faults at chainage 2498-2517 m, Pont Ventoux, Italy. The following is a summary. 1. Clay compromises grippers. 2. Plentiful water erodes a shaft and loosens blocks. 3. Tunnel stability reduces. 4. Cutter-head is repeatedly blocked. See Fig. 19 for specific details experienced during a 5 months period

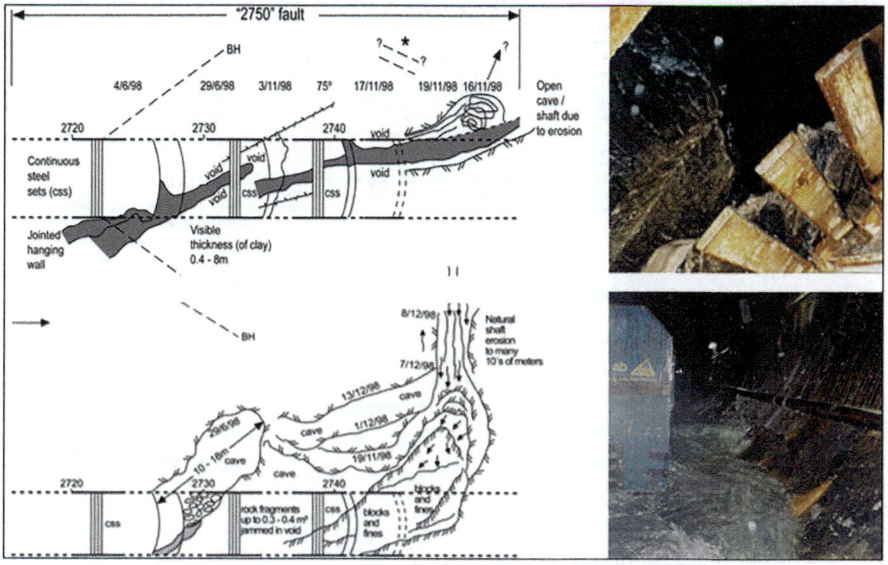

Fig. 19 This sub-parallel fault was encountered for some 30 m and delayed the project by 5 months. A 'natural' shaft developed above the cutterhead. It was eroded by water and falling blocks. The final result was a bypassed TBM (see one of the distant crosses in Fig. 3a). (Figures, photos from Karl Gunnar Holter and NB. The author superimposed the dated over-break records of KGH)

In carefully instrumented uniaxial compression tests, acoustic emission is also seen to start at the stress/strength ratio of '0.4'. The reasons are actually the arithmetically equal ratio of tensile strength and Poisson's ratio: $0.4 \times \text{UCS} \approx \sigma_t/\nu$, (Fig. 25). This more correct explanation of fracture initiation was discovered by Baotang Shen (Shen and Barton 2018). Shen was the developer of the fracture mechanics based FRACOD code which is seen applied to highly 'stressed' (or strained) TBM simulations in Fig. 25 (Old Beaumont TBM) and in Fig. 26 (1000 m deep simulations in granite) showing the 'positive' effect of jointing in dissipating the highest stress and strain.

Roughly speaking, strain-induced tensile fracturing will start in a TBM driven in massive rock when the estimated tangential stress reaches $4 \times$ to $5 \times \sigma_t$, since Poisson's ratio is often 0.2–0.25. If stresses are much higher, propagation in unstable shearing (rock burst) may occur (Shen and Barton 2018).

8 Conclusions

- A principal finding from case records, both open-gripper, double-shield and world record performances is that, after the learning curve there is deceleration of advance rate (AR) due to reduction of utilization (U) as challenges appear as the

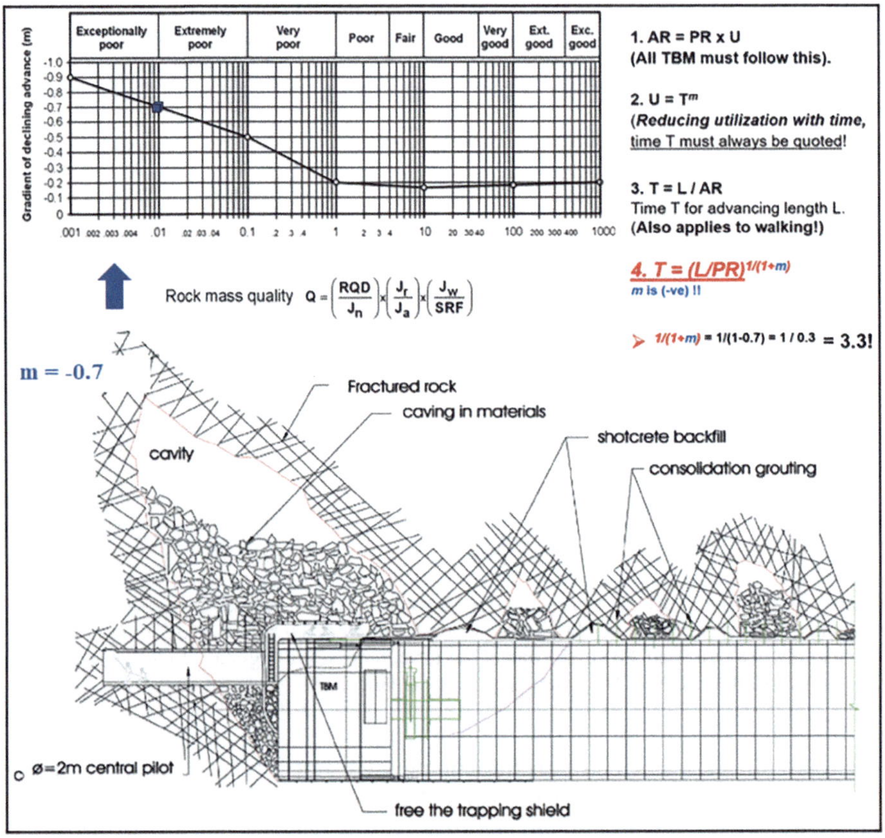

Fig. 20 Top-left is the suggested link between the regular Q-value and the deceleration gradient (−)m. This was published in Barton (2000). With the benefit of projects such as Follobanen (Norway), Guadarrama (Spain) and Gastau (Brazil) driven in particularly high-Q, high-UCS rocks a 'U-shaped' trend curve could now be envisaged, climbing on the right side from approx. Q = 40 when Q_C-values are 'too high'. The equations are explained in the text that follows

tunnel gets longer (factors such as difficult faults, excessive water, pre-injection needs, abrasive hard rock, soft-mixed-face, hard-mixed-face, increased cutter ring and bearing replacement). U needs to be recast as T^m where T is total hours (24, 168, 720, etc.) and m is the negative deceleration gradient.

- There is a need to estimate rock mass strength to compare with cutter force in TBM prognoses. Sometimes PR may reduce despite increased cutter force. The number of joint sets J_n and joint properties J_r and J_a in the Q-rating obviously affect over-break and local cutter action, and the potential for over-boring, so they are also needed and are important parts of Q_{TBM}.
- Descriptions of the rock mass are limited to each 24 h (maintenance shift) in double-shield with PC-elements. So double-shield case records with logging at

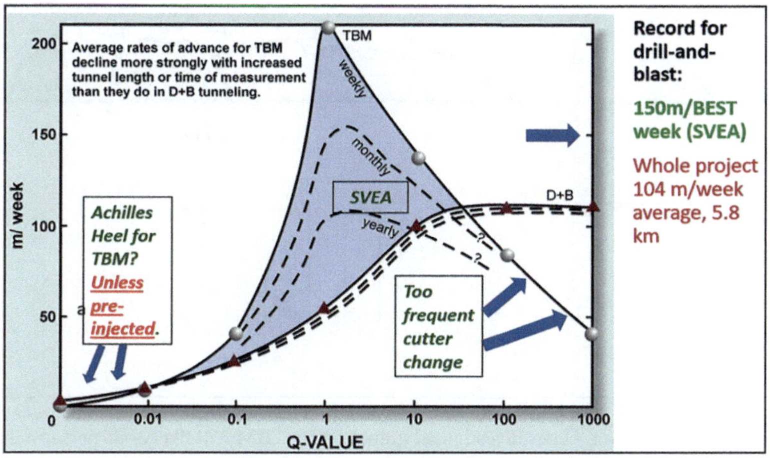

Fig. 21 Comparison of a moderately fast TBM project (5 km in 1 year) showing dependency on ideally jointed conditions if to remain faster than drill-and-blast (Barton 2000 with additions). The boxes emphasize the two delaying factors for TBM where drill-and-blast may be faster. Note the world record SVEA data for 5.8 km of D + B in coal-measure rocks on Svalbard

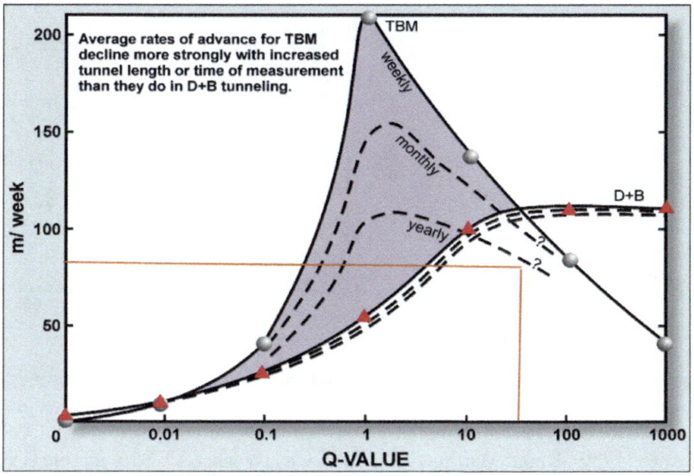

Fig. 22 The added 'rectangle' assumes a Q-value of ≈ 40, intersecting an approx. 2 years predicted result and with mean ≈ 80 m/week on the vertical axis. Mean completion time for the four 9 km tunnels: 110 weeks from Table 3 (Check: 0.49 m/h = 168 × 0.49 = 82 m per week average, 82 × 110 ≈ 9.02 km. Mean tunnel length = 9.02 km)

only 15, 25, or 35 m, etc., intervals may be unreliable in relation to continuous open-gripper rock mass descriptions. Critics (of other's TBM prognoses) should be careful using such sporadic 'data'.

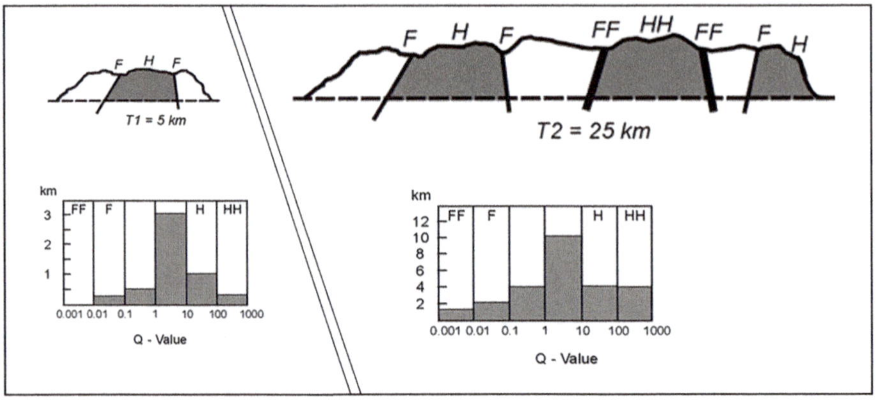

Fig. 23 **Left**: lower cover, better investigated tunnel where TBM will likely out-perform drill-and-blast by a considerable margin (except for the delivery/assembly time delay). **Right**: a much longer, deeper tunnel with the likelihood of a number of major faults (FF) and harder more massive rock (HH). The 'extremes' at either end of the Q_{TBM} scale (Fig. 6) may delay the one (or two) TBM more than expected

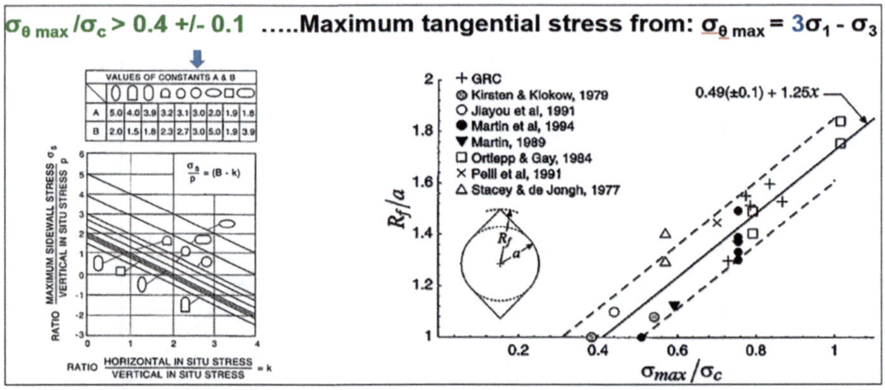

Fig. 24 Due to their cylindrical/circular shape and 'slow' method of excavation, TBM tunnels maximize the likelihood of Kirsch-based elastic theory applying. The above example giving a 100 MPa estimate for $\sigma_{\theta\,max}$ is the elementary first step ready for comparison with UCS or σ_c. Left: Hoek and Brown (1980). Right: Martin et al. (1999). Clearly when UCS for marble is 90–120 MPa, and the tangential stress is at least as high as 100 MPa, then major problems must be expected. The author recommended removal of the TBM at Jinping II, but this did not occur until after the major fatalities in the pilot bore, with final completion of all tunnels by drill-and-blast. Remarkably there are now 'high-stress' (rock-burst) research laboratories parallel to the completed tunnels, where cover exceeds 2 km

Table 5 The empirically based 'acceleration' of SRF after the stress/strength ratio has exceeded ≈ 0.4 (Grimstad and Barton 1993). Higher SRF gives lower Q-value, hence heavier support, especially based on fibre-reinforced energy absorbing shotcrete, and bolts that can yield, such as grouted D-bolts

(b) Competent rock, rock stress problem		σ_c/σ_1	σ_θ/σ_c	SRF
H	Low stress, near surface, open joints	> 200	< 0.01	2.5
J	Medium stress, favorable stress condition	200–10	0.01–0.3	1
K	High stress, very tight structure. Usually favorable to stability, may be unfavorable for wall stability	10–5	0.3–0.4	0.5–2
L	Moderate slabbing after > 1 h in massive rock	5–3	0.5–0.65	5–50
M	Slabbing and rock burst after a few minutes in massive rock	3–2	0.65–1	50–200
N	Heavy rock burst (strain-burst) and immediate dynamic deformations in massive rock	< 2	> 1	200–400

- Pre-injection helps to improve most of the six Q-parameters, so the deceleration gradient (-m) is made less negative. This means fewer long delays for the TBM, despite the time needed for the pre-injections.
- A simple formula that describes why there may be major delays in low Q-value fault zones has been available for more than two decades. Improvement of the effective ultra-poor quality of fault zones with flushing and cement injection ahead of the face using multiple boreholes can help here. The Q-value is effectively improved.
- Pre-injection will also be needed to control against groundwater drawdown and settlement damage. The problem is that TBM may get delayed before the water-tight lining reaches the leaking zone, so avoiding damage caused by ground-water drawdown is helped by pre-injection after discovery of permeable zones by multiple, continuous probe drilling.
- Do not 'automatically' choose TBM for long tunnels. A 'hybrid' solution might have advantages. Drill-and-blast the deeper and less investigated sections, if intermediate access is possible, while waiting for TBM delivery for fast excavation of the better investigated sections. Help maintain the good reputation of TBM. There have been many bad decisions and final forced hybrid with TBM destroyed or remaining in the mountain, and regrettable multiple fatalities in well-known cases such as in Taiwan, China, Pakistan. In the case of Taiwan it was just the severely faulted rock mass, not great depth.
- The assumption of 'stress-induced' fracturing when the estimated maximum tangential stress reaches approx. 0.4 × UCS in deep hard-rock TBM tunnels

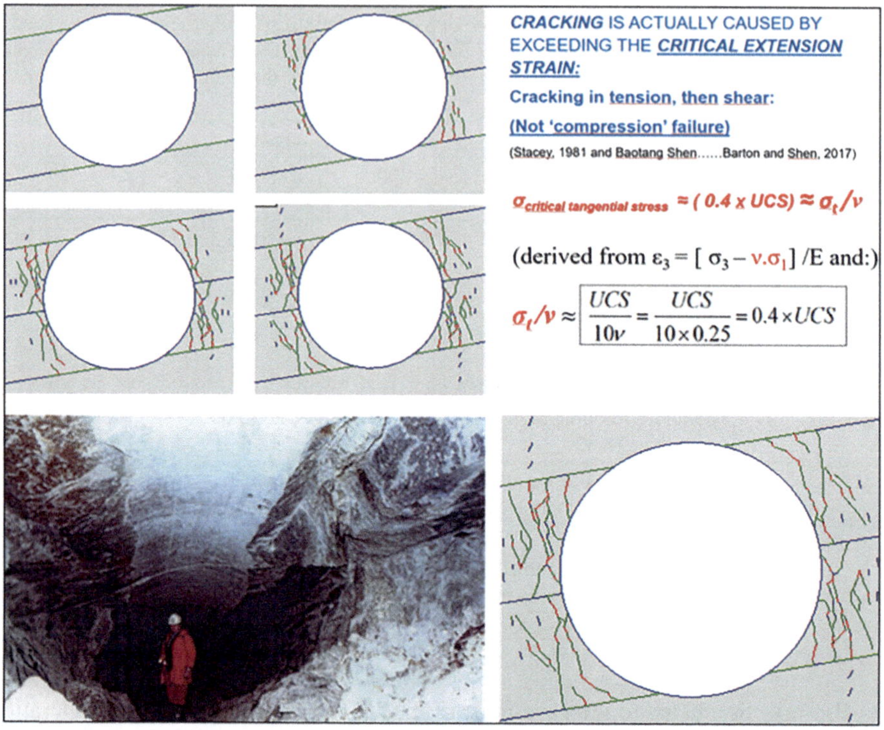

Fig. 25 The explanation for extension strain caused extension fracturing (red) with propagation in shear (green) from FRACOD modelling by Baotang Shen. The rule-of-thumb that fracturing (and AE in a laboratory test) start when the tangential or major stress reaches 0.4 × UCS (± 0.1) is more correctly given as tensile strength divided by Poisson's ratio. The photograph is of the Old Beaumont TBM pilot tunnel from 1880, driven about 1 km, 110 years before the nearby Channel Tunnel in the same chalk marl. Here the tunnel has curved under a 70 m high cliff, which was enough to increase the tangential stress by several MPa, which was sufficient to fracture the weak 4–9 MPa rock (Barton and Shen 2017)

is standard practice, also for SRF estimation in drill-and-blast tunnels in massive rock. Recent work by fracture mechanics expert Baotang Shen shows that extension strain caused by Poisson's ratio and the anisotropic stress near the tunnel wall actually causes (extensional) tensile fracturing, nothing to do with 'compressive' failure. Unstable propagation in shear might follow.

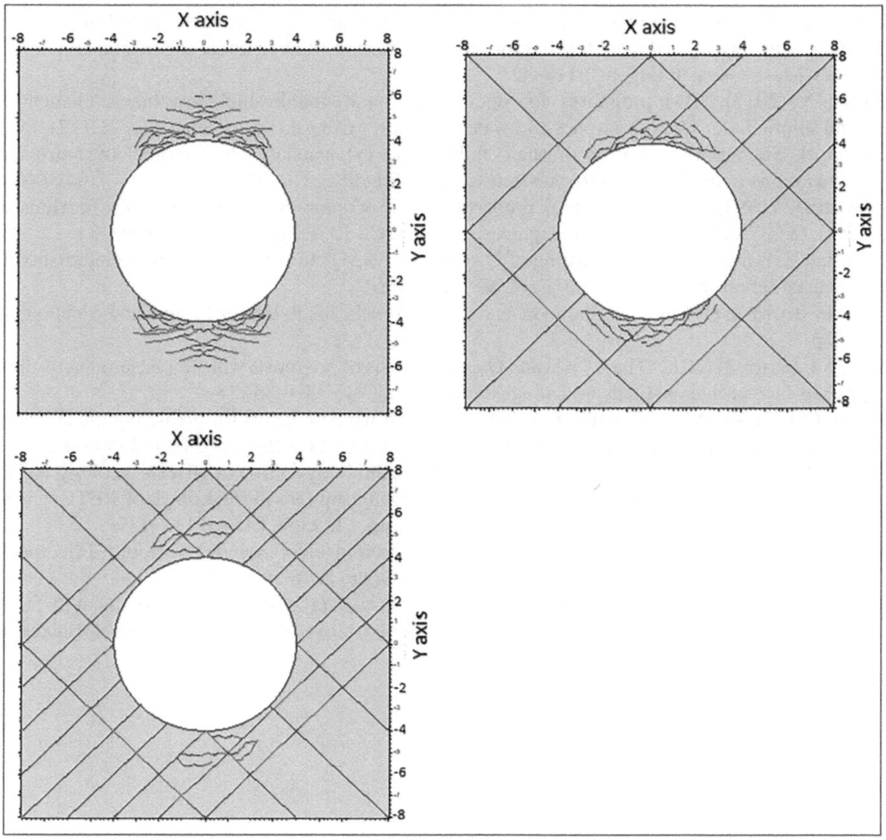

Fig. 26 FRACOD modelling by Shen of hypothetical 1000 m deep TBM tunnel in massive unjointed, and partially jointed rock masses with the same strength properties, but with simple Mohr–Coulomb 'c and ϕ' jointing. Minor displacement on the joints dissipates to some extent the highest stresses. The damage caused by drill-and-blast as compared to the 'non-damaging' TBM is the reason for the occasional need to not rely on TBM when tunnels are exceptionally deep and the rock is not strong enough

References

Barton N (1995) The influence of joint properties in modelling jointed rock masses. In: Keynote lecture, 8th ISRM congress, Tokyo, vol 3. Rotterdam, Balkema, pp. 1023–1032

Barton N (2000) TBM tunnelling in jointed and faulted rock. Balkema, Rotterdam, 173p

Barton N (2007) Near-surface gradients of rock quality, deformation modulus, Vp and Qp to 1km depth. First Break, EAGE 25:53–60

Barton N, Gammelsæter B (2010) Application of the Q-system and QTBM prognosis to predict TBM tunnelling potential for the planned Oslo-Ski rail tunnels. Nordic Rock Mechanics Conf., Kongsberg, Norway

Barton N (2012) Reducing risk in long deep tunnels by using TBM and drill-and-blast methods in the same project—the hybrid solution. Keynote lecture. Risk in underground construction. J Rock Mech Geotech Eng 4(2):115–126

Barton N (2013) TBM prognoses for open-gripper and double-shield machines, challenges and solutions for weakness zones and water. Fjellsprengningskonferansen, Oslo, 21.1–21.17

Barton N, Shen B (2017) Risk of shear failure and extensional failure around over-stressed excavations in brittle rock. J Rock Mech Geotech Eng 9(2):210–225

Fawcett DF (1993) The effects of rock properties on the economics of full face TBMs. In: Hudson et al. (eds) Comprehensive rock engineering, vol 4, Ch 10. Pergamon, UK, pp 293–311

Grimstad E, Barton N (1993) Updating of Q system for NMT. In: Proceedings of the international symposium on sprayed concrete, Fagernes, pp 46–66

Hoek E, Brown E (1980) Underground excavations in rock. Institution of Mining and Metallurgy, 527p

Macias J, Barton N (2022) The NTNU and Q_{TBM} methods of prognosis with emphasis on hard-rock mixed-face problems. Fjellsprengningskonferansen, Oslo 35(1–35):31

Martin CD, Kaiser PK, McCreath DR (1999) Hoek-Brown parameters for predicting the depth of brittle failure around tunnels. Can Geotech J 36:136–151. https://doi.org/10.1139/t98-072

Shen CP, Tsai HC, Hsieh YS, Chu B (1999) The methodology through adverse geology ahead of Pinglin large TBM. In: Orlando FL, Hilton, Samuelson (eds) Proceedings of RETC, Ch 8. Society for Mining, Metallurgy, and Exploration, Inc., Littleton, CO, pp 117–137

Shen B, Barton N (2018) Rock fracturing mechanisms around underground excavations. J Geomech Eng Korea Sci **16**(1):35–47 https://doi.org/10.12989/gae.2018.16.1.035

Sundaram NM, Rafek AG (1998) The influence of rock mass properties in the assessment of TBM performance. In: Moore, Hungr (eds) Proceedings of 8th IAEG congress, Vancouver. Balkema, Rotterdam, pp 3553–3559

University of Tehran Model, a Geological-Based Model for Rock Mass Boreability Classification and TBM Performance Prediction

Jafar Hassanpour and Jamal Rostami

1 Introduction

Over the past few decades, researchers have developed various models to predict the performance of tunnel boring machines (TBMs) in rock. The most important ones are included in this book. These models, categorized into theoretical and empirical types, are widely used at different stages of mechanized tunneling projects to analyze TBM performance. Although they effectively determine penetration and advance rates, the advancements in manufacturing technology and diverse geological conditions necessitate the development of new models based on updated data.

Over the past two decades, many long and deep tunnels have been excavated in different geological zones of Iran using mechanized technology. The information collected from these projects has been invaluable for developing comprehensive databases to evaluate machine performance and create new models. Given the vastness and geological diversity of Iran, along with the presence of numerous tunnels, this database features highly varied geological conditions, which is a significant advantage.

Recently, the authors of this chapter have developed a new empirical model (Hassanpour et al. 2009, 2010, 2011, 2015, 2021, 2023a; b; Goodarzi et al. 2021; Pourhashemi et al. 2021a, b, c; Samadi and Hassanpour 2025) using this valuable data, which will be detailed in this chapter. Since a large part of the research carried out to develop this new model was conducted as research projects at the School of

J. Hassanpour (✉)
University of Tehran, Tehran, Iran
e-mail: hassanpour@ut.ac.ir

J. Rostami
Colorado School of Mines, Golden, CO, USA
e-mail: rostami@mines.edu

J. Hassanpour and J. Rostami (eds.), *Performance of Tunnel Boring Machines (TBM) in Rock*, Springer Series in Geomechanics and Geoengineering,
https://doi.org/10.1007/978-3-032-20424-0_5

Geology, University of Tehran, this model is named after the "University of Tehran" and is abbreviated as the "UT Model" in this book.

A unique feature of the UT model is its emphasis on the geological characteristics of the rock mass. During its development, efforts were made to consider certain geological parameters that influence the cutting process but are difficult to quantify. Consequently, the data in the developed database were analyzed from an engineering geology perspective before model development, resulting in an initial classification based on direct observations of the cutting process in various geological conditions.

Based on the above explanations, the methodology used in this research to analyze TBM performance and develop a novel empirical prediction model consists of five distinct stages:

- Collecting raw data from completed or ongoing tunneling projects;
- Screening the data to create a comprehensive database;
- Classifying the data based on geological conditions affecting rock mass boreability;
- Analyzing the classified data to develop empirical equations (Developing UT model);
- Validating the proposed model.

2 Collecting Data from Tunneling Projects

Obtaining suitable data on geological conditions and actual machine operational parameters is crucial for the investigation of TBMs performance and the development of new empirical models. These data need to be both quantitatively adequate and qualitatively reliable, devoid of flaws and errors as far as possible.

As mentioned, a notable number of long and deep tunnels have been excavated using mechanized method in soft to hard rock conditions across different geological zones in Iran in the past two decades. The locations of these tunnels are presented in the simplified geological map of Iran (Fig. 1).

Fortunately, many of these projects provided sufficient information on machine performance and the geological conditions of the tunnel routes. Due to the authors' direct involvement with several of these projects, a comprehensive database was established. Table 1 lists the projects from which the authors personally gathered data during their construction.

Table 1 shows that all of these tunnels, totaling 122.7 km in length, are used for inter-basin water transfer. The diameters of the tunnels in Iran range from 4.5 to 6.7 m, all excavated and completed by shield machines capable of installing concrete segments. The only exception is the Manapouri second tailrace tunnel (MSTT) in New Zealand, which was excavated using an open/gripper-type machine with a diameter of 10 m, due to the project's geology located in hard and stable rock.

During data collection phase, geological data and machine performance details were collected through tunnel face observations and measurements (Fig. 2), in situ tests, laboratory tests on collected samples, shift/daily report reviews, data analysis

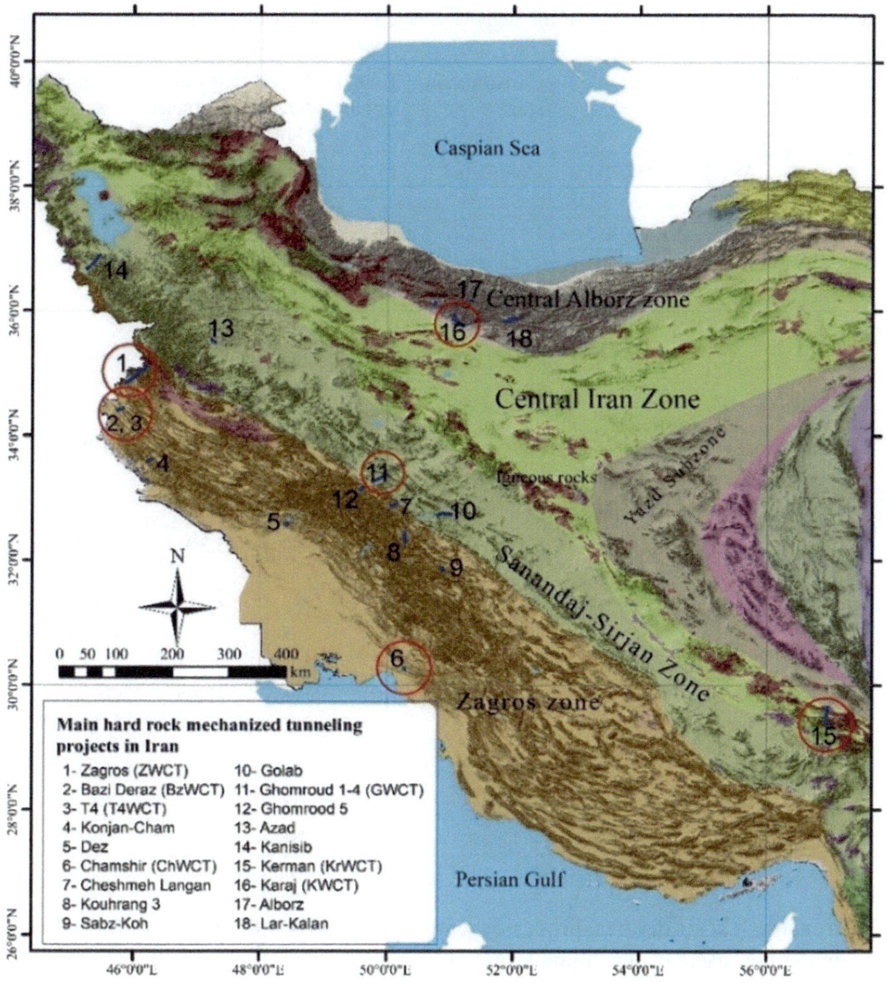

Fig. 1 The locations of completed and ongoing mechanized rock tunnels in various geological zones of Iran since the 1990s (Hassanpour et al. 2023a)

from the TBM acquisition system, examination of as-built maps, and analysis of geological and geotechnical reports from preconstruction and construction phases. This data was then systematically organized, classified, and documented to analyze machine performance and develop empirical models (see next sections).

Table 2 presents the geological characteristics of the selected tunnels. As shown in Table 2 and Fig. 1, these tunnels are situated in four distinct geological zones of Iran, each with unique characteristics.

Classifying tunneling project sites by geological zones is essential, as each zone has unique characteristics that significantly influence construction conditions. The geological phenomena and processes encountered in each zone highlight these

Table 1 General characteristics of the tunnels selected for establishing database and developing UT model

No.	Project name	Abbr.	Construction stage	Tunnel length (km)	Bored length (km)	Tunnel diameter (m)	Machine type
1	Karaj WCT, lot 1	KWCT-1	Completed	15.7	15.7	4.66	Double shield
	Karaj WCT, lot 2	KWCT-2		13.7	13.7		
2	Ghomrood WCT, lots 3, 4	GWCT-3,4		18	18	4.52	
3	Zagros WCT, lot 2	ZWCT-2		26	26	6.73	
4	Bazideraz WCT	BzWC-T		8.6	8.6	6.86	EPB-HR
5	T4 WCT	T4WC-T		4.2	4.2		
6	Chamshir WCT	ChWC-T	Stopped	7.7	2.5	5.30	Single shield
7	Kerman WCT, southern lot	KrWCT-S	Under construction	19	5.5	5.28	Double shield
8	Manapouri second tailrace tunnel (New Zealand)	MSTT	Completed	9.8	9.8	10.05	Gripper type / Open
Sum				**122.7**	**104.0**		

differences. These diverse processes can significantly affect the performance of the machines used in the tunnels, which is the primary focus of this chapter. The variability of geological conditions in these tunnels is very important because the geological diversity of the established database will directly influence the quality and applicability of the developed prediction models.

From an engineering geology perspective, a wide range of features can be identified in the rock masses surrounding the selected tunnels, including very fractured, jointed/blocky, and intact rock. Variations from soft to extremely strong rocks, as well as from non-abrasive to highly abrasive rocks, are evident along the tunnel paths. Various types of foliated and non-foliated rocks have been encountered, along with rocks exhibiting different bed thicknesses and varying intensities of folding and faulting.

The significant variability in the geological and geotechnical properties of these rocks has resulted in diverse data being recorded in databases. As stated above, this diversity is vital for the development of comprehensive models that aid in predicting machine performance under varying geological conditions.

Fig. 2 Collecting required data during construction and pre-construction phases; **a–c** Tunnel face observation in limestones and marly limestones of ZWCT-2 project, **d** intact core obtained from a borehole in KWCT-1 project, **e, f** investigation of rock mass condition behind the shield in two sections of GWCT-3,4 project (Hassanpour et al. 2023a)

3 Main Database

As previously stated, after thorough evaluation, eight tunneling projects were selected based on their substantial and relevant data (see Tables 1 and 2). Table 3 details the data from these projects used to build the databases and develop and validate the UT model.

The databases established in this study are designed to evaluate the impact of engineering geological and operational parameters on machine performance (Hassanpour 2009) and cutter consumption (Hassanpour et al. 2014; Hassanpour 2018), develop and validate new empirical models (Hassanpour et al. 2009, 2010, 2011, 2015, 2016, 2021, 2023a; b; Goodarzi et al. 2021; Pourhashemi et al. 2021a, b, c), correct common prediction models (Hassanpour et al. 2024), and assess the correlation between engineering rock mass properties (Hassanpour et al. 2022).

As shown in Table 3, 199 tunnel sections from various projects were selected, and their corresponding data were documented in the databases used for developing TBM performance prediction models. Typically 5 to 20 m long, each section corresponds to one shift or day and includes several installed segment rings.

Table 2 Geological characteristics of the tunnelling projects selected for establishing database and developing UT model (Hassanpour et al. 2023a)

No.	Project	Geological formations	Lithology	Geological zone	Rock strength	Typical structure
1	KWCT-1, KWCT-2	Karaj formation	Pyroclastic and volcanic rocks such as tuff, tuffite, agglomerate, andesitic tuff and monzodiorite	Central Alborz (CA) zone	Moderately strong to very strong	Blocky and very blocky rocks accompanied by fractured fault zones
2	GWCT-3,4	Jurassic metamorphic rocks (low to medium grade) and Cretaceous limestones	Slate, phyllite, schist and limestone	Sanandaj-Sirjan (SS) zone		Schistose and foliated rocks accompanied by fault zones
3	ZWCT-2	Pabdeh, Gurpi and Ilam formations	Carbonate-Argillaceous rocks such as shale, marlstone, limy shale, argillaceous limestone	Zagros simply folded (ZSF) subzone	Soft to very strong	In surface outcrops: Bedded and jointed; In depth: rocks containing clay minerals are often continuous and isotropic
4	BzWCT	Bakhtiari Fm., Fars group, Asmari Fm	Marlstone, siltstone, sandstone, limestone, gypsum, anhydrite			
5	T4WC-T	Aghajari Fm	Sandstone, quaternary alluvium, siltstone and claystone		Soft (weak) to moderately strong	
6	ChWCT	Aghajari Fm., Bakhtiari Fm	Claystone, marlstone, conglomerate		Soft to very strong	
7	KrWCT-S	Eocene volcanic rocks	Volcanic, pyroclastic and plutonic rocks such as Andesite, Basalt, andesitic tuff, Granodiorite	Urumieh-Dokhtar (UD) zone	Strong to extremely strong	Massive to blocky accompanied by fractured fault zones
8	MSTT	Paleozoic high grade metamorphic rocks	Calcsilicate, metadolorite, meta-andesite, paragneiss, and granitic gneiss	Fiordland complex	Strong to extremely strong	Massive

Table 3 List of tunnels with available data used for developing and validating the UT model (Hassanpour et al. 2023a)

No.	Project	Total length (km)	TBM performance database			Cutter wear database	
			Length of tunnel used for development of model (km)	No. of selected tunnel sections (datasets)	Length of tunnel used for validation of the model (km)	Length of tunnel used for development of the model (km)	No. of tunnel sections (datasets)
1	KWCT-1	15.7	8.7	41	–	16.0	15
	KWCT-2	13.7	–	–	13.5	13.5	12
2	ZWCT-2	26.0	5.6	37	10.0	5.3	2
3	GWCT-3,4	18.0	18.0	31	–	18.0	12
4	BZWCT	8.6	5.5	10			
5	T4WCT	4.2	1.5	5			
6	ChWCT	7.7	2.5	5			
7	KrWCT-S	19	5.5	21			
8	MSTT	9.8	10.0	49	–	10.0	11
	Total	122.7	57.3	199	23.5	62.8	52

In these selected tunnel sections required geological/geotechnical data were obtained directly from the tunnel face and muck materials observations and measurements and by investigating drilled borehole logs and core boxes and related laboratory tests on core samples and mucks (Fig. 2), as well as as-built maps.

The established databases include engineering geological parameters and actual operational and performance parameters of the TBMs employed in the selected projects. To facilitate analysis, the databases have been divided into three distinct parts:

- Part 1: Actual operational parameters of machines and calculated results of machine performance metrics.
- Part 2: Directly measured geotechnical and geological parameters for the rock mass in each tunnel section.
- Part 3: Results of the calculations of machine performance parameters using existing prediction models.

In the first part of the databases, machine parameters, including averages of machine thrust, and rotation speed, torque, and power of the cutterhead, were computed and documented for each selected section. Furthermore, factors such as the length of the excavated section and the time taken to complete the excavation of the section (boring time) were also recorded in this part. Using this data, parameters associated with machine performance, such as averages for penetration rate, cutterhead penetration, field penetration index, and specific energy, were computed and logged with the aid of the relationships outlined in Chap. 1.

The second part of these databases includes geotechnical and engineering geological metrics for each selected section. Data on intact rock properties and the characteristics of discontinuities are essential in this part of the databases. Using this information, key parameters such as rock mass engineering classification systems (e.g., RQD, RMR, Q, GSI) and rock mass strength parameters were calculated for each section.

In the third part of these databases, the necessary input data for the CSM, NTNU, and Q_{TBM} models were calculated or estimated using actual data collected from the selected tunnel sections. Subsequently, machine performance parameters, such as the penetration rate, were determined and documented using these models.

The excavation of tunnel projects using shielded machines and the immediate lining with precast segmental linings posed challenges in accurately identifying geological features along the entire length of the tunnel. For this research, only tunnel sections with reliable geological data were included in the databases. These sections were selected based on the following criteria:

- Access to the tunnel face at both ends for assessment, sampling, and direct measurements;
- Proximity to a geotechnical borehole;
- Engineering judgment that supported reliable extrapolation of rock mass characteristics from surface outcrops to tunnel depth.

Additionally, these tunnel sections displayed the following characteristics:

- No geological issues were reported, including tunnel collapses, blocky rock masses, high water inflow, or cutterhead clogging;
- Geological characteristics and machine operational parameters showed minimal fluctuations, and the machine's performance remained unaffected by any specific or unidentified issues;
- The disc cutters were in good condition with appropriate levels of wear.

4 Engineering Geological Classification of Recorded Data

As explained in previous sections, diverse rock units with various lithologies can be observed along the selected projects' paths, featuring almost all major rock groups, including:

- Detrital sedimentary rocks like conglomerate, sandstone, siltstone, mudstone, and shale;
- Chemical sedimentary rocks like different types of limestone;
- Evaporitic rocks like gypsum and anhydrite;
- Volcanic rocks such as andesite, basalt, and dacite;

- Pyroclastic rocks such as tuff, andesite/dacite tuff, tuffite and agglomerate;
- Igneous intrusive rocks like granite, granodiorite, monzonite, and monzodiorite;
- Metamorphic rocks like slate, phyllite, various types of schists, gneiss, granitic gneiss.

Each rock unit has distinct characteristics and reacts differently during mechanized excavation, influenced by several factors, some of which remain unknown. Therefore, grouping these units based on similar behaviors enhances analyses accuracy by accounting for hard-to-quantify unknown parameters. In this study, an attempt was made to categorize the rock units according to their engineering geological characteristics—such as rock strength, plastic or brittle behavior, and rock mass structure, including foliation and schistosity—focusing on boreability and behavior in mechanized tunneling. Using UNESCO's definitions[1] (IAEG 1976) and considering the study's scale, all rock units in the database were classified into 6 lithological types (LT), as summarized in Table 4. These LTs have been analyzed in various projects, revealing significant differences in their behavior during TBM excavation. Pictures of some of these LTs are shown in Fig. 3. Moreover, Table 4 outlines the projects and geological formations for each LT. Data specific to each identified LT has been utilized independently to develop models tailored for distinct geological conditions.

---

[1] In the field of engineering geology, the classification system put forth by the Commission on IAEG-UNESCO 1976 is utilized for the purpose of delineating engineering geological units. This classification scheme delineates four distinct of engineering geological units based on their levels of homogeneity: Engineering Geological Type (ET), Lithological Type (LT), Lithological Complex (LC), and Lithological Strings (LS).

The Engineering Geological Type exhibits the highest degree of homogeneity in terms of its physical and mechanical properties. It is imperative that this unit demonstrates uniformity in its lithological characteristics and physical state. These units can be characterized based on specific statistical values obtained through the determination of physical and mechanical properties from various samples.

The Lithological Type (LT) represents a larger engineering geological unit composed of multiple engineering geological types. This unit features a homogeneous composition, structure, and fabric, yet displays variations in its lithological characteristics and physical and mechanical properties. Consequently, it is challenging to attribute reliable average values of the physical and mechanical properties to the entirety of the unit. Instead, a broad overview of the general properties can only be provided by considering limited variations in values.

The Lithological Complex (LC) consists of a variety of lithological types that share a common geological genesis and have been formed under similar tectonic settings. While the spatial arrangement of lithological types within a lithological complex is uniform and distinct, the group as a whole may not exhibit uniformity in both geological characteristics and physical properties. It is not feasible to establish comprehensive definitions for the physical and mechanical properties of an entire lithological group, and such properties can only be ascribed to the individual lithological types within the group.

A Lithological String (LS) is composed of multiple lithological groups that have generally experienced analogous paleo-geographic and tectonic conditions. Within a lithological string, there exist common lithological features that impart a shared uniformity and distinguish each sequence from others. For a lithological sequence, only rudimentary engineering geological properties can be delineated.

Table 4 Categorizing rock units to different lithological types based on their corresponding engineering geological properties (Hassanpour et al. 2023a)

No.	Lithological type	Main rock groups	Lithologic variations	Typical structure	Rock mass behavior	Project/ formation
1	LT-I	Pyroclastic and volcanic rocks	Tuff, andesite, agglomerate	Medium bedded to massive + 2–3 sets of joints	Isotropic, discontinuous	Most members of Karaj formation in KWCT
2	LT-II	Moderately strong to strong sedimentary rocks	Sandstone, dolomite, carbonate	Thin to medium bedded + 2 sets of perpendicular joints	Isotropic, discontinuous	Parts of Gurpi and Pabdeh formations in ZWCT-2, cretaceous carbonate rocks in GWCT-3,4
3	LT-III	Low to medium grade metamorphic rocks	Shale, slate, phyllite, schist	Foliated (one predominant discontinuity set) + joint sets	Anisotropic, discontinuous	Jurassic metamorphic rocks in GWCT3,4, and shales of Karaj formation in KWCT
4	LT-IV	Moderately strong to very strong intrusive rocks	Granite, granodiorite	Jointed (2–3 joint sets)	Isotropic, discontinuous	Intrusive rocks in KrWCT-S and MSTT
5	LT-V	Soft sedimentary rocks (< 50 MPa)	Claystone, marlstone, shale, siltstone	In depth, lacking effective discontinuities	Isotropic, continuous	Soft sedimentary rocks from Neogene formations of SFZ
6	LT-VI	Very strong to extr. strong intrusive rocks (> 150 MPa)	Granite, granodiorite	Massive	Isotropic, continuous	Intrusive rocks in KrWCT-S

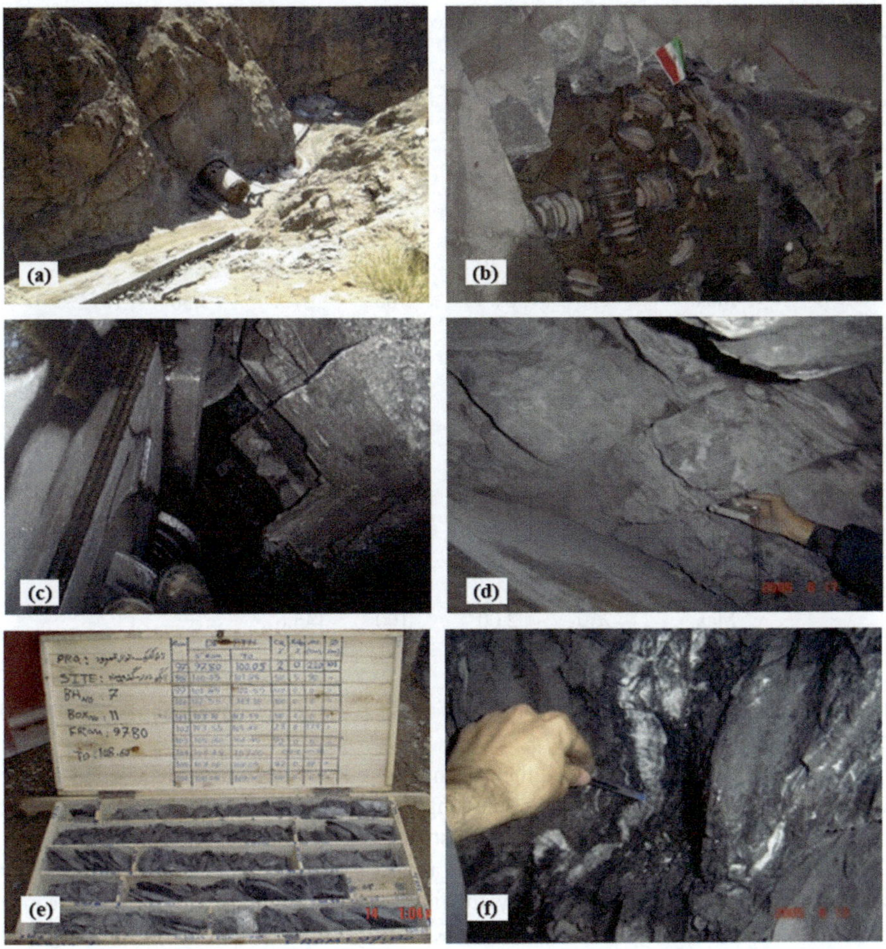

Fig. 3 Images of lithological types of: **a** LT-I (tuff thick layers); **b** LT-II (medium bedded limestone and sandstone); **c** LT-III (Slate and shale), in various tunneling projects of Iran (Hassanpour et al. 2023a)

5 Statistical Distribution of Parameters Recorded in the Database

As mentioned in previous sections, 199 tunnel sections were carefully selected for analysis. Various engineering geological, operational, and machine performance parameters were measured and recorded within these selected sections, providing the necessary data for this study. The results of the statistical analyses may help clarify the application ranges of the empirical relationships introduced in the following sections of this chapter. Table 5 presents statistical parameters for LT-I to LT-IV. Additionally,

Table 5 Statistical metrics for 158 datasets related to LT-I to LT-IV (Hassanpour et al. 2023a)

Lithological type	Statistical measures	UCS (MPa)	Spacing (m)	RQD (%)	FPI (kN/mm/rev)	ROP (m/h)	Cutter load (kN)
LT-I	Ave	72	0.28	56	14.6	3.87	148.5
	Min.	30	0.10	10	6.5	2.12	94.7
	Max.	150	0.80	100	22.8	5.40	196.5
LT-II	Ave	53	0.24	40	16.9	2.46	118.2
	Min.	20	0.10	95	8.9	1.00	76.4
	Max.	150	0.50	66	27.2	3.50	161.2
LT-III	Ave	79	0.29	57	23.2	3.87	114.9
	Min.	20	0.03	10	2.8	1.38	57.4
	Max.	170	0.80	100	70.7	5.88	208.3
LT-IV	Ave	145	0.87	94	71.6	1.11	223.9
	Min.	50	0.20	70	18.40	0.48	139.3
	Max.	250	2.00	100	145.6	1.84	248.1
LT-I to LT-IV	Ave	92	0.45	70	34.5		
	Min.	20	0.03	10	2.8		
	Max.	250	2.00	100	145.6		

Fig. 4 shows the frequency distribution of these parameters based on the combined data from all four LTs.

As shown in Table 5 and Fig. 4, the rock masses associated with the recorded data in this portion of database exhibit a diverse range of characteristics, from fractured, weak rocks to strong, jointed/blocky, and very strong, massive/intact rocks. For example, the UCS variations in these 158 datasets range from 20 to 250 MPa, while RQD varies from 10 to 100%. Additionally, the field penetration index (FPI) ranges from a minimum of 2.75 to a maximum of 145.6 kN/cutter/mm/rev. The variation ranges of parameters for the two other LTs (LT-V and LT-VI) will be discussed in the following sections of this chapter.

6 Specifications of the Machines Used in the Selected Projects

Table 6 outlines the main specifications of the machines used in the selected projects, while Fig. 5 showcases images of these machines. As shown in Table 6, the machines used in the KWCT, ZWCT-2, GWCT-3, GWCT-4, and KrWCT-S projects are all categorized as double (telescopic) shield machines. The machines in the KWCT and GWCT-3,4 projects were particularly similar, as they had nearly identical diameters. In contrast, the ChWCT project utilized a second-hand, refurbished single-shield machine.

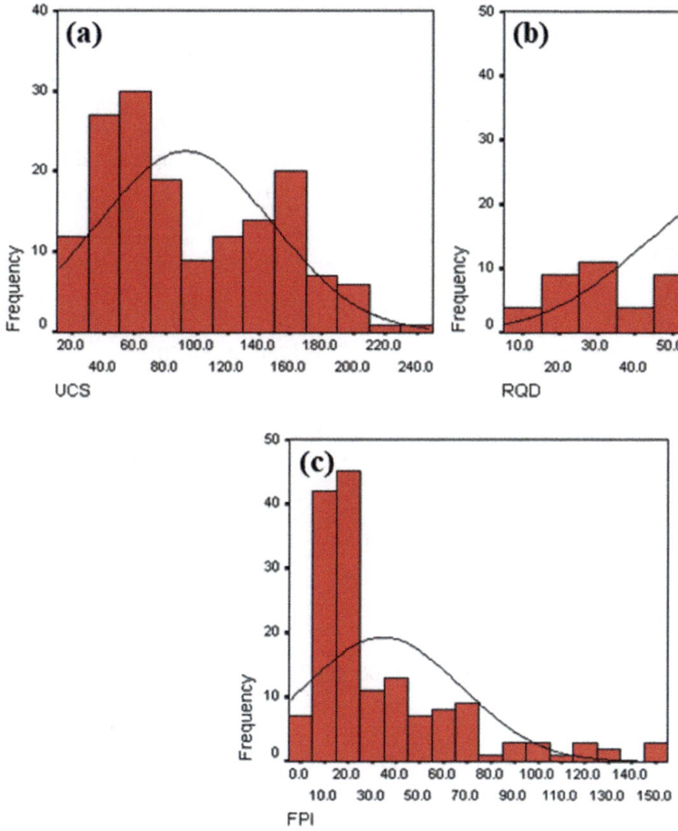

Fig. 4 Frequency distribution of **a** UCS, **b** RQD and **c** FPI for 158 sections of 4 LTs (LT-I to LT-IV) (Hassanpour et al. 2023a)

Analysis The machine used in the MSTT project was notably different in type, diameter, and design parameters. This open/gripper machine had a diameter of about 10 m, more than double the size of the machines employed in the KWCT and GWCT-3,4 projects, making it distinct in power capacity, maximum thrust, and number of disc cutters. In contrast, the machines employed in the BzWCT and T4WCT projects were of the EPB-hard rock type. Although their cutting mechanisms were similar to those of other machines, they differed significantly in their methods of collecting and transporting muck materials from the tunnel face.

In summary, despite some fundamental differences in the types of machines used in the selected projects, there are notable similarities, particularly in the type and configuration of the disc cutters on the cutterheads, which facilitates the model development and validation process in this chapter. All machines had a flat cutterhead with disc cutters measuring 432 mm (17 inches) in diameter and 18 to 19 mm in edge

Table 6 Main specifications of machines employed in the selected projects

Parameter	ZWCT-2	GWCT-3,4	KWCT	KrWCT-S	MSTT	ChWCT	T4WCT/ BzWCT
Manufacturer	Herr.[a]	Wirth	Herr.	Herr.	Robbins	Herr.	Lovat
Machine type	Double shield				Open	Single sh	EBP-HR
Mach. Dia. (m)	6.73	4.525	4.665	5.275	10.05	5.3	6.86
Cutter Dia.(mm)	432 (17 in)						
No. of cutters	42	35	31	37	68	34	46
Cutter load (kN)	250				280	250	
Cutter width (mm)	18–19 mm						
Max. thrust (kN)	28,134 in 350 bar	18,000	16,913	20,000		9350	26,500
Power (kw)	2100	1250	1120	1750	3463	2000	1600
RPM	0–11	0–12	0–11	0–10.9		0–9	Max 2.77
Torque (kN.m)	4450 in RPM 9	802 in 90% effi	1029 in RPM 11		9859	2500 in RPM 6.5	9640 in RPM 1.5
Course length of thrust jacks (mm)	1700	1400	1400	1700		1700	2250
Screw conveyor capacity (m^3/h)	690	250	200	500		300	520
Weight (ton)	573		170	300			

[a]Herrenknecht

width, with a cutter load capacity of 250 kN. The average spacing between the disc cutters for all machines was between 70 and 85 mm.

7 Analysis of Data Recorded in Database

As highlighted in Chap. 1, three key parameters—engineering geological parameters, machine operating parameters, and specifications for the cutterhead and cutting tools—significantly influence machine performance, particularly the penetration rate. This section will detail how these operating and geological parameters affect machine performance, utilizing data related to the first four LTs (LT-I to LT-IV), which comprise 158 tunnel sections and form the core of the database. The third group of parameters, which pertains to cutterhead and cutting tool specifications, is

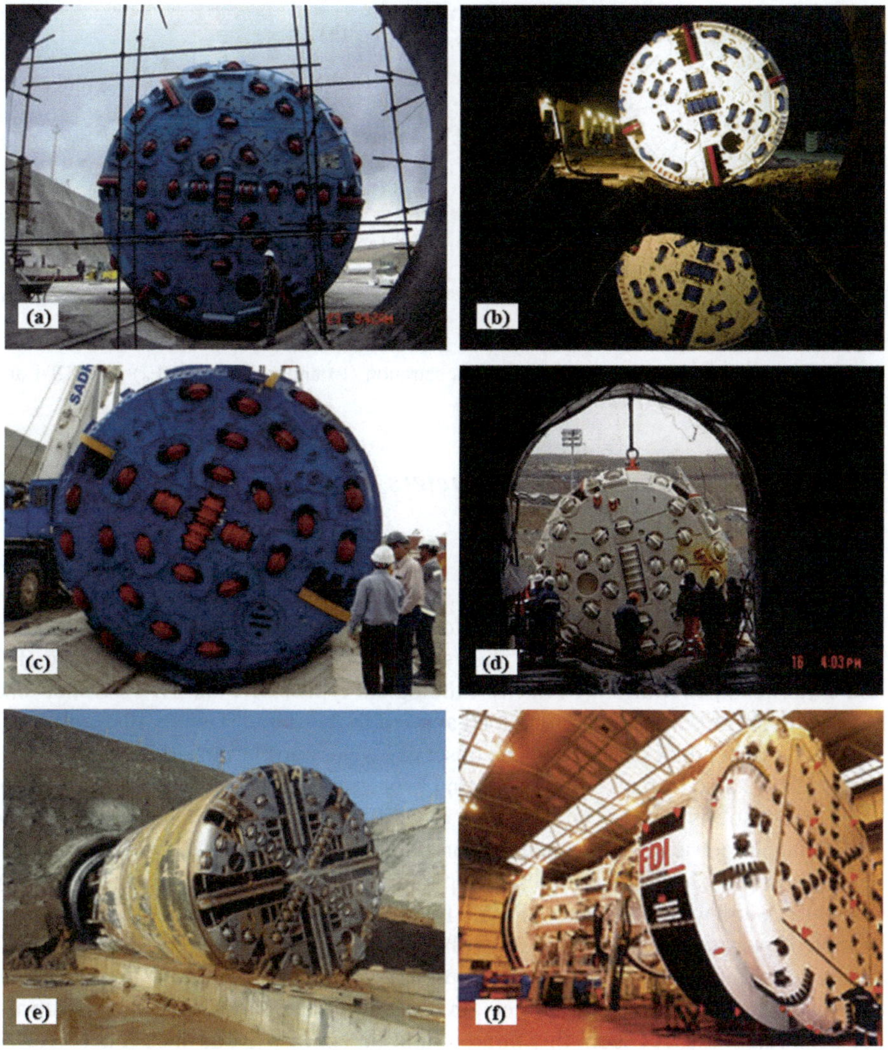

Fig. 5 View of tunnel boring machines in the selected projects, **a** double shield machine in the ZWCT-2 project, **b** double shield machine in the KWCT project, **c** double shield machine in the KrWCT-S project, **d** double shield machine in the GWCT-3,4 project, **e** EPB-Hard rock machine in the BzWCT project, and **f** open-type machine in the MSTT project

nearly identical across the selected projects and will, therefore, be excluded from the analyses.

Obviously, investigating the influence of various engineering geological parameters on TBM performance offers valuable insights and enhances our understanding of rock mass behavior during the cutting process.

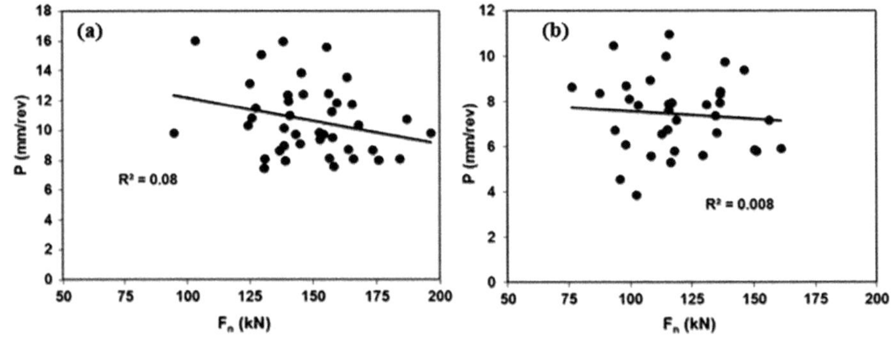

Fig. 6 The relationship between cutterhead penetration (P) and cutter load (F_n) in: **a** LT-I and **b** LT-II lithological types (Hassanpour 2009)

7.1 Impact of Operating Parameters on Machine Performance

Key operating parameters of the machine, particularly cutterhead thrust that translates to cutter load (F_n) and cutterhead rotation speed (RPM), significantly affect TBM performance. For example, Fig. 6 shows the relationship between cutter load (F_n) and cutterhead penetration (P) in two lithological types, LT-I and LT-II. Cutter load (F_n), which represents the average load applied to the tunnel face by disc cutters, is a crucial independent parameter. The observed weak correlations indicate that additional factors also impact the rock-cutting process beneath the disc cutters. Similar weak correlations have been noted in other lithological types between machine performance metrics, such as penetration rate (PR), specific energy (SE), and field penetration index (FPI), and operating parameters such as cutterhead torque (T_q), Power (P_o), and rotation speed (RPM). Therefore, any correlation between these two groups of parameters is incomplete without considering geological factors. Meanwhile, one should note that the negative trends between P and F_n in Fig. 6 is due to the fact that in the same tunneling project, in harder rock the machine uses higher thrust but penetrates less. As such, this figure might be confusing but once it is corrected for rock strength it would show the right trend, which is increasing P with F_n in the same rock type/strength.

7.2 Impact of Engineering Geological Parameters on Machine Performance

Many researchers have explored the impact of engineering geological parameters on machine performance using actual data from various projects. Findings indicate that the most significant geological factors are the strength of the intact rock and the degree of fracturing of rock mass (Hassanpour et al. 2009, 2010, 2011). Often, these two parameters alone suffice to assess geological influences on machine performance. Thus, compound indicators like rock mass strength parameters and engineering rock mass classification systems can be used to predict machine performance if appropriate parameters with suitable weights are applied. In this section, the relationships between these parameters and machine performance are analyzed, with particular emphasize on rock strength, degree of fracturing, orientation of critical discontinuities, and composite parameters such as rock mass classification systems and rock mass strength parameters.

It should be noted that while there are many geotechnical parameters that can impact the machine, it is merely impossible to capture all the related interactions. This is mainly due to lack of consistent information gathering from the field. As a result, the parameters used in the analysis and therefore in the prediction models are those who are 1) the most important, and 2) those that are widely available in the tunneling projects. As it turns out, the most frequently available rock information is the UCS and it is the most important rock parameter that controls cutter penetration. Similarly, RQD is the most widely available measure of rock mass properties and as a result is used in many of the formulas for estimation of TBM performance.

7.2.1 Intact Rock Strength

Among various intact rock properties, UCS is a crucial indicator for predicting machine performance. In many projects, rock strength closely correlates with performance parameters, often serving as the primary input in prediction models. Extensive research highlights that as rock strength increases, the penetration rate of cutting tools typically decreases (Fawcett 1993; Barton 1996; Nelson et al. 1983; Rostami 2013).

Figures 7, 8, 9 and 10 depict the relationships between UCS and various performance metrics (PR, P, SE, and FPI) for different LTs. These graphs demonstrate that among the performance parameters, FPI exhibits the strongest correlation with UCS. This is likely because FPI is a composite parameter that takes into account the effects of cutter load (F_n) and RPM. Similar trends can also be observed in other engineering geological parameters. Thus, FPI was selected as an appropriate parameter to explore the impact of various factors on machine performance and to develop new empirical models, which will be discussed in the following sections of this chapter.

7.2.2 Rock Mass Fracturing Degree

This study also investigates the relationship between rock mass fracturing degree and performance parameters. The rock mass fracturing degree (or frequency of discontinuities in rock) is among the crucial characteristics that significantly affect machine performance. The chipping process under disc cutters is greatly influenced by the number and spacing of fractures in the rock.

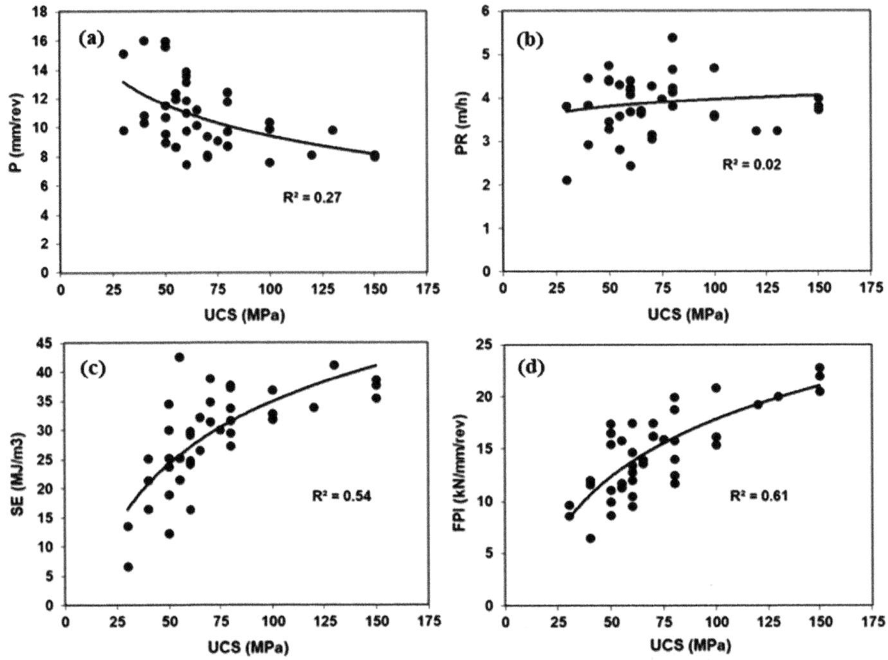

Fig. 7 Relationship between UCS and performance parameters of the machine for LT-I lithological type (Hassanpour et al. 2010)

The degree of rock mass fracturing can be defined using various parameters, including spacing of critical discontinuity set, rock quality designation (RQD; Deere et al. 1967), and the joint count number (Jv; Palmström 1995).

Figures 11, 12, 13 and 14 illustrate the relationships between spacing (of the most critical set) and RQD_0 (measured parallel to the tunnel axis), and FPI (as the selected performance parameter) for four LTs (LT-I to LT-IV). The graphs show a relatively strong correlation between these parameters and machine performance, with FPI decreasing significantly as the number of rock discontinuities increases.

7.2.3 Orientation of Discontinuities

The results of this study showed a weak correlation between discontinuity orientation and machine performance parameters like FPI (Fig. 15). This finding is consistent with observations from similar projects, which indicate that establishing a clear relationship between discontinuity orientation and machine operational parameters is generally difficult.

The impact of discontinuity orientation on the cutting process and penetration rate is not consistent. For discontinuities oriented at angles between 0 and 60 degrees to the tunnel axis, the PR initially rises before declining at angles more than 60

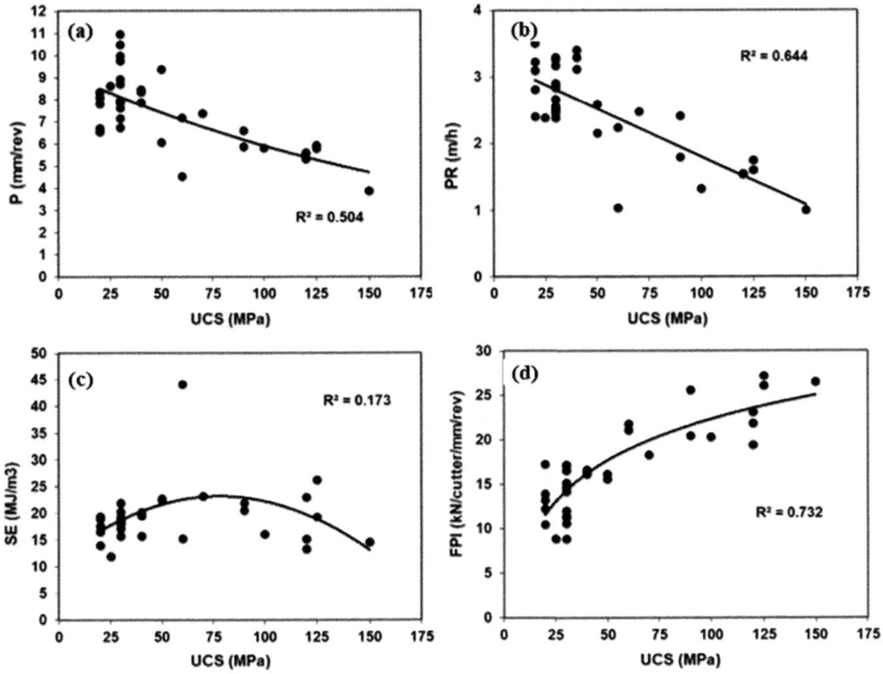

Fig. 8 Relationship between UCS and performance parameters of the machine for LT-II lithological type (Hassanpour et al. 2009)

degrees, particularly when there is only one major joint set in the rock mass (such as foliated and bedded rocks). Additionally, the influence of discontinuity orientation also depends on joint spacing.

In blocky and layered rock formations with two or three nearly perpendicular sets of discontinuities (isotropic rock masses), the impact of orientation is minimal. Orientation becomes significant only in cases where one or two sets of discontinuities with limited spacing and distinct characteristics exist, such as in foliated or schistose rock masses (anisotropic rock masses).

7.2.4 Compound Rock Mass Parameters

In this study, the impact of various rock mass classification systems (such as Q and RMR) and strength parameters (like RMCI and Q_c) on machine performance metrics (such as FPI) has been investigated. These parameters are composite metrics derived from simpler properties of intact rock and discontinuities, allowing multiple factors affecting the cutting process to be considered simultaneously in the analyses.

The results are presented in Figs. 16, 17, 18 and 19 for the four LTs (LT-I to LT-IV), and indicate a strong correlation between rock mass classification systems

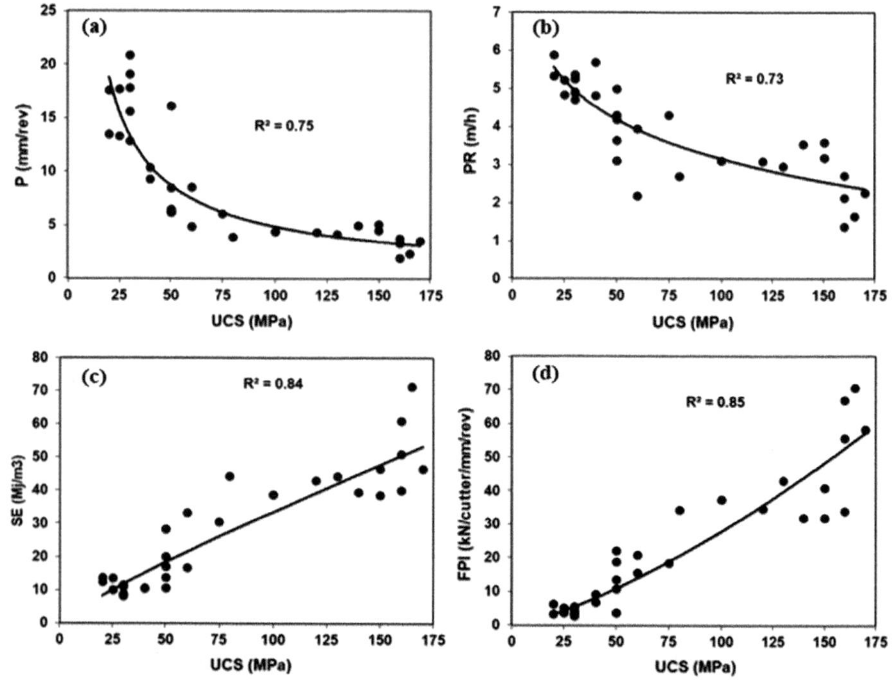

Fig. 9 Relationship between UCS and performance parameters of the machine for LT-III lithological type (Hassanpour et al. 2021)

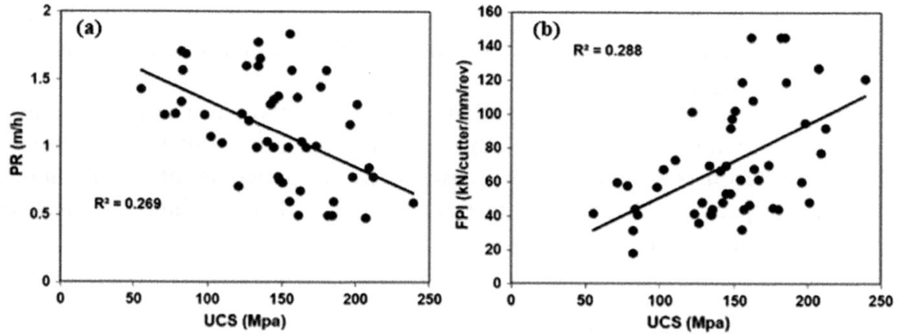

Fig. 10 Relationship between UCS and performance parameters of the machine for LT-IV lithological type

and strength parameters with FPI. Notably, rock mass cuttability index (RMCI)[2] demonstrates an even stronger correlation with FPI. As a composite parameter calculated from UCS and RQD, RMCI's robust relationship with machine performance

[2] $RMCI = UCS \times (RQD/100)^{2/3}$.

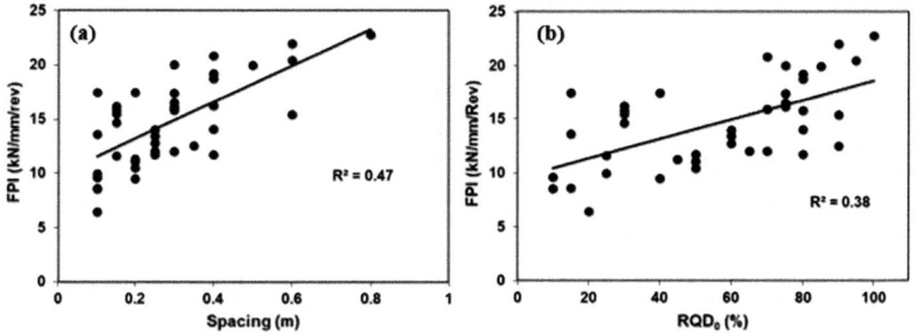

Fig. 11 The relationship between FPI and **a** spacing of discontinuities, **b** RQD_0, for LT-I lithological type (Hassanpour et al. 2010)

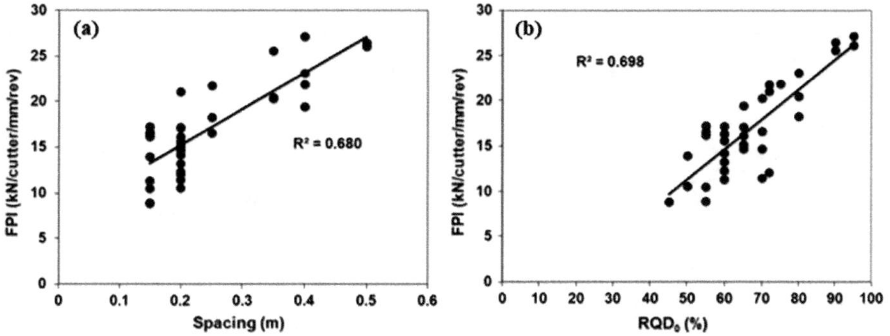

Fig. 12 The relationship between FPI and **a** spacing of discontinuities, **b** RQD0, for LT-II lithological type (Hassanpour et al. 2009)

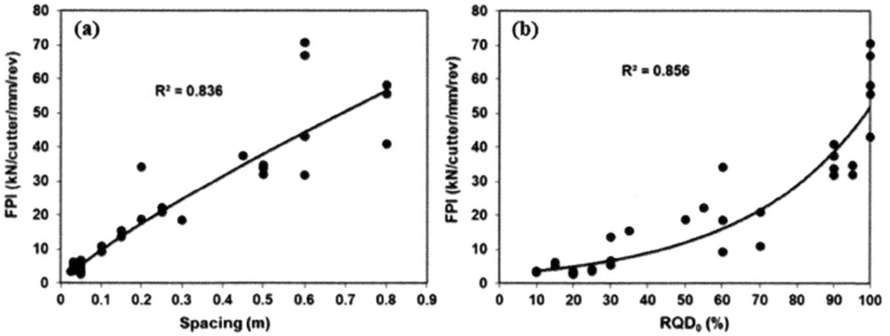

Fig. 13 The relationship between FPI and **a** spacing of discontinuities, **b** RQD_0, for LT-III lithological type (Hassanpour et al. 2021)

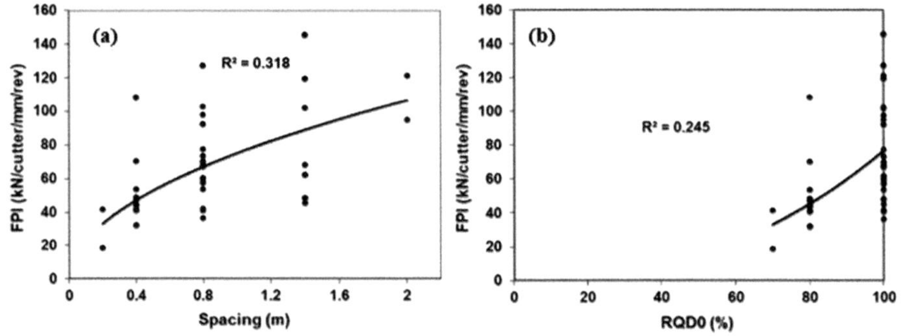

Fig. 14 The relationship between FPI and **a** spacing of discontinuities, **b** RQD_0, for LT-IV lithological type

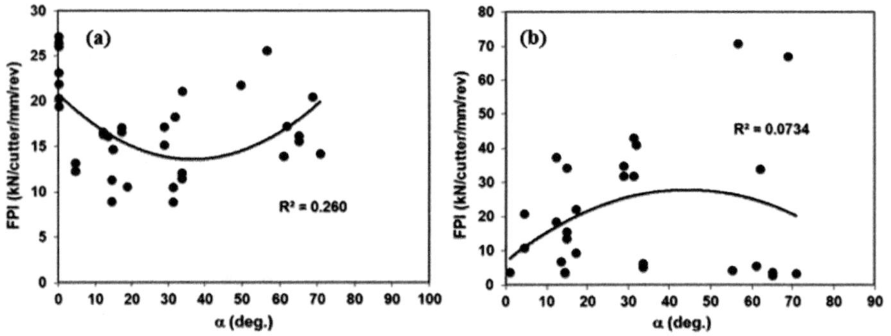

Fig. 15 The relationship between orientation of discontinuities and FPI for: **a** LT-II and **b** LT-III lithological types

metrics leads to its use in subsequent sections of the chapter for developing empirical models to predict machine penetration rate. As FPI is also a composite parameter, the relationship between RMCI and FPI can be considered relatively holistic.

It should be noted that in many studies the relationship between FPI and standard rock mass classification system ratings (Q or RMR) are not very strong due to the fact that these rating were developed for ground support design and evaluation and not excavation. The good correlation between FPI and RMCI is that this parameter is specifically introduced as a measure to represent rock mass behavior in excavation applications and as such, it shows better correlation than RMR or Q.

7.2.5 Analyzing Combined Data

As previously mentioned, this study examines how engineering geological parameters affect machine performance and establishes a foundation for a comprehensive model applicable across various geological conditions and machine types. To

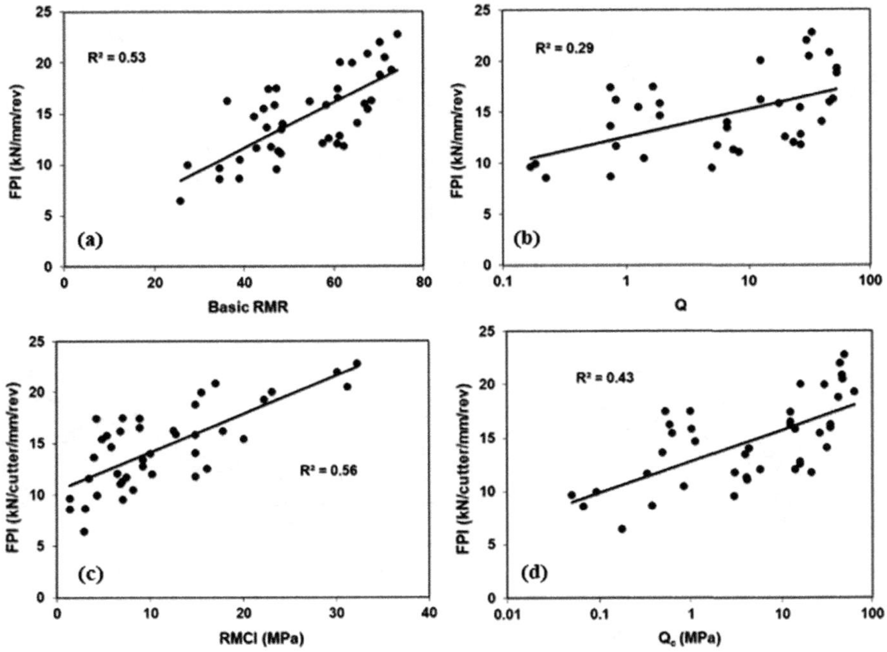

Fig. 16 Correlation between rock mass parameters and FPI for LT-I lithological type (Hassanpour et al. 2010)

achieve this, data from four LTs were combined and reanalyzed. Figure 20 illustrates the correlation between the FPI and engineering geological parameters using the combined data from the four LTs. As shown, the correlation coefficients between the engineering geological parameters and FPI have decreased compared to LT-III but increased relative to the other LTs. Notably, even with the combined data, FPI has maintained its strong correlations with engineering geological parameters. Among the rock mass parameters, the RMCI has shown a particularly strong correlation with FPI.

8 Developing Empirical Relationships for Estimating PR

To develop the initial version of the UT model, data were analyzed from four LTs related to three Iranian projects (KWCT-1, GWCT-3,4, and ZWCT-2), as well as the MSTT project in New Zealand. As detailed in Table 3, a total of 158 datasets were examined, comprising 41 sections from KWCT-1, 37 from ZWCT-2, 31 from GWCT-3,4, and 49 from the MSTT project. This version was published as part of the first author's doctoral dissertation and post-doctoral research, along with related

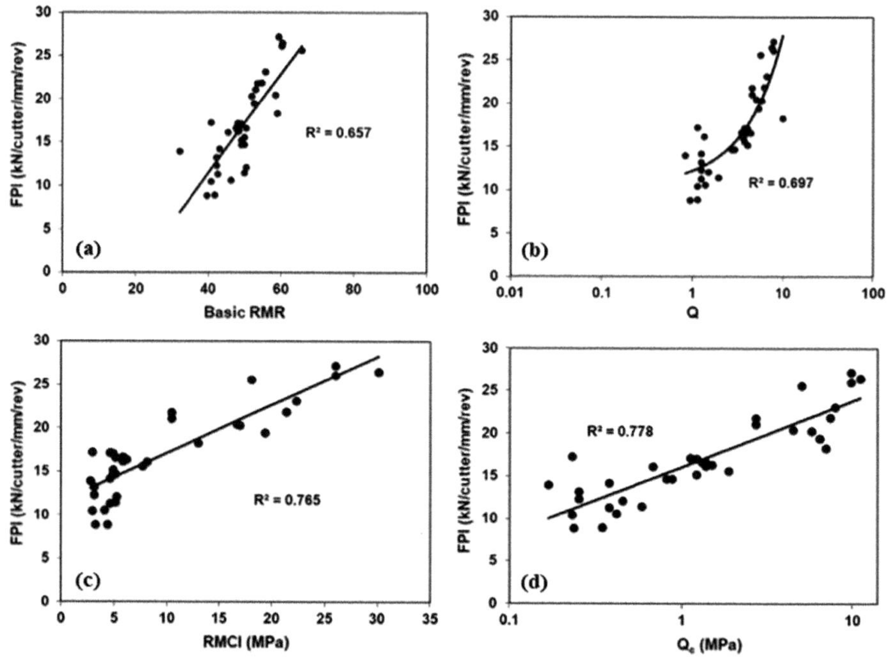

Fig. 17 Correlation between rock mass parameters and FPI for LT-II lithological type (Hassanpour et al. 2009)

articles, between 2009 and 2011(Hassanpour 2009; Hassanpour et al. 2009, 2010, 2011).

In the following years, efforts focused on improving the model by enhancing the database and expanding the geological conditions with data from new tunneling projects. In recent years, specific relationships for two other lithological types, LT-V and LT-VI, each having distinct boreability/chipping conditions, were developed. These lithological types include soft rocks (UCS < 50 MPa) and very strong rocks (UCS > 150 MPa). These relationships were respectively developed by Goodarzi et al. (2021) and Pourhashemi et al. (2021a, b, c) in their PhD theses. In the following sections, after a detailed explanation of the main model, the developed relationships for these lithological types will also be presented.

To complete the UT model, sections on estimating disc cutter wear (Hassanpour et al. 2014; Hassanpour 2018), utilization factor, and advance rate of machine were added, using data from various projects.

It must be noted that one of the most successful empirical models for TBM performance prediction has been the Field Penetration Index (FPI) models. FPI models were introduced by Nelson et al. (1983) and have been used off and on with rock properties such as Total hardness (TH), but given the limited availability and subjectivity of TH in various projects, it enjoyed limited acceptability and use. It was the subsequent attempts to link FPI with other rock and rock mass properties that initiated a new

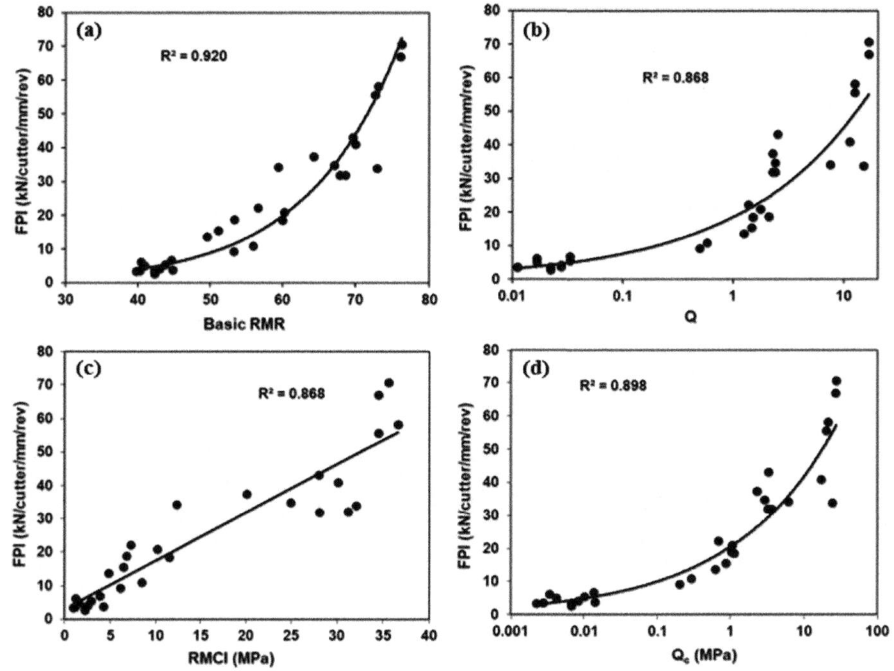

Fig. 18 Correlation between rock mass parameters and FPI for LT-III lithological type (Hassanpour et al. 2021)

wave of use and adoption of this method by researchers. This is because these models take into account machine size through RPM and then normalize penetration with Thrust (which is the most important operational parameter controlled by the operator). This allows for comparing the normalized penetration of the cutter with rock and rock mass properties.

8.1 Initial Version of the Model

Single and multivariable regression techniques were employed to establish empirical relationships among three main parameter sets: operational and performance parameters of the machine and rock mass parameters. The results of these analyses are as follows:

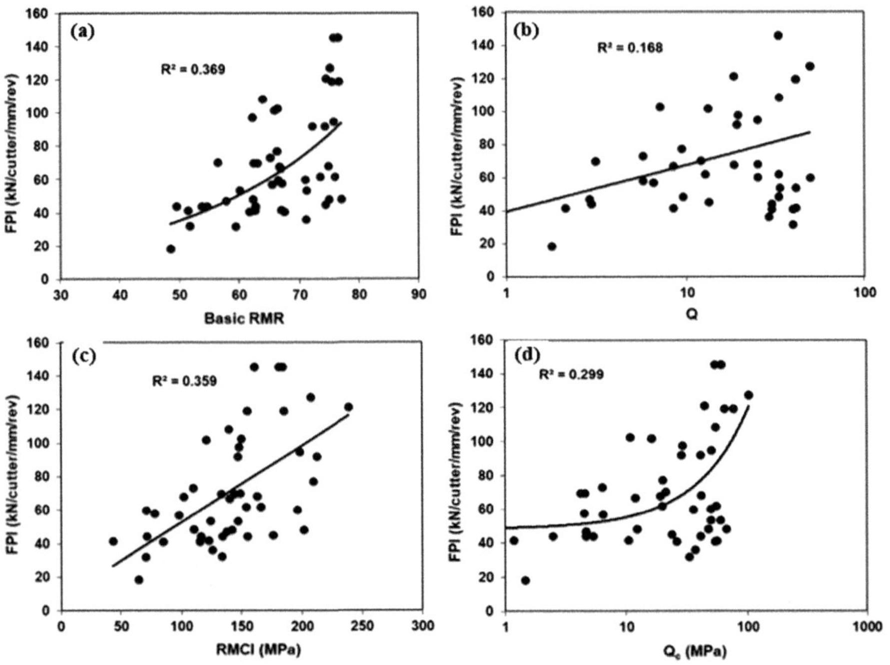

Fig. 19 Correlation between rock mass parameters and FPI for LT-IV lithological type (Hassanpour et al. 2023a)

8.1.1 Single Regression Analyses

As previously mentioned, FPI, which is a composite performance parameter, demonstrates stronger correlations with engineering geological parameters across all LTs (LT-I to LT-IV) compared to other machine performance indicators. Therefore, it is reasonable to utilize FPI for developing empirical relationships.

Table 7 (sections a to d) presents the results of the correlations between FPI and various engineering geological parameters, along with the related equations for the four LTs (Eqs. 1–42).

According to Table 7, the RMCI shows a strong correlation with FPI (Eqs. 12, 24, 35, and 42). RMCI is a composite parameter derived from two key rock mass parameters: UCS (as a key index of intact rock strength) and RQD (indicative of rock mass fracturing degree). As previously noted, FPI is also a compound machine performance parameter that combines two key components: cutter load (F_n) and cutterhead penetration or P (as outlined in Eq. 1.22). Importantly, all these geological (UCS and RQD) and machine (F_n) parameters are available even in the early stages of tunnel design, making Eqs. 12, 24, 35, and 42 suitable for estimating P under similar geological conditions and machine characteristics.

To develop a more universally applicable model that covers a broader range of geological conditions and machine specifications, the data from all of the four LTs

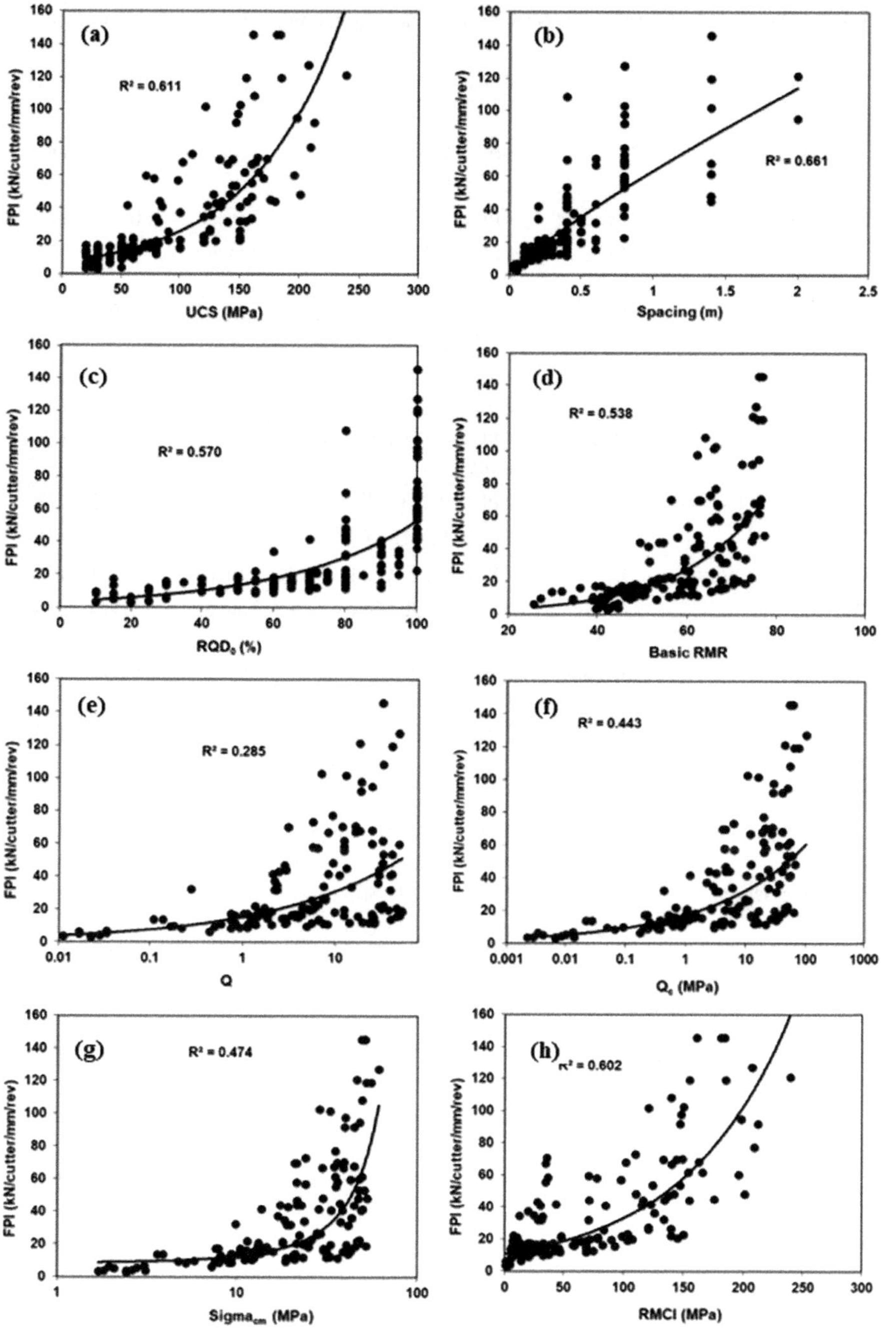

Fig. 20 The correlation between engineering geological parameters and FPI after combining data from four LTs (Hassanpour 2009)

Table 7 The results of correlation between FPI and engineering geological parameters in four lithological types

Parameter		Equation	Regr. type	R^2	Eq. no
(a) LT-I					
UCS (MPa)		FPI = 7.887 Ln (UCS) – 18.497	Log	0.612	(1)
Fracturing degree	Spacing	FPI = 16.721 Sp + 9.882	Linear	0.472	(2)
	RQD_0	FPI = 0.091 RQD_0 + 9.528	Linear	0.378	(3)
	$K_{s\text{-}tot}$	FPI = -9.664 Ln ($K_{s\text{-}tot}$) + 17.781	Log	0.401	(4)
Rock mass classification systems	RMR_{Basic}	FPI = 0.222 BRMR + 2.755	Linear	0.532	(5)
	RMR'_{89}	FPI = 0.178 RMR'_{89} + 4.005	Linear	0.449	(6)
	GSI	FPI = 9.273 $e^{0.008GSI}$	Exp	0.412	(7)
	Q	FPI = 11.718 $Q^{0.098}$	Power	0.369	(8)
Rock mass strength parameters	Q_c	FPI = 0.157 Q_c + 12.239	Linear	0.493	(9)
	$Sigma_{cm}$	FPI = 0.190 $Sigma_{cm}$ + 9.731	Linear	0.475	(10)
	UCS_{rm}	FPI = 11.990 UCS_{rm} 0.118	Power	0.496	(11)
	RMCI	FPI = 0.080 RMCI + 10.412	Linear	0.556	(12)
(b) LT-II					
UCS (MPa)		FPI = 6.280 Ln (UCS) – 6.733	Log	0.697	(13)
Fracturing degree	Spacing	FPI = 36.795 Sp + 8.150	Linear	0.639	(14)
	RQD_0	FPI = 0.293 RQD_0 + 2.420	Linear	0.636	(15)
	$K_{s\text{-}tot}$	FPI = -7.284 Ln ($K_{s\text{-}tot}$) + 19.793	Log	0.344	(16)
Rock mass classification systems	RMR_{Basic}	FPI = 0.474 BRMR – 6.314	Linear	0.571	(17)
	RMR'_{89}	FPI = 0.412 RMR′89 – 6.113	Linear	0.688	(18)
	GSI	FPI = 0.405 GSI – 3.208	Linear	0.669	(19)
	Q	FPI = 1.597 Q + 11.113	Linear	0.624	(20)
Rock mass strength parameters	Q_c	FPI = 1.285 Q_c + 13.428	Linear	0.718	(21)
	$Sigma_{cm}$	FPI = 0.583 $Sigma_{cm}$ + 8.034	Linear	0.718	(22)
	UCS_{rm}	FPI = 3.419 Ln (UCS_{rm}) + 13.756	Log	0.809	(23)

(continued)

Table 7 (continued)

Parameter		Equation	Regr. type	R^2	Eq. no
	RMCI	FPI = 0.116 RMCI + 11.849	Linear	0.748	(24)
(c) LT-III					
UCS (MPa)		$FPI = 0.054\ UCS^{1.354}$	Power	0.861	(25)
Fracturing degree	Spacing	$FPI = 68.268\ Sp^{0.849}$	Power	0.916	(26)
	RQD_0	$FPI = 2.749\ e^{0.029\ RQD}$	Exp	0.870	(27)
Rock mass classification systems	RMR_{Basic}	$FPI = 0.162\ e^{0.080\ BRMR}$	Exp	0.919	(28)
	RMR'_{89}	$FPI = 0.307\ e^{0.062\ RMR'\ 89}$	Exp	0.871	(29)
	GSI	$FPI = 0.918\ e^{0.054\ GSI}$	Exp	0.883	(30)
	Q	$FPI = 18.502\ Q^{0.388}$	Power	0.928	(31)
Rock mass strength parameters	Q_c	$FPI = 20.462\ QC^{0.310}$	Power	0.940	(32)
	$Sigma_{cm}$	$FPI = 1.877\ {Sigma_{cm}}^{0.931}$	Power	0.940	(33)
	UCS_{rm}	$FPI = 6.796\ {UCS_{rm}}^{0.563}$	Power	0.905	(34)
	RMCI	FPI = 1.445 RMCI + 2.911	Linear	0.868	(35)
(d) LT-IV					
UCS (MPa)		$FPI = 26.103\ e^{0.006UCS}$	Exp	0.301	(36)
Fracturing degree	Spacing	$FPI = 74.679\ Sp^{0.51}$	Power	0.382	(37)
	RQD_0	$FPI = 0.002\ RQD^{2.35}$	Power	0.336	(38)
Rock mass classif. Sys	RMR_{Basic}	$FPI = 5.786\ e^{0.036BRMR}$	Exp	0.393	(39)
	Q	$FPI = 40.599\ Q^{\ 0.175}$	Power	0.180	(40)
Rock mass strength param	Q_c	$FPI = 0.706\ Q_c + 48.299$	Linear	0.299	(41)
	RMCI	$FPI = 25.311\ e^{0.007RMCI}$	Exp	0.385	(42)
(e) all of the lithological types					
UCS (MPa)		$FPI = 6.883\ e^{0.013\ UCS}$	Exp	0.611	(43)
Fracturing degree	Spacing	$FPI = 63.267\ Sp^{\ 0.847}$	Power	0.661	(44)
	RQD_0	$FPI = 3.490\ e^{0.027\ RQD}$	Exp	0.570	(45)
Rock mass classification systems	RMR_{Basic}	$FPI = 1.032\ e^{0.055\ BRMR}$	Exp	0.538	(46)
	GSI	$FPI = 4.619\ e^{0.023\ GSI}$	Exp	0.401	(47)
	Q	$FPI = 15.309\ Q^{0.304}$	Power	0.285	(48)
Rock mass strength parameters	Q_c	$FPI = 17.389\ {Q_c}^{0.269}$	Power	0.443	(49)
	$Sigma_{cm}$	$FPI = 8.317\ e^{0.042\ Sigmacm}$	Exp	0.474	(50)
	RMCI	$FPI = 10.525\ e^{0.011\ RMCI}$	Exp	0.602	(51)

have been combined and reanalyzed. Table 7 (section e) summarizes the results, including relevant equations. The RMCI showed a strong correlation with FPI for the combined data (Eq. 51). Therefore, Eq. 51 can be considered as a valid relationship for more general geological conditions and machine specifications.

8.1.2 Multivariable Regression Analysis

Multivariable regression analyses were also employed to explore empirical relationships between FPI and common geological parameters. To apply this statistical method, four rock mass parameters, including UCS, discontinuity spacing,[3] RQD, and orientation of discontinuities (α angle[4]) were used as independent variables, and FPI was treated as the dependent variable. The impact of each variable on FPI was evaluated using forward stepwise regression analyses. For all LTs, the best linear fit with a 95% confidence level between Ln (FPI) as the response parameter and UCS and RQD as predictors was identified, while spacing and orientation of discontinuities were excluded after testing various combinations of parameters. As a result, new predictive models (Eqs. 52–55) were developed for the four lithological types LT-I to LT-IV, respectively:

$$\text{LT}-\text{I}: \ FPI = \exp(0.005UCS + 0.002RQD + 2.129) \tag{52}$$

$$\text{LT}-\text{II}: \ \text{FPI} = \exp\left(0.004\ \text{UCS} + 0.008\ \text{RQD} + 2.077\right) \tag{53}$$

$$\text{LT}-\text{III}: \ \text{FPI} = \exp\left(0.004\ \text{UCS} + 0.023\ \text{RQD} + 1.003\right) \tag{54}$$

$$\text{LT}-\text{IV}: \ \text{FPI} = \exp\left(0.005\ \text{UCS} + 0.020\ \text{RQD} + 1.644\right) \tag{55}$$

$$\text{All LTs}: \ FPI = \exp(0.008UCS + 0.015RQD + 1.384) \tag{56}$$

Table 8 provides a summary of the results of the statistical analyses performed to develop Eqs. 52–56.

Figure 21 compare the calculated FPI values (derived from Eqs. 52–56) with the actual values from the database. The graphs indicate that, while all calculated values are generally close to the actual ones, the FPIs for lithological types LT-I to LT-III align more closely with the actual values than those for LT-IV.

To facilitate the easy use of the established empirical equations and to effectively compare different rock masses regarding their boreability conditions some simple charts have been introduced. These charts enable quick estimation of the FPI values

---

[3] Spacing of most favorable discontinuity set.

[4] $\alpha = \text{Arcsin}\,(\text{Sin}\ \alpha_f.\ \text{Sin}\ (\alpha_t - \alpha_s))$; where α_t = tunnel azimuth, α_s = Strike of discontinuity plane and α_f = Dip of discontinuity plane.

Table 8 The results of multivariable regression analyses

Eq. no.	Lithological type	R	R^2	$R^2{}_{Adj}$	Std. error
(52)	LT-I	0.740	0.547	0.523	0.1993
(53)	LT-III	0.939	0.882	0.874	0.3779
(54)	LT-II	0.815	0.665	0.645	0.1755
(55)	LT-IV	0.684	0.468	0.445	0.3463
(56)	All LTs	0.888	0.788	0.785	0.4119

based on variations in rock mass parameters, specifically UCS and RQD. The charts are illustrated in Figs. 22, 23, 24 and 25. As shown, different geological units in the selected tunnel projects display distinct ranges of FPI and boreability conditions based on their strength and degree of fracturing.

Sensitivity analyses in Fig. 26a–d show that in LT-I rocks, the FPI value is more responsive to UCS variations, indicating a strong effect on TBM performance. In LT-II and LT-III, FPI is more sensitive to RQD. These results emphasize the models' high sensitivity to geological conditions and the importance of developing tailored models for specific geological settings.

A general chart for predicting rock mass boreability, is shown in Fig. 27. This chart offers a comprehensive model and a quick method for estimating boreability across various rock masses. The yellow hatches on the chart indicate rock masses for which boreability estimates using this general model are not recommended, according to Hassanpour et al. (2011). This highlights the need for further research in these areas. A revised chart will be presented in subsequent sections of this chapter to address certain deficiencies based on data from recent projects.

Figure 27 categorizes rock masses into six classes (B-0 to B-V) based on properties such as rock strength and rock mass structure regarding boreability. Boreability indicates the ease/difficulty of excavating rock with disc cutters, which is vital for predicting machine performance. Thus, classifying rock masses by boreability is essential for these studies. Detailed explanations of the various classes of rock mass boreability and their attributes will be provided later in this chapter.

8.2 *Updated Version of the UT Model*

The application of developed empirical relationships across recent projects has highlighted the need to better address previously overlooked geological conditions to refine and expand the model. While these relationships cover a wide range of rock types (LT-I to LT-IV), certain geological conditions still require additional considerations.

The models previously introduced may be inaccurate for two types of rocks: (1) continuous soft rocks, and (2) very strong, massive rocks. In both cases, the chipping process beneath the disc cutters remains incomplete. In the first category, intact

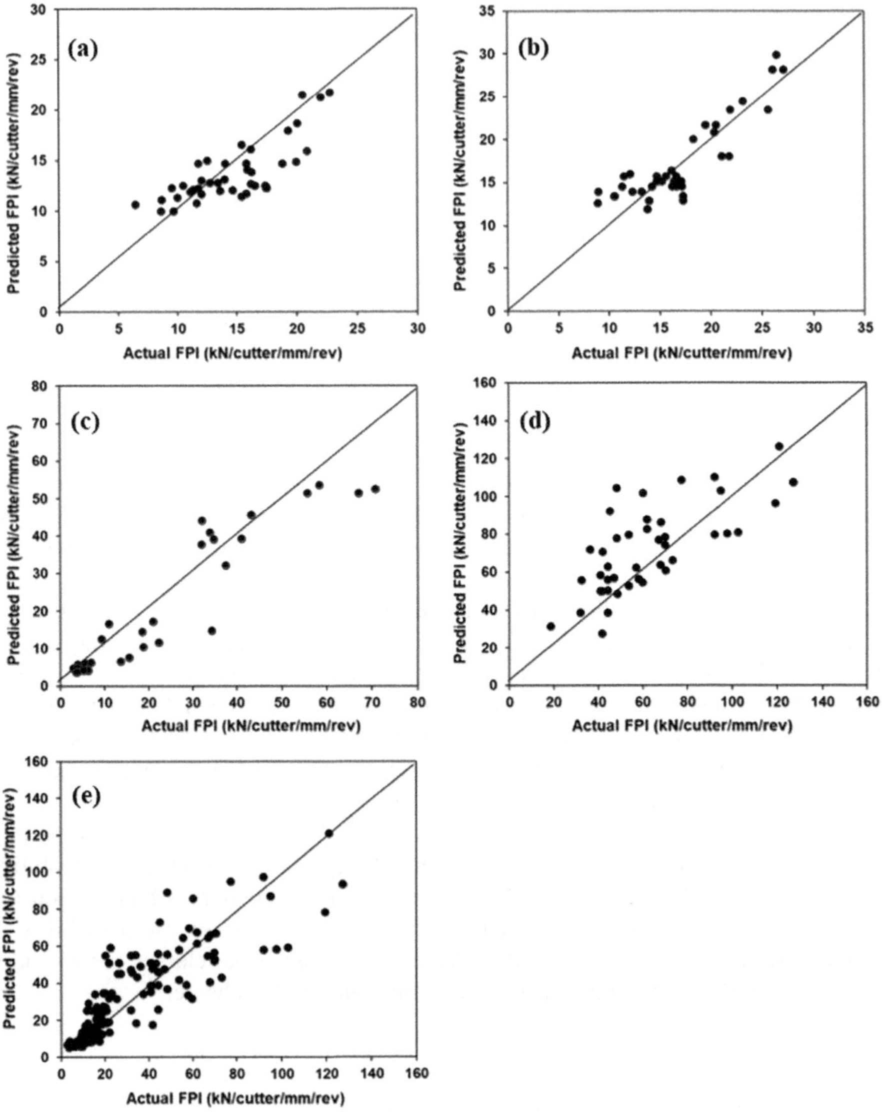

Fig. 21 Comparison of actual FPI with calculated FPI using the following equations: **a** Eq. 52 (LT-I); **b** Eq. 53 (LT-II); **c** Eq. 54 (LT-III); **d** Eq. 55 (LT-IV); **e** Eq. 56 (all LTs); (Hassanpour et al. 2009, 2010, 2011)

soft rocks exhibit plastic deformations rather than brittle behavior (Hassanpour et al. 2023b; Goodarzi et al. 2021). As clay minerals increase in the mineralogical composition of the rock and rock strength decreases, plastic behavior becomes dominant, diverging from brittle characteristics (Kanji 2014). In the second, the high strength of the rock prevents proper cutter penetration, leading to predominant grinding and

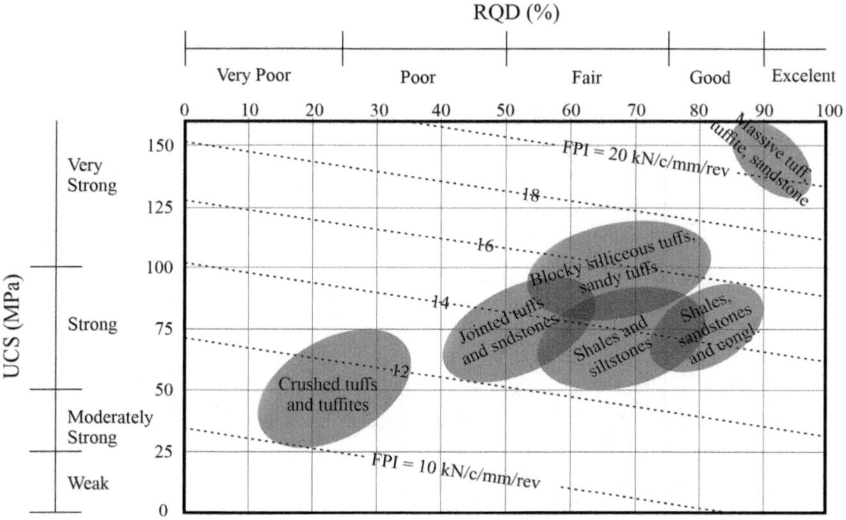

Fig. 22 FPI prediction chart for LT-I lithological type, along with variation ranges of FPI in various geological units (Hassanpour 2009)

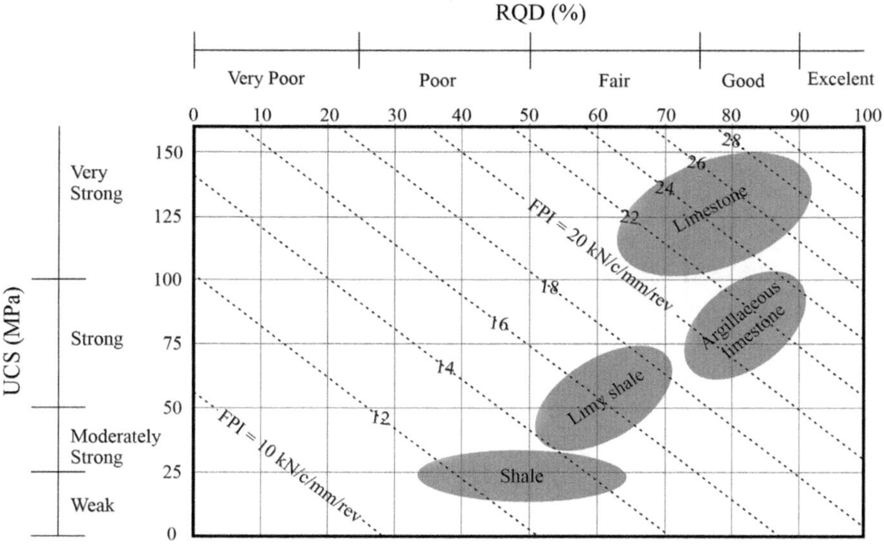

Fig. 23 FPI prediction chart for LT-II lithological type, along with variation ranges of FPI in various geological units (Hassanpour et al. 2009)

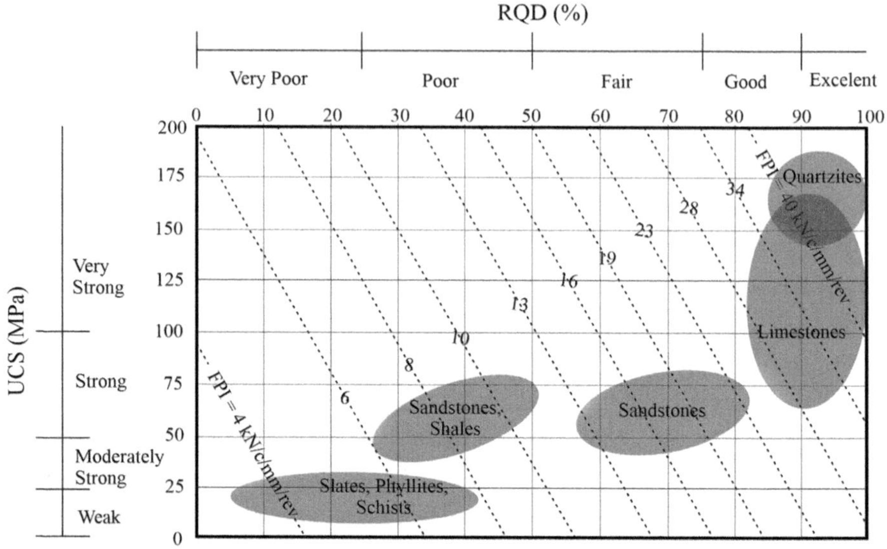

Fig. 24 FPI prediction chart for LT-III lithological type, along with variation ranges of FPI in various geological units (Hassanpour 2009; Hassanpour et al. 2021)

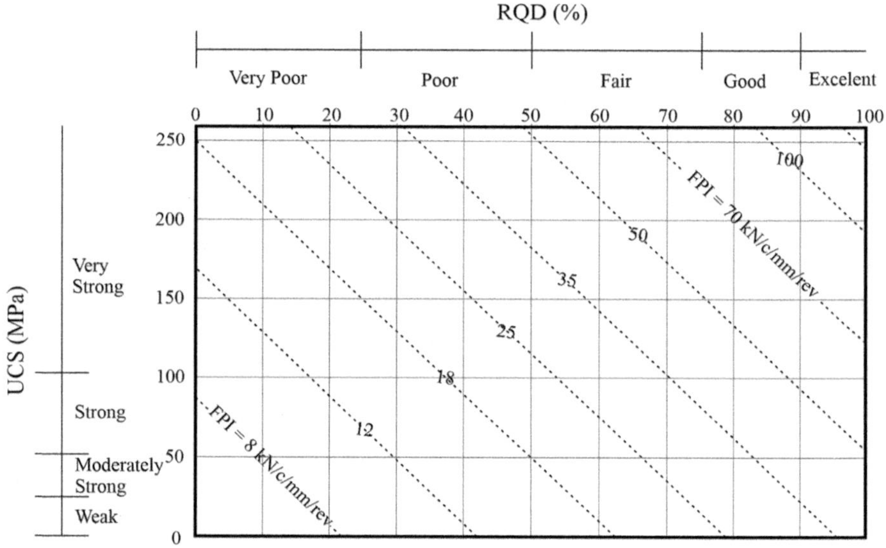

Fig. 25 FPI prediction chart for LT-IV lithological type, along with variation ranges of FPI in various geological units

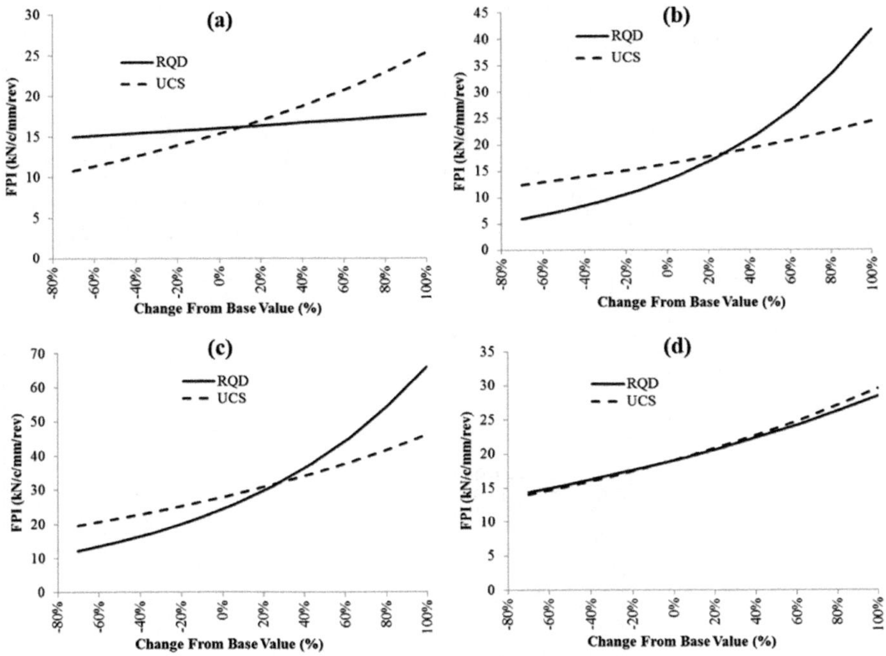

Fig. 26 Sensitivity analysis of the FPI to input parameters of empirical equations for four LTs; **a** LT-I, **b** LT-II, **c** LT-III, **d** LT-IV (Hassanpour et al. 2021)

a drastic decrease in the penetration rate of the machine (Pourhashemi et al. 2021a, b, c).

Consequently, efforts have been made to create separate empirical relationships for two additional lithological types, LT-V and LT-VI (soft rocks and very strong, massive rocks, respectively). These lithological types are unique in mechanized tunneling projects due to the inefficiencies in the rock chipping process with disc cutters. The following sections will present the developed empirical relationships for these two specific LTs.

8.2.1 Soft Sedimentary Rocks (LT-V)

The main characteristic that sets soft rocks apart in mechanized tunneling is their plastic behavior, which inhibits crack propagation during excavation and prevents complete rock cutting process under the disc cutter.

Many of the soft sedimentary rocks, especially those composed mainly of clay minerals (such as claystones, marlstones, shales, etc.), exhibit "non-structurally controlled" behavior in deep tunnels and below the groundwater level, and are completely continuous and intact (Hassanpour et al. 2019a). In these rocks whose behavior gradually tends to be more in plastic mode, shearing is more dominant than

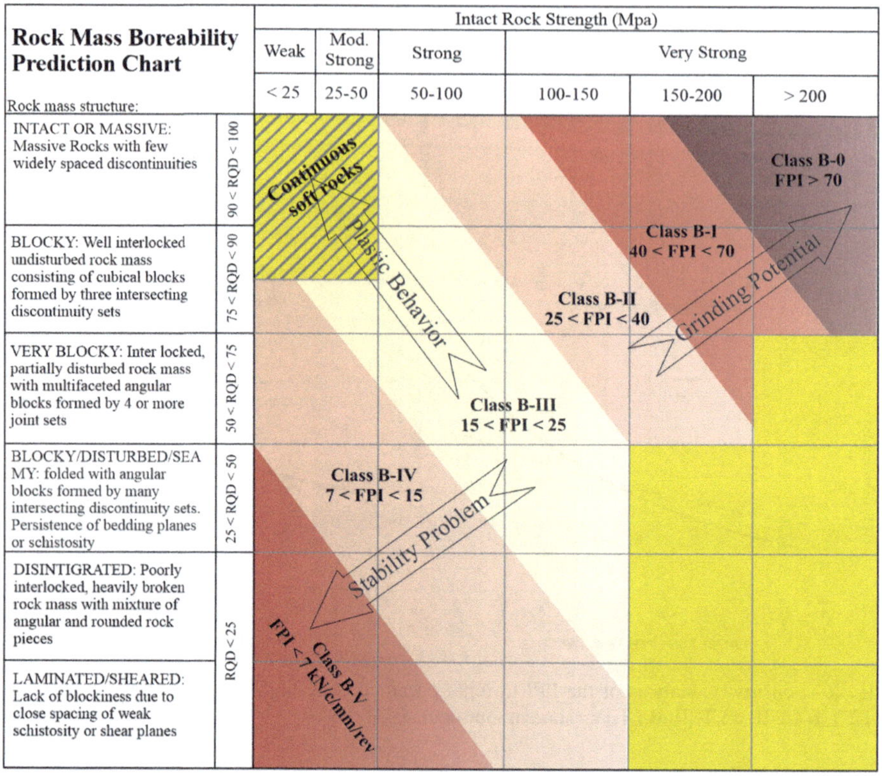

Fig. 27 The general rock mass boreability classification chart (Hassanpour et al. 2011; Goodarzi et al. 2021)

the development of tensile cracks (which are active in brittle rocks). Therefore, the chipping process, described in Chap. 1, is not applicable to these cases. In reality, the mechanical rock cutting process includes, to varying degrees, tensile and shear types of failure modes, depending on rock properties, cutter and cutting geometry (Copur et al. 2003).

In this case, a portion of the machine energy is spent on rock deformations that do not play much of a role in the chipping process. Due to the low brittleness of these rocks, tensile cracks do not sufficiently propagate and therefore the chipping process is incomplete. In addition, the plastic deformation under the discs blades moves the rock material instead of initiating the fractures. As a result, ridges of rock may remain between the two cutters, which may cause extra damage in the cutterhead and increase in torque of machine, due to contact between the ground and the cutterhead structure.

In addition, due to the presence of more clay minerals, there is also the issue of stickiness and potential for clogging, which makes the boring and progress of the machine more difficult. In such cases, the machine penetration rate may increase due to deeper penetration, but some problems such as extraordinary jamming of the disc cutters, abnormal cutter wear, secondary wear (high abrasion in other parts of the machine such as the cutterhead structure) and clogging can cause long stoppages and lead to reduced utilization and advance rate of the machine.

In soft rocks with high plastic behavior, different types of drag bits can be used instead of the disc cutters. In this case, shearing is dominant in the cutting process. According to Sutcliffe (1996), rocks with UCS up to 90 MPa can be mined using drag bits. However, the experience of the authors shows that the efficiency of the drag type tools is greatly reduced in rocks with UCS > 50 MPa. However, disc cutters are commonly used in soft rocks with good results and long service life. It is inevitable to make use of this tool, especially in grounds characterized by different rock conditions including sequences of soft and plastic rocks and hard and brittle rocks. Therefore, it is necessary to modify the existing prediction models to suit the application of the equations in soft rock or to develop new models for such conditions. This will help predict performance in soft rocks more accurately.

Although intact rocks have been classified according to their strength in different scales and terms by different researchers, there is a general agreement that the upper limit of UCS in very low strength rocks is 25 MPa and in low strength rocks is 50 MPa (Deere and Miller 1966; Bieniawski 1973). On the other hand, the lower limit of the strength of soft rock, distinguishing it from soils, is more difficult to establish. Many criteria have been proposed, but UCS = 0.4 MPa (Terzaghi and Peck 1967) and UCS = 1 MPa (ISRM 1979; Bieniawski 1973) are more acceptable limits for this delineation. In this study, it seems that selecting 1 and 50 MPa as the lower and upper boundaries, which covers almost all rocks that include the definition of soft rocks in geology, is a suitable choice. This range is equivalent to three classes of very low, low, and medium–low strength rocks, according to ISRM (1979) classification and is considered based on the difference in the behavior of these rocks in mechanized tunneling and the machine performance in this range of the rock strength as well as the experience of the authors. Although there are no sharp boundaries, experience has shown that this is the typical threshold for rocks exhibiting plastic behavior, especially in continuous or intact sedimentary rocks with high clay minerals content, where the chipping process is not complete beneath the disc cutters.

(a) Establishing a special database for soft rocks

This section aims to examine the boreability characteristics of continuous soft sedimentary rocks and to develop a new model that provides a more accurate estimation of TBM performance in this particular lithological type, which falls into the indicated area of Fig. 27 (hatched yellow color area). To reach this goal, the data obtained from selected sections of four water conveyance tunnels (ZWCT-2, BzWCT, T4WCT and ChWCT), excavated in soft sedimentary rocks of the Zagros simply folded zone (ZSF) were collected, screened, and analyzed (Goodarzi et al. 2021). The characteristics of these four tunnels are listed in Tables 1 and 2.

Figure 28 shows the location of these tunnels in the simplified geological map of the ZSF zone in Iran. ZSF zone covers the entire western to southwestern and south-central parts of Iran and is considered as one of the main geological zones

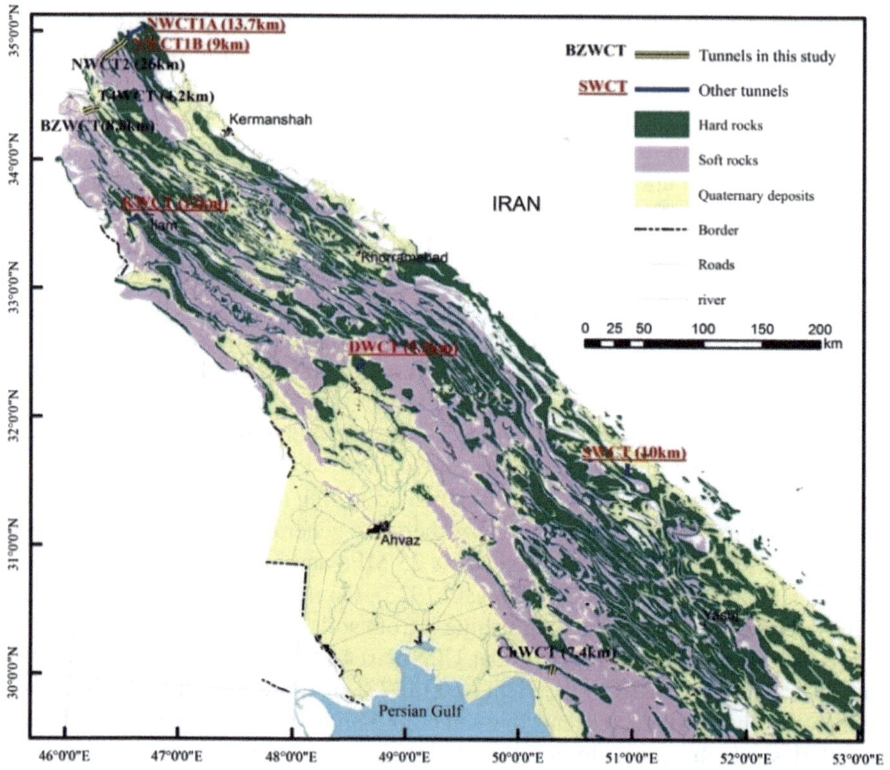

Fig. 28 Geographic situation of projects in the simplified geological map of the Zagros zone (Goodarzi et al. 2021)

in the country. This zone contains about 10 km thick of sedimentary rocks most of which can be classified as soft rocks. As shown in Fig. 28, major parts of the selected tunnels were excavated in soft rock formations.

Table 9 presents the geological formations of the ZSF zone along the tunnels alignment. A total of 30 tunnel Sections (5–20 m in length) composed of soft sedimentary rocks were selected for analysis in the four tunnels under investigation. These sections include various soft sedimentary rocks such as marlstone, mudstone, gypsum, anhydrite, sandstone, siltstone, shale, and limestone, all part of the Neogene-age sedimentary formations within the ZSF zone. The notable formations identified are Bakhtiari (Bk), Aghajari (Aj), Gachsaran (Gs), Asmari (As), Pabdeh (Pd), and Gurpi (Gu). Figure 29 displays some images of these formations across various projects, while Fig. 30 showcases photos specifically of the Pabdeh and Gurpi formations within the BZWCT and ZWCT-2 projects. Notably, the Pabdeh and Gurpi formations make up a significant portion of the studied sections in the BZWCT and ZWCT-2 projects.

Detailed studies showed that there are various rock units with different qualities along studied parts of these tunnels, varying from very poor and very soft to very good and strong rock masses. These rock units are listed in Table 10 for each tunnel

Table 9 Distribution of geological formations along the studied tunnels (Goodarzi et al. 2021)

Formation	Main lithology	Age	Projects			
			ZWCT-2	BzWCT	T4WCT	ChWCT
Quaternary Dep.	Unconsolidated soil	Quaternary			***	
Bakhtiari (Bk)	Conglomerate, sandstone, siltstone	Pliocene-Pleistocene				1
Aghajari (Aj)	**Marlstone***, **mudstone**, sandstone	Upper Miocene		3	5	4
Gachsaran (Gs)	**Marlstone, shale**, gypsum, anhydrite	Lower Miocene		3		
Asamri (As)	Limestone, marly limestone, dolomite	Oligocene-Miocene				
Pabdeh (Pd), Gurpi (Gu)	**Shale**, limestone, **mudstone**	Paleocene	10**	4		
Ilam (Il)	Limestone	Upper cretaceous				

*Bolded lithologies are incompetent soft rocks
**Numbers in the cells show the number of selected tunnel sections in each formation, for each tunnel
***Gray cells represent formations which exist along each tunnelling project

Fig. 29 Pictures of outcrops of formations in different projects: **a** Aghajari formation in BzWCT, **b** Pabdeh Formation in BzWCT, **c** Gurpi Formation in ZWCT-2, **d** Aghajari Formation in ChWCT (Goodarzi et al. 2021)

Fig. 30 Pictures of Pabdeh-Gurpi formation at tunnel face, **a** BzWCT project, **b** ZWCT-2 project (Goodarzi et al. 2021)

separately. As shown, some rocks like limestones, conglomerates and sandstones are competent and more brittle rocks with well-developed joint systems (Hard, discontinuous rock masses) and some of them like mudstones, marlstones and shales are incompetent rocks with more plastic behavior and weathered features at outcrops (Soft, continuous rock masses). The main petrographic, physical and mechanical characteristics of these rock units are summarized in Table 10. Based on the information in Table 10, the rock units can be divided into three groups as shown in Table 11.

The aim of this study is to investigate the performance of the machine in the rocks of third group, which can be categorized as LT-V lithological type and fall into the indicated area of Fig. 27 (hatched yellow color area), with characteristics as listed below:

- Their mineralogical content is rich in clay minerals;
- They are sensitive to water presence and behave very differently in saturated state than dry state
- They exhibit ductile or plastic behavior in uniaxial loading tests, especially when saturated;
- During loading by disc cutters, the rocks in this group usually deform and crack propagation and chip formation in rocks are incomplete and ineffective processes;
- Their UCS is typically between 10 and 50 MPa;

Table 10 Main characteristics of the identified rock units in the selected tunneling projects

Project	Lithology	Formation	Main minerals	UCS (MPa)	Class
NWCT2A	Limestone	IL	Calcite	75–125	Hard, discontinuous
	Arg. limestone	Pd-GU	Calcite, clay minerals	50–100	Hard, discontinuous
	Limy shale	Pd-GU	Clay minerals, calcite, quartz	20–50	Soft, continuous
	Shale	Pd-GU	Clay minerals, quartz	15–30	Soft, continuous
BZWCT& T4WCT	Limy shale	Pd-GU	Clay minerals, calcite, quartz	25–50	Soft, continuous
	Shale	Pd-GU	Clay minerals, quartz	10–30	Soft, continuous
	Limestone	As	Calcite, dolomite	70–120	Hard, discontinuous
	Gypsum and anhydrite	Gs	Gypsum and anhydrite	10–30	Soft, continuous
	Mudstone and marlstone	Aj	Clay minerals, calcite	5–25	Soft, continuous
	Sandstone, siltstone	Aj	Calcite, shell fragments, clay minerals	10–30	Soft, discontinuous
	Sandstone, conglomerate	Bk	Calcite, dolomite, rock fragments	80–150	Hard, discontinuous
T4WCT	Mudstone and marlstone	Aj	Clay minerals, calcite	5–25	Soft, continuous
	Sandstone, siltstone	Aj	Calcite, shell fragments, clay minerals	10–30	Soft, discontinuous
ChWCT	Mudstone and marlstone	Aj	Clay minerals	5–25	Soft, continuous
	Sandstone, siltstone	Aj	Calcite, shell fragments, clay minerals	10–30	Soft, discontinuous
	Sandstone, conglomerate	Bk	Calcite, dolomite, rock fragments	80–150	Hard, discontinuous

- Despite some discontinuity planes in surface outcrops, they do not exhibit "structurally controlled" behavior in deep tunnels and in saturated state, instead seem to behave as continuous media. During tunnel mapping, only very few ineffective discontinuities can be observed and measured in the tunnel face.

Table 11 Classification of identified rock units in the selected tunneling projects (Goodarzi et al. 2021)

Group	Lithology	Main minerals	Typical behavior in UCS test	Range of UCS (Mpa)	Structural condition in deep tunnels	Description	Formations
1	Limestone, argillaceous limestone, dolomite, and siliceous sandstone, conglomerate	Calcite, dolomite, quartz, clay minerals	Brittle	40–120	Discontinuous (Bedded and jointed)	Hard, Discontinuous	Asmari, Bakhtiari
2	Siltstone, calcareous weak sandstone	Calcite, dolomite, quartz, clay minerals	Brittle-Ductile	10–50	Discontinuous (Bedded and jointed)	Soft, Discontinuous	Some parts of Aghajari Fm
3	Marlstone, mudstone, shale, limy shale, gypsum, anhydrite, some siltstones	Clay minerals, calcite or (gypsum and anhydrite)	Ductile	10–50	Continuous	Soft, Continuous	Main parts of Pabdeh-Gurpi and Aghajari Fms

As stated above, the discontinuities do not impact the mechanical behavior and boreability characteristics of the rock mass in tunnel depth due to the plasticity properties and presence of clay minerals in the mineralogical composition of these rocks. Consequently, determining discontinuity properties, such as the number of discontinuity sets, spacing, persistence, and other properties of discontinuity surfaces, is inconsequential for these rocks at tunnel depth. This implies that, in the establishment of database and development of empirical relationships, parameters related to discontinuities are not relevant, and only parameters associated with the intact rock, such as UCS, are influential. According to the above explanation, the database developed in this study only includes the intact rock strength (UCS) and the operating and performance parameters of the machine, averaged for each selected tunnel section. The statistical distribution of more important recorded parameters in the database utilized for developing a specialized prediction model for soft rocks is shown in Fig. 31 and summarized in Table 12, including descriptive statistical characteristics of these data.

(b) Developing empirical equations

The relationship between three parameters of rock strength (UCS), cutter load (F_n), and cutterhead penetration (P) was analyzed in 30 selected tunnel sections using data recorded in the database. Figure 32a and Eq. (57) depicts the relation between the cutter penetration and cutter load at the 30 selected sections. As can be seen, there is a strong relationship between cutter load (F_n) and penetration (P). In this graph, the size of circles represents the average UCS at each tunnel section. Figure 32b illustrates that the relationship between penetration and cutter load significantly improves by categorizing the tunnel sections into two groups based on their UCS: 10–25 MPa and 25–50 MPa, leading to the development of Eqs. 58 and 59.

$$10 < UCS < 50\,MPa : F_n = 106.72Ln(P) - 100.42\,R^2 = 0.7612 \tag{57}$$

$$25 < \mathrm{UCS} < 50\,\mathrm{MPa} : \mathrm{F_n} = 153.37\,\mathrm{Ln}\,(\mathrm{P}) - 188.86\,\mathrm{R}^2 = 0.9356 \tag{58}$$

$$10 < \mathrm{UCS} < 25\,\mathrm{MPa} : \mathrm{F_n} = 127.38\,\mathrm{Ln}\,(\mathrm{P}) - 166.88\,\mathrm{R}^2 = 0.808 \tag{59}$$

These equations can also be used inversely to calculate the penetration for 17-inch disc cutters with cutter spacing ranging between 80 and 85 mm. Obviously, the estimated penetration should be checked for the availability of machine torque and power.

Multivariable regression technique was also used to find a correlation between cutter load (F_n) and UCS as independent variables (or predictors) and the natural logarithm of penetration or Ln(P) as the dependent variable (or response parameter). After evaluating different statistical models, the best relationship was obtained with $R^2 = 0.914$, as follows:

$$\mathrm{Ln}\,(\mathrm{P}) = 0.006\,\mathrm{F_n} - 0.016\,\mathrm{UCS} + 1.833 \tag{60}$$

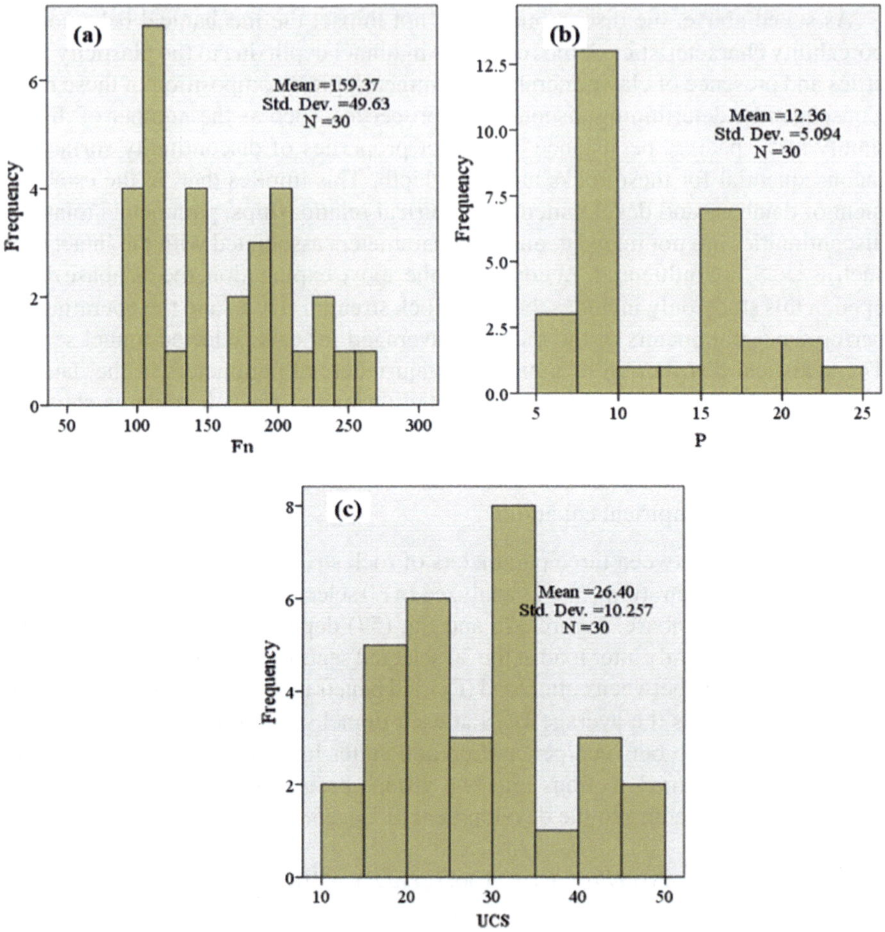

Fig. 31 Frequency histograms of the parameters recorded in the soft rocks database, **a** cutter load (F_n); **b** cutterhead penetration (P); **c** rock strength (UCS) (Goodarzi et al. 2021)

Table 12 Statistical characteristics of parameters recorded in the database of soft rocks

Parameters	Ave	Std. Dev	Med	Min.	Max.	Var
UCS (MPa)	26.4	10.3	25.0	10.0	50.0	105.2
F_n (kN)	159.4	49.6	142.4	98.1	257.1	2463
P (mm/rev)	12.4	5.1	9.9	6.1	23.8	25.9

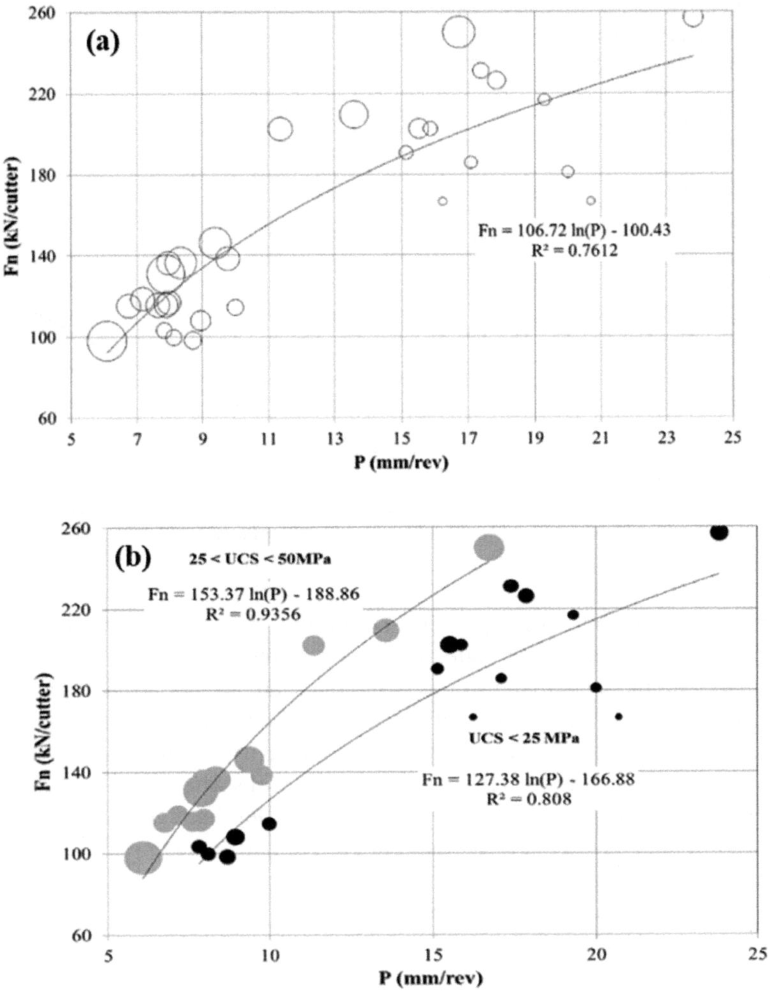

Fig. 32 The relationship between average cutter load and penetration in selected tunnel sections; **a** all data; **b** categorized data (Goodarzi et al. 2021)

or:

$$P = \text{Exp}\,(0.006\,F_n - 0.016\,\text{UCS} + 1.833) \quad (61)$$

Equations 57–61 can be used to calculate the performance of the machine in continuous, soft sedimentary rocks. The penetration in these types of rocks can be calculated by identifying only two parameters of UCS and cutter load. To facilitate using these equations a graph is also offered (based on Eq. 61) in Fig. 33. If the rock

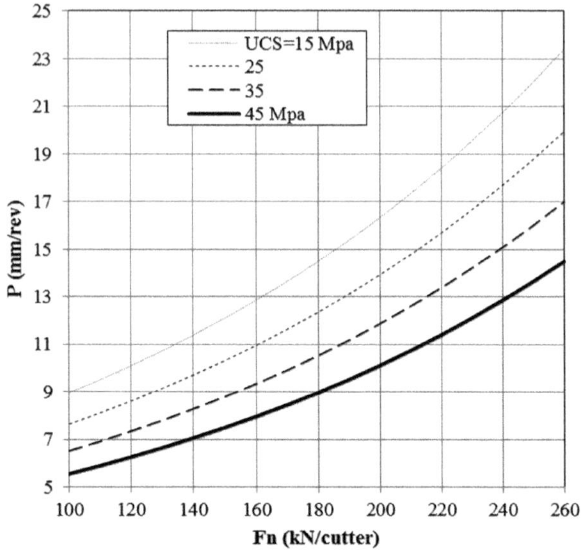

Fig. 33 New prediction model for estimating cutter penetration in continuous soft sedimentary rocks (Goodarzi et al. 2021)

strength (UCS) and one of the parameters of cutter load or penetration are known, the other parameter can be estimated using this graph.

It must be emphasized that the proposed models are the simple empirical relationships between three parameters of cutter load (as an operating parameter), penetration (as a performance parameter) and the rock strength (as a geological parameter). These relationships are applicable for the continuous soft rocks with the UCS in the range of 10 to 50 MPa, and for a machine with 17-inch diameter disc cutters (with a constant cross section) and a cutter spacing ranging between 80 and 85 mm.

The newly developed equations can be compared with models that were previously prepared for hard rocks. For this comparison, Eq. 53, which pertains to LT-II rocks, has been selected. This equation is also relevant to sedimentary rocks with a broader range of strength, which closely aligns with the new equation, making them complementary to each other.

For an accurate comparison, it's essential to consider wide discontinuity spacings (RQD = 100) for the rocks in Eq. 53. The results of this comparison, based on various values of UCS, are presented in Fig. 34. In this graph, thicker lines correspond to Eq. 53, while thinner lines relate to Eq. 61. Although the curves for UCS = 50 MPa appear similar in both equations, the difference between the two models becomes significant as rock strength decreases. For instance, the discrepancy between the models is greatest at UCS = 10 MPa, reaching up to 90%. This suggests that the new model, which pertains to continuous soft sedimentary rocks, differs substantially from the model for hard sedimentary rocks.

Therefore, it is recommended to use Eq. 61 for soft sedimentary rocks with high clay minerals content, in deep tunnels and below groundwater level, and Eq. 53 for hard sedimentary rocks (with UCS > 50 MPa). Considering the common boundary

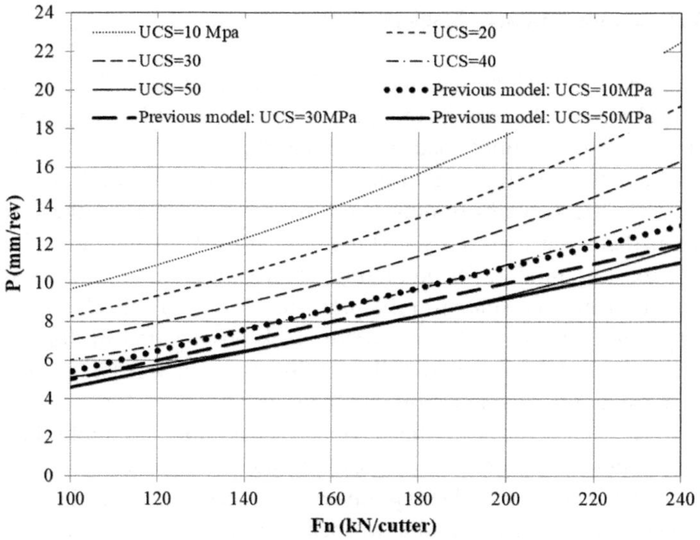

Fig. 34 Comparison of the new model (Eq. 61) with the previous model (Eq. 53) for massive and continuous sedimentary rocks with UCS < 50 MPa (Goodarzi et al. 2021)

of UCS = 50MPa for the two models, Fig. 35 can be proposed as a complete chart to predict the performance of the TBMs equipped with 17-inch disc cutters in massive and continuous sedimentary rocks.

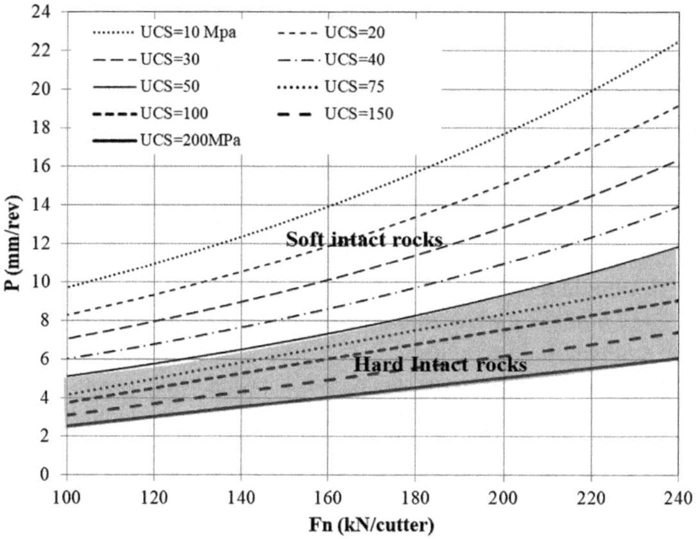

Fig. 35 TBM performance prediction chart in hard and soft, massive and continuous, sedimentary rocks (Goodarzi et al. 2021)

8.2.2 Very Strong, Massive Rocks (LT-VI)

The penetration of a machine into the rock is crucially tied to the process of chip formation, and any hindrances in this process can lead to disruptions in machine performance. When the cutter load is less than the critical load, the predominant behavior of the rock is grinding, resulting in the production of a high volume of fines (Dardashti et al. 2024). Transitioning from the grinding stage to the formation of chips requires the cutter load to exceed the critical load, which varies for different rocks. Many existing models for predicting cutter penetration into the rock and the empirical relationships presented in the previous section of this chapter (Sect. 8.1) are only applicable under conditions where crack propagation in the rock and the chipping process are fully developed (Fig. 1.26). In other words, the cutter load is sufficient for the required penetration into the rock.

In very strong, massive rocks, the capacity of the cutters to apply the necessary load for sufficient penetration into the rock is limited, and, in many cases, the cutter load is less than the critical load (Hassanpour et al. 2019b). Under these conditions, existing models are unsuitable, and new, specific models had to be developed. This section offers relationships between rock mass characteristics and machine performance parameters using data from the first 5 km of the KrWCT-S project, where a double-shield machine was employed to excavate in strong to extremely strong igneous and pyroclastic rocks. The objective is to establish a new predictive model for estimating the machine's penetration rate under this specific condition (Pourhashemi et al. 2021a).

(a) Project description

Kerman Water Conveyance Tunnel (KrWCT) is about 38 km long and is located in southern Iran (see Fig. 1). It is one of the longest tunnels currently being constructed in a volcanic and plutonic zone worldwide. The tunnel has been divided into northern (L = 19.1 km) and southern (L = 18.7 km) lots, which are under construction using two TBMs with diameters of 5.275 and 4.625 m, respectively. Only the data from the first 5 km of the southern lot (KrWCT-S) have been used in the analyses of this study, wherein grinding has occurred in many tunnel sections.

A double shield machine, manufactured by Herrenknecht, was employed for excavating KrWCT-S project (Fig. 5c). The cutterhead, as shown in Fig. 5c, features 37 disc cutters (29 single + 4 double cutters) with an average spacing of 72 mm. Each disc cutter is 432 mm (17 inches) in diameter. Other main technical specifications of the TBM are summarized in Table 6.

From a geological perspective, the KrWCT project site is situated within the Urumieh-Dokhtar zone (UDZ). This zone features Tertiary volcanic rocks that extend approximately 2000 km in length and 50–200 km in width, running from northwest to southeast across Iran. Magmatic activity in this area began in the Late Paleocene and reached its peak during the Late Eocene. The zone is composed of various volcanic, pyroclastic, and plutonic rocks. Therefore, it was anticipated that the tunnel will encounter different types of rocks, as illustrated in Fig. 36a.

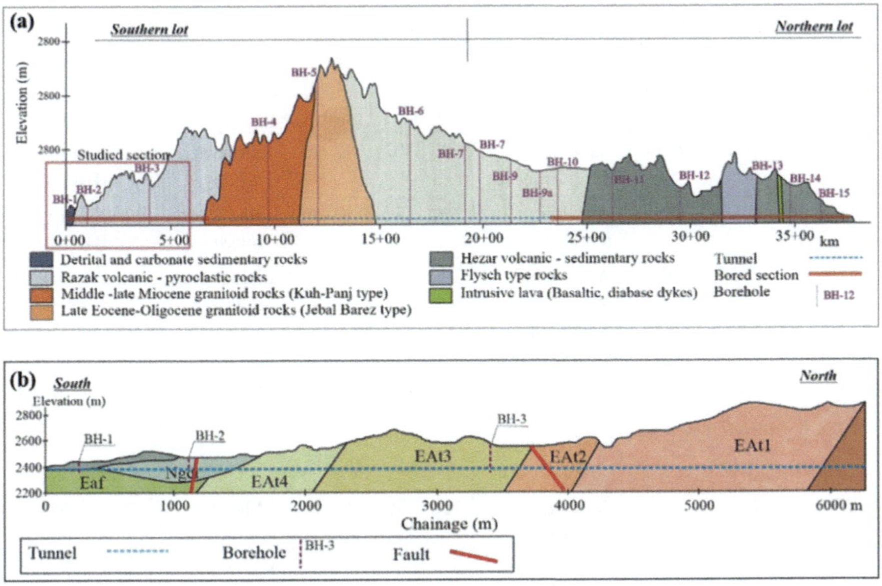

Fig. 36 **a** Geological section along the KrWCT project (northern and southern lots); **b** engineering geological section along the first 5 km of the KrWCT-S lot (Pourhashemi et al. 2021a)

Based on an engineering-geological point of view, the first 5 km of the southern lot of the project can be divided into 5 rock units with different engineering geological characteristics (SOI 2016) as listed in Table 13. Moreover, Fig. 36b illustrates the engineering geological section and the spatial distribution of the identified units along this section of the tunnel.

As shown in Table 13, the majority of rock units encountered in the tunnel, particularly the EAT1 to EAT3 units, typically exhibit high strength and low fracturing. Therefore, the rock mass boreability conditions in most sections of this tunnel were anticipated to be challenging. In this study, firstly, the model presented in Fig. 27 (Hassanpour et al. 2011) was used to evaluate rock mass boreability conditions. According to this classification, rocks located closer to the upper right corner of the chart, specifically in the B-0 and B-I classes, show a greater potential for grinding. Since the rock masses identified in the studied section of the KrWCT-S project were primarily categorized as B-0, B-I, and B-II classes (as indicated in Fig. 37 and Table 13), grinding was expected to be a common occurrence in this section of the tunnel.

Additionally, the analysis of the muck materials revealed that while the chipping phenomenon was observed in most sections of the tunnel (see Fig. 38a), grinding was also the dominant process in many areas. As illustrated in Fig. 38b–d, in the sections where grinding occurred, the grooves formed beneath the cutters were shallow, and the particle size of muck was small in size, consisting primarily of fine grains. This

Table 13 Typical engineering geological properties of the identified rock units along the first 5 km of the KrWCT-S project (SOI 2016)

Rock unit	Chainage (m)	Lithology	UCS (MPa)	RQD (%)	Structure	Boreability class
Ngcl	500–1300	Sandstone, conglomerate	70–100	60–80	Seamy to very blocky	B-II to B-III
EAT4	1300–2120	Andesite, agglomerate	100–150	60–80	Blocky to massive	B-II to B-III
EAT3	2120–3530	Andesite, andesitic basalts and Granodiorite	120–200	90–100	Blocky to massive	B-I to B-II
EAT2	3530–4100	Andesite, andesitic basalts and Granodiorite	180–300	90–100	Massive	B-0 to B-I
EAT1	4100–...	Andesite, andesitic basalts, tuff and rhyolite	180–230	90–100	Massive	B-0 to B-II

is a sign of low thrust/cutter load compared to the threshold normal loading that is needed for efficient chip formation and brittle rock fragmentation.

(b) Grinding conditions in the tunnel

Graphs (a) to (f) of Fig. 39 how the variations of different machine parameters (including boring time, RPM, cutter load, penetration, and S/P ratio) in each stroke along the tunnel. As shown in Fig. 39 cutterhead rotation speed and cutter load (as operational parameters) and boring time, penetration, and S/P ratio (as performance parameters) have large fluctuations in the bored section of tunnel. The variations in the performance parameters are due to the changes in geological and operational parameters. Therefore, once the geological conditions are varied, the operator tries to keep the performance parameters at optimum values by changing the operating parameters. Furthermore, the operator needs to monitor the machine's limitations, such as the maximum cutter load and cutter head speed, to prevent the damage to various parts of the machine.

In this project, by increasing the rock strength, the operator has also attempted to increase the penetration rate by increasing the thrust and rotation speed of the cutterhead. However, this approach was not successful at all times. In many sections, limitations prevented sufficient penetration into the rock, resulting in penetration rates that did not meet the desired levels. In these sections, as shown in the graphs presented in Fig. 39, the chip formation process remained incomplete and the machine worked like a stone mill where penetration was less than 3 mm, boring time was longer than 75 min per stroke, FPI was higher than 50 kN/c/mm/rev, and cutter load was about the maximum disc capacity (about 230 kN).

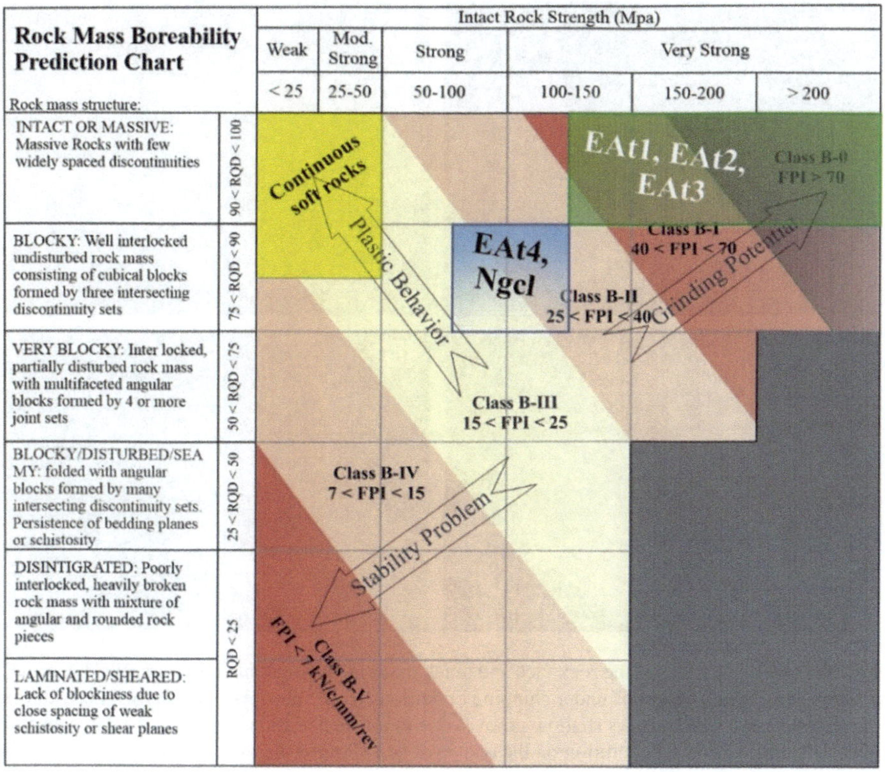

Fig. 37 Variation ranges of FPI and boreability classes for the project's identified rock units based on the UT model's boreability classification chart (Pourhashemi et al. 2021a)

Table 14 indicates that the machine encountered stronger rock masses toward the end of the bored section. This trend is also reflected in the FPI values. As illustrated in Fig. 39d, which shows the variations of FPI along the tunnel alignment, there is a significant increase in FPI values within the two rock units, EAT1 and EAT2.

In Fig. 40, values for cutter load are plotted against penetration for various rock units. Additionally, Table 14 presents the results from the statistical analyses of actual machine operating and performance data, which were conducted separately for each rock unit. The blue hatch in the upper left corner of each graph designates rocks with little cutter penetration despite the maximum force being applied to the cutter. As expected, the stronger rock units are more likely to be located in this area. The hatched area shown in these graphs can be assumed to be consistent with rocks with high potential for grinding during rock excavation by machine. As can be seen, this phenomenon has occurred especially in EAT1 and EAT2 rock units. Furthermore, the results shown in Table 14 indicate that the average FPI and cutter load in EAT1 and

Fig. 38 Different views of the rock face and muck material during tunnelling in KrWCT-S project; **a** normal size muck materials under chipping conditions on the conveyor belt: **b** rock mass in front of the cutterhead which shows shallow grooves due to low penetration in rock, **c, d** small size muck materials under grinding condition on the conveyor belt (Pourhashemi et al. 2021a)

EAT2 were higher compared to the other units. Therefore, the average penetration was lower than that of the other rock units. It was mentioned earlier that these two rock units are categorized as B-0 and B-I, and sometimes B-II classes (Fig. 37 and Table 13).

(c) Establishing a special database for very strong rocks

In this study, a number of tunnel sections (mostly in stronger rocks of EAT1, EAT2, and EAT3 rock units) with the availability of accurate geological data and dominance of the grinding phenomenon were selected to establish a suitable database. To ensure that the grinding phenomenon is predominant, in addition to checking muck materials, a great deal of attention was also paid to the recorded penetrations by PLC and depth of grooves created by disc cutters on the tunnel face. The frequency histogram of UCS, cutter load (F_n), and penetration (P) in the database are given in Fig. 41.

It must be emphasized that in all of the selected tunnel sections, the quality of rock masses was good (massive rocks with RQD > 90) and intact rock strength or UCS ranges between 150 and 300 MPa (Fig. 41a). Actually, in most of these sections no discontinuity was observed during rock face observations or they were widely spaced. Therefore, they didn't have important effect on cutting process and boreability characteristics of the rock masses.

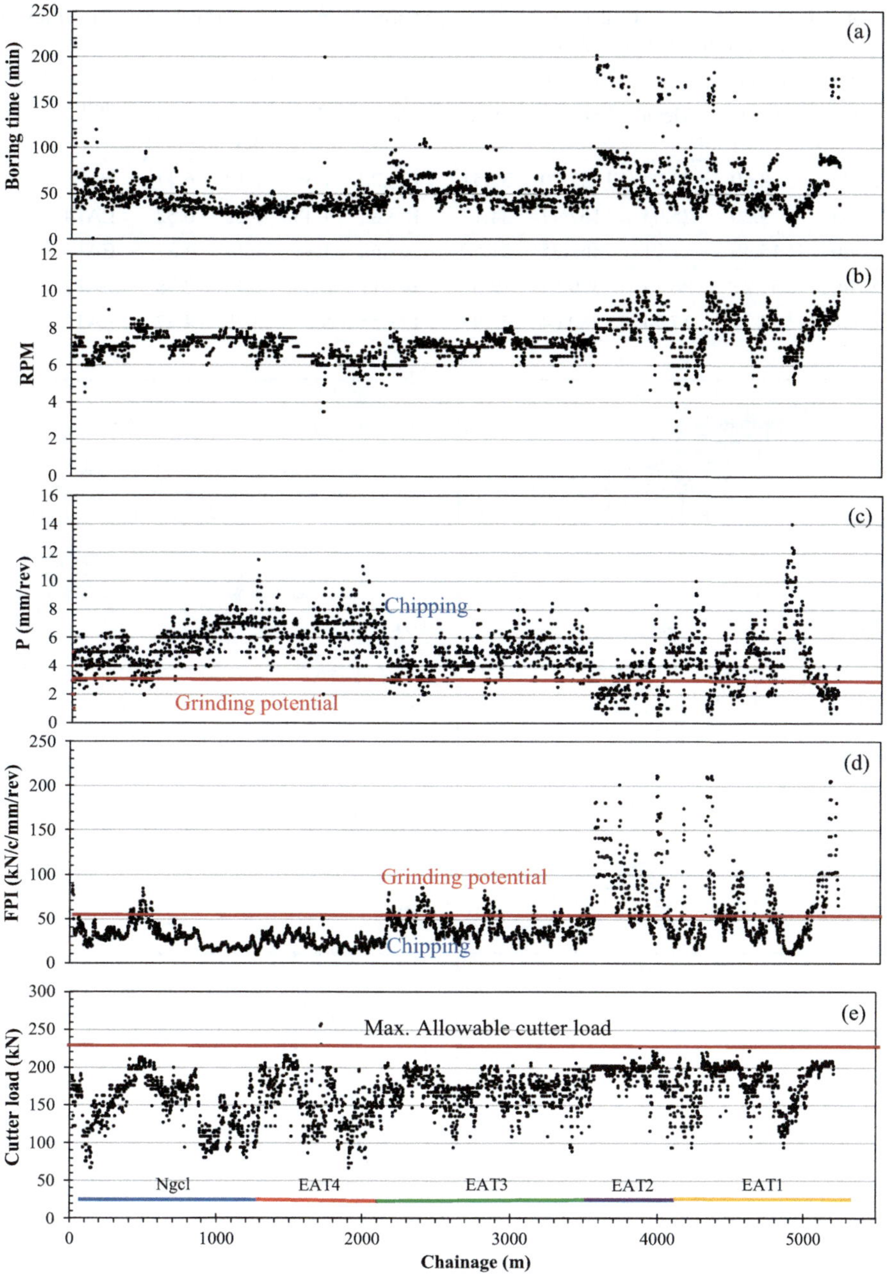

Fig. 39 Variations of different operating and performance parameters of each stroke along tunnel alignment: **a** boring time of each stroke, **b** Cutterhead RPM, **c** Cutterhead penetration, **d** FPI, **e** cutter load (Pourhashemi et al. 2021a)

Table 14 Descriptive statistical analysis of main performance and operating parameters in different rock units (Pourhashemi et al. 2021a)

FPI (kN/c/mm/rev)			Cutter load (kN)			Penetration (mm/rev)			Geological unit
Var	St. dev	Mean	Var	St. dev	Mean	Var	St. dev	Mean	
73.99	8.60	23.00	1057.20	32.51	138.20	1.36	1.17	6.37	Ngcl
84.55	9.20	25.04	1167.47	34.17	153.82	1.48	1.22	6.52	EAT4
237.01	15.40	41.01	462.35	21.50	168.60	1.41	1.19	4.57	EAT3
2624.27	51.23	86.12	380.52	19.51	191.11	2.05	1.43	2.98	EAT2
1796.82	42.39	57.74	780.04	27.93	178.55	5.06	2.25	4.42	EAT1

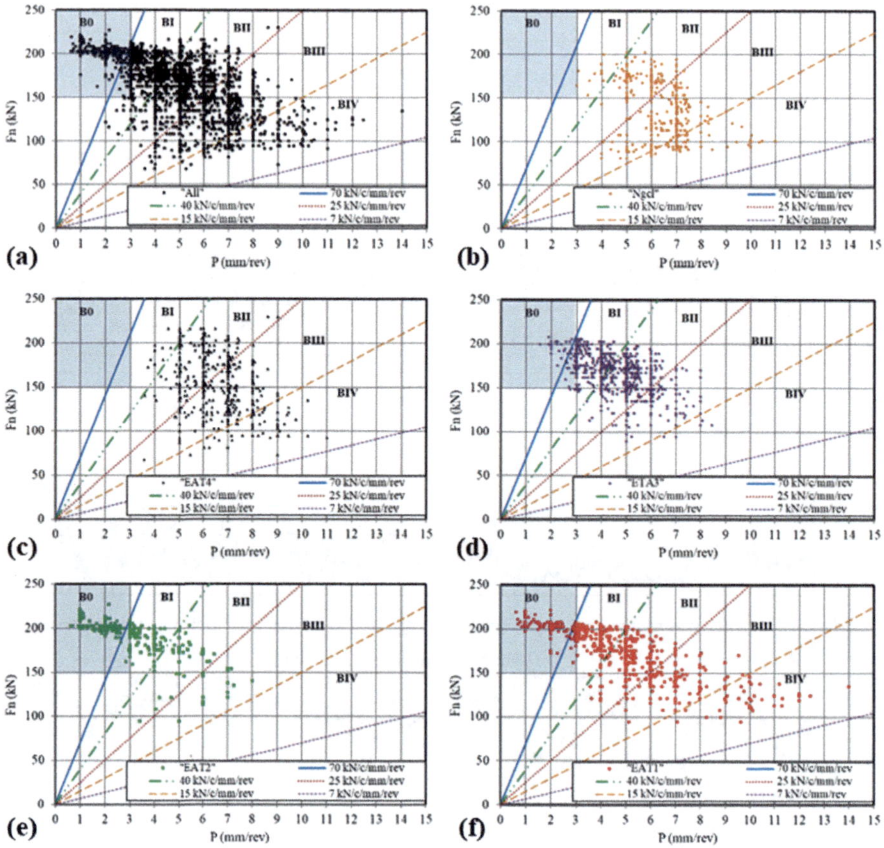

Fig. 40 Relationship between cutter load and penetration in each stroke for different rock units: **a** all rock units, **b** Ngcl, **c** EAT4, **d** EAT3, **e** EAT2, **f** EAT1; The hatched area in each graph corresponds to the grinding condition (Pourhashemi et al. 2021a)

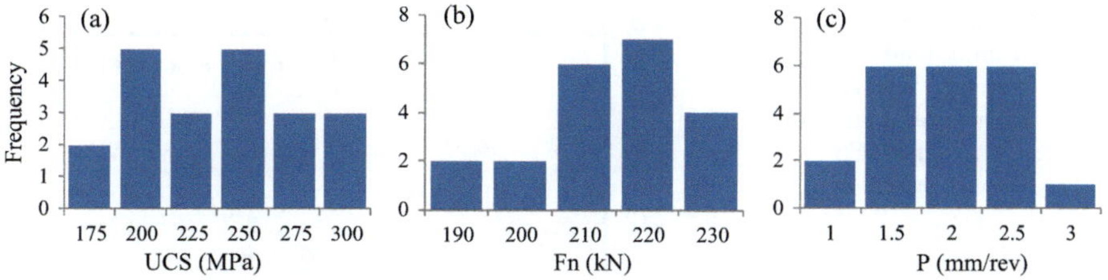

Fig. 41 Frequency histogram of: **a** UCS, **b** cutter load and **c** cutterhead penetration in the database (Pourhashemi et al. 2021a)

(d) Developing empirical equations

In this study, simple and multivariable regression analyses were performed on data recorded in established database. In simple regression analyses, to obtain better results, the recorded data were divided into three groups based on the rock strength (Table 15) estimated in each tunnel sections and analyses were performed on three group separately. As shown in Table 15 and Fig. 42 three linear equations (Eqs. 62–64) have been developed to relate cutter load (F_n) and penetration (P) for each group, which can be employed as predictive tools for estimating cutter load required to reach the desired penetration once grinding is the predominant phenomenon rather than chipping. They can also be used conversely for estimation of cutterhead penetration in very strong rocks when a given load is applied on disc cutters.

In this study, multivariable regression technique was also used to find a correlation between cutterhead penetration (P) and UCS as independent variables (or predictors) and the cutter load (F_n) as the dependent variable (or response parameter). After evaluating different statistical models, the best relationship was obtained with $R^2 = 0.773$, as follows:

$$F_n = 0.278 \text{ UCS} + 6.378 \text{ P} + 135.602 \tag{65}$$

or:

$$P = 0.157 F_n - 0.044 \text{ UCS} - 21.261 \tag{66}$$

Table 15 Summary of regression models between cutter load (F_n) and penetration (P) and proposed empirical equations developed for predicting TBM performance in grinding condition (Pourhashemi et al. 2021a)

Eq. no.	R^2	Equation	UCS range (MPa)	Group
(62)	0.758	$F_n = 15.67P + 161.4$	150–200	1
(63)	0.788	$F_n = 10.74P + 193.49$	200–250	2
(64)	0.782	$F_n = 12.25P + 206.83$	250–300	3

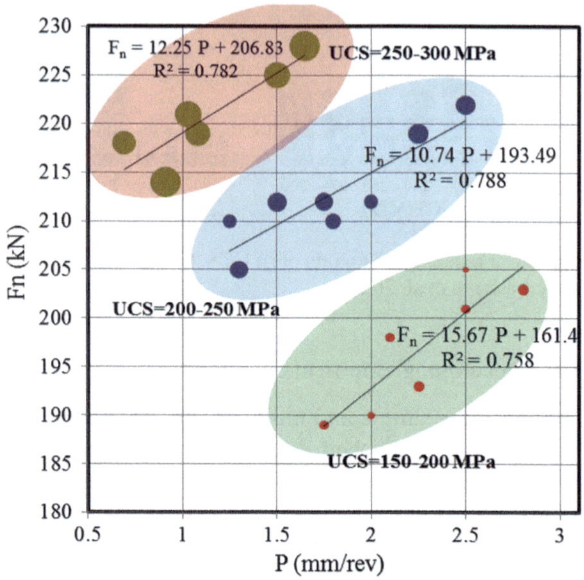

Fig. 42 Relationship between cutter load and penetration for each category of rocks (the size of circles is proportional to the UCS values) (Pourhashemi et al. 2021a)

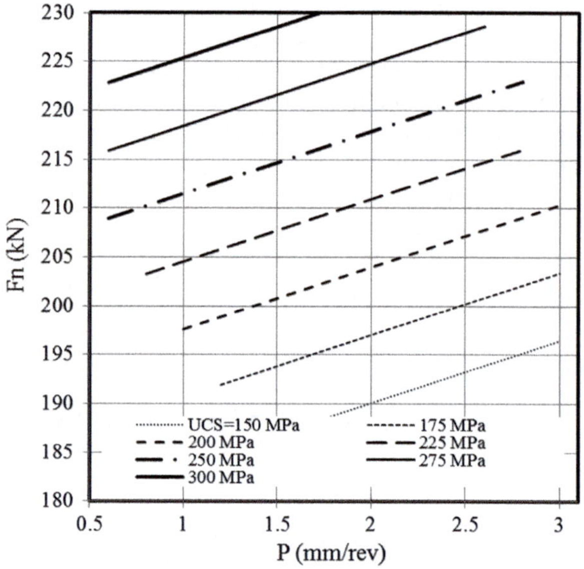

Fig. 43 New prediction model for estimating cutter penetration in grinding conditions (Pourhashemi et al. 2021a)

Equations 65 and 66 can also be used to calculate the performance of the machine in grinding conditions. To facilitate using these equations a graph was also prepared and presented in Fig. 43. If the rock strength (UCS) and one of the parameters of cutter load or cutterhead penetration are known, the other parameter can be estimated using this graph.

The developed model has specified the relationship between penetration and cutter load for rocks with different strengths. However, this model can only be used for the above-mentioned conditions, namely very strong rocks (UCS = 150–300 MPa), with massive structure (RQD > 90%). The developed model is only applicable to 17-inch-diameter disc cutters with a maximum load capacity of 230–250 kN.

9 Rock Mass Boreability Classification

During the development of the UT model, the authors proposed a rock mass boreability classification chart (see Fig. 27). This initial chart was based on data from four initially identified LTs. In this section, a revised version of the chart, which includes new experiences and additional data from new LTs, is presented (Fig. 44). In the updated chart, the six rock mass boreability classes originally presented in Fig. 27 are retained, and a new class, B-VI, has been added. The general characteristics of the seven rock mass classes are detailed in Table 16 and Fig. 45.

Table 16 illustrates that the boreability conditions of the rock mass improve from the top to the bottom, whereas the stability conditions of the rock mass deteriorate for tunnel construction. Both factors are essential for the effective operation of tunnel boring machines (TBMs). Therefore, evaluating the challenges associated with using a TBM (known as TBM excavatability) necessitates considering both boreability and stability conditions.

To gain a better understanding of machine performance in rock masses with different boreability classes, Table 17 provides an example that estimates the performance of a Tunnel Boring Machine (TBM) in six assumed rock units, labeled “A” to “F” (indicated by stars in Fig. 44). Each unit corresponds to a specific boreability class, ranging from B-0 to B-V.

The FPI values for rock units “A” to "F" were estimated based on their engineering geological characteristics, as illustrated in Fig. 44. The machine penetration rate was then calculated using the available equations from Chap. 1 (Eqs. 1.22 and 1.12), with an average cutter load of 200 kN and a cutterhead rotation speed of 7 rpm. Assuming a utilization factor of 25%, the theoretical advance rate of the machine in rock units “A” to “F” ranges from 2.5 to 73.6 m/day.

Achieving the calculated high penetration and advance rates for rock mass classes B-IV and B-V is often impractical due to issues such as unstable tunnel walls, face instability, and difficulties in collecting and transporting excavated materials, as indicated in Table 16. Consequently, operators are required to reduce the thrust force and the rotation speed of the cutterhead. In these situations, the operator maintains a constant penetration rate to manage ground challenges and the limitations of the machinery. Using more realistic assumptions for the cutter load and RPM can lead to more accurate penetration and advance rates in unstable ground, classes B-IV and B-V. The results of these calculations are presented in part (b) of Table 17. As shown, the highest advance rates are observed in rock units C, D, and E, which are classified as B-II, B-III, and B-IV, respectively. These rock mass classes provide

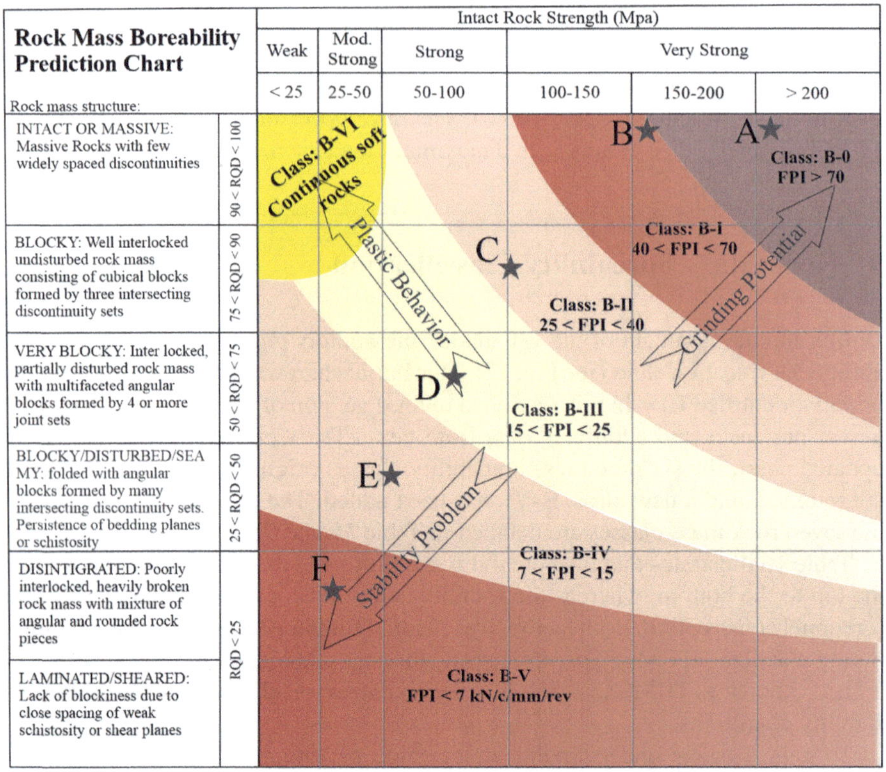

Fig. 44 The latest version of the UT model's rock mass boreability prediction chart, which is used for FPI prediction and the identification of rock mass boreability classes under general geological conditions (Hassanpour et al. 2023a)

optimal conditions for mechanized tunneling regarding rock mass boreability and tunnel stability. According to Table 16, these conditions are rated as good to very good for tunnel boring machines.

10 Cutter Consumption

In the UT model, empirical equations developed based on data from Iranian mechanized tunneling projects are used to evaluate cutter consumption, crucial for evaluating machine performance and calculating the utilization factor. For this purpose, tunnels alignment was divided into different sections with the same engineering geological properties including abrasivity of intact rock, strength, and degree of fracturing of the rock mass. These sections more or less coincided with engineering geological units identified and mapped during the construction phase. This was

Table 16 The proposed system for classification of rock mass boreability (Hassanpour et al. 2023a)

Class	FPI (kN/mm/rev)	Descr. of boreability	Stability condition	Descr. of TBM excavatability	LT	Example
B-0	> 70	Tough	Completely stable	Tough	LT-VI	Very strong and extremely strong, massive quartzitic veins, intrusive and metamorphic rocks
B-I	40–70	Fair-tough	Stable	Fair	LT-IV	Massive igneous and metamorphic rocks
B-II	25–40	Good-fair	Minor instabilities	Good	LT-I	Blocky and jointed tuffs, Tuffites, limestone
B-III	15–25	Good	Only local structural instabilities	Very good	LT-II	Alternations of sandstones, limestone and shale
B-IV	7–15	Very good	Some major instabilities	Good	LT-III	Alternations of thin bedded shale and sandstone beds
B-V	< 7	Excellent	Collapse, gripper problems, squeezing , etc	May be problematic	–	Highly foliated and schistose metamorphic rocks (slate, Phyllite, graphite schist), Shale, marlstone, thick fault zones
B-VI		Very good	Clogging, instabilities in the tunnel face	Good (above the groundwater)	LT-V	Clayey rock (mudstone, siltstone)

followed by estimating the average values of recorded geological parameters and disc cutter life for each tunnel section.

Disc cutter life can be expressed in many different ways. A frequently used unit is the rolling distance (km) before a disc has to be replaced. In this concept the penetration rate and the position of each cutter should be known. Also, disc life can be expressed as the length of tunnel mined or volume of rock excavated per cutter, shown as H_m and H_f, respectively (Bruland 1998):

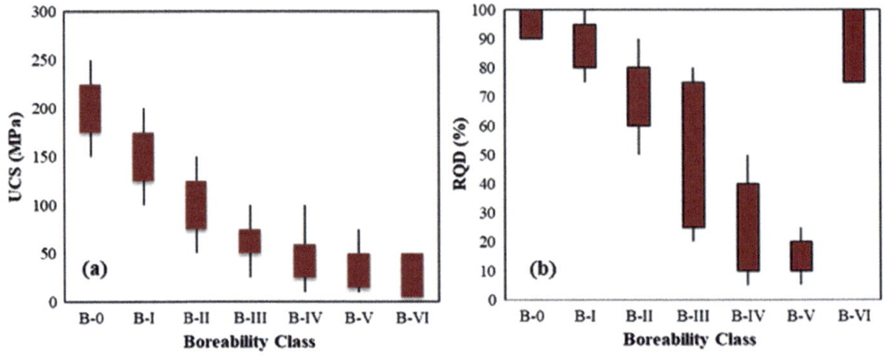

Fig. 45 The variation ranges of UCS (**a**) and RQD (**b**) in different boreability classes based on data recorded in the database

Table 17 Calculating the machine performance in rock masses with different boreability classes

Rock mass	A*	B	C	D	E	F
Boreability class	B-0	B-I	B-II	B-III	B-IV	B-V
UCS (MPa)	210	160	100	75	50	30
RQD (%)	100	100	80	60	40	20
(a)						
FPI (kN/mm/rev)	96	64	29	18	11	7
F_n (kN/cutter)	200					
RPM (rev/min)	7					
Penetration per revolution (mm/rev)	1	3.11	6.78	11.18	18.44	29.20
Penetration rate (m/h)	0.42	1.31	2.85	4.70	7.74	12.27
Utilization factor (%)	25					
Advance rate (m/day)	2.5	7.8	17.1	28.2	46.5	73.6
(b)						
FPI (kN/mm/rev)	96	64	29	18	11	7
F_n (kN/cutter)	220	180	180	180	80	50
RPM (rev/min)	7				4	3
Penetration per revolution (mm/rev)	2.0	2.8	6.10	10.06	7.37	7.14
Penetration rate (m/h)	0.84	1.18	2.56	4.23	1.77	1.29
Utilization factor (%)	25					
Advance rate (m/day)	5.0	7.1	15.4	25.4	10.6	7.7

*For rock mass A, assuming grinding conditions, Figs. 42 and 43 have been utilized

$$\mathrm{H_m} = \frac{\mathrm{L}}{\mathrm{N_{TBM}}} \tag{67}$$

$$H_f = \frac{H_m \cdot \pi \cdot D_{TBM}^2}{4} \tag{68}$$

$$W_m = \frac{1}{H_m} \tag{69}$$

where $\mathrm{H_m}$ (in m/cutter) is the length of tunnel bored, and $\mathrm{H_f}$ (in m^3/cutter) is the volume of rock excavated per each cutter change, respectively. Wm (in cutter/m) is the number of cutters needed to excavate each unit length of tunnel. $\mathrm{D_{TBM}}$ is TBM diameter, $\mathrm{N_{TBM}}$ is total number of disc cutters changed and L (in m) is tunnel length excavated per full dressing of the head. This does not mean that all the cutters on the cutterhead (from center to gage) were replaced, instead means that the total number of cutter replacements are the same as $\mathrm{N_{TBM}}$, granted that cutters in the gage area are replaced at faster rate.

During development of models, in order to estimate actual cutter life along the tunnels, a procedure explained by Bruland (1998) was used. In this procedure, instantaneous cutter life parameters ($\mathrm{H_{mi}}$ and $\mathrm{H_{fi}}$) are calculated for sections along the tunnel by summation of cutter life at each section between cutter change positions.

As shown in Table 3, a total of 52 zones (sections) in four projects were identified and related cutter life and rock characterizations were compiled in the database. A series of statistical analysis was performed on data recorded in the database to develop relationships between rock properties and cutter wear parameters. These analyses yielded a set of empirical equations to estimate disc cutter life in different geologies. An in-depth discussion on the development of these models can be found in Hassanpour et al. (2014, 2015), Hassanpour (2018) and Karimi et al. (2024).

To develop simple empirical equations for cutter life prediction, the relationships between actual cutter life or $\mathrm{H_f}$ (m3/cutter) and different ground parameters including intact rock properties (UCS, quartz content, and Vicker's Hardness Number or VHNR) and fracturing degree or rock mass properties (joint count number or $\mathrm{J_V}$, RQD, Basic RMR, and GSI) were evaluated. In this case after checking different statistical models, very good correlations between $\mathrm{H_f}$ as response parameter and UCS and VHNR as predictors, were obtained. Table 18 lists the summary results of correlations and proposed equations developed based on data from each LT introduced in Table 4. A general equation was also proposed by combining data from all lithological types.

In a given project, rock mass properties have direct influence on ease or difficulty (rate) of boring and abrasivity of ground. Usually stronger and less fractured rock masses are more difficult to cut by disc cutters and require use of higher thrusts to achieve a certain level of penetration. In such ground conditions disc cutter life is also reduced dramatically. Therefore, lower values of $\mathrm{H_f}$ are usually recorded in strong and massive rock masses with high percentages of abrasive minerals (and vice versa). So, $\mathrm{H_f}$ can be selected as an index for categorizing rock abrasivity (Table 19).

Table 18 Equations developed by the authors for cutter life estimation of each tunnel project separately and the combined equation for all projects

No.	Equation	R^2	LT	Reference	Eq. no.
1	$H_f = -2.669\ VHNR - 7.891\ UCS + 3430.955$	0.693	LT-I	Hassanpour et al. (2014)	(70)
2	$H_f = -2.013\ VHNR - 8.074\ UCS + 2859.350$	0.785	LT-III	Hassanpour (2018)	(71)
3	$H_f = -0.553\ VHNR - 2.302\ UCS + 938.887$	0.444	LT-IV	Hassanpour et al. (2015)	(72)
4	$H_f = -2.544\ VHNR - 8.331\ UCS + 3288.248$	0.771	General	Hassanpour et al. (2015)	(73)

Table 19 Summary of ground conditions for various abrasivity classes (Hassanpour et al. 2015)

Abrasivity class	H_f (m^3/cutter)	Rock mass abrasivity	Example
A-0	< 100	Extremely abrasive, High cutter wear	Very strong quartzitic veins, intrusive and metamorphic rocks with high quartz contents
A-I	100–500	Very abrasive	Strong to very strong granites and gneisses
A-II	500–1000	Abrasive	Strong siliceous tuffs and quartz sandstones
A-III	1000–1500	Moderately abrasive	Moderately strong to strong well-cemented sandstones
A-IV	1500–2000	Slightly abrasive	Shales, strong limestones and dolomites
A-V	2000–2500	Not very abrasive	Moderately strong limetones, dolomites
A-VI	> 2500	Non-abrasive, almost no cutter wear	Marls, mudstones, claystone, weak limestones

A prediction chart based on the general equation (Eq. 73) was developed and shown in Fig. 46, enabling quick estimation of cutter consumption and the number of disc cutters needed for completing tunnel excavation by TBM, during the preliminary design of the project.

11 Utilization Factor

As already mentioned at the Chap. 1, the TBM penetration rate alone does not determine the actual machine advance rate over a particular period. The actual advance rate is deeply affected by other factors such as rock support, cutter replacement, machine

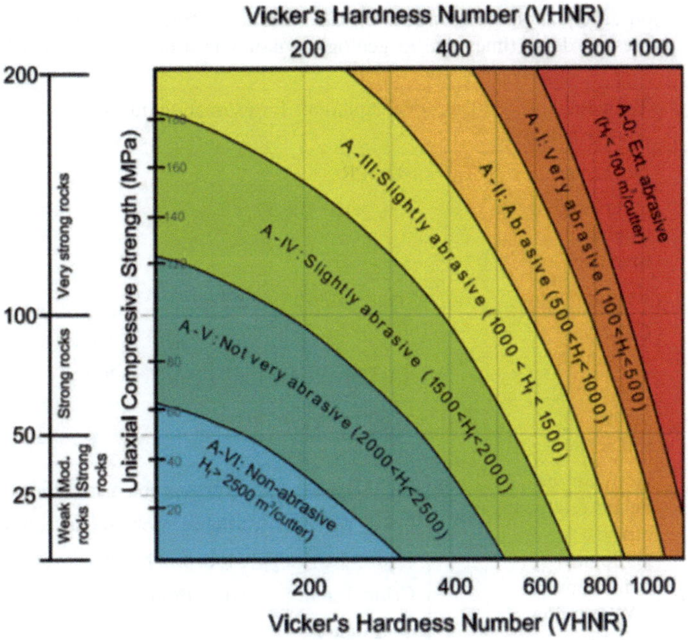

Fig. 46 General cutter life prediction chart (Hassanpour et al. 2015)

repairs, efficiency of the back-up system to remove the muck from the excavation face and to deliver construction material and supplies, etc. (Maidl et al. 2008). In relation to all this, it is also clear that adverse geological conditions may slow down all the tunneling activities thus leading to smaller TBM advance rates.

The combined effect of these factors is captured by the machine utilization factor, which is the percentage ratio of net boring time to the total tunneling activity time, including delays (Eq. 74).

$$U = \frac{100 \cdot T_B}{T_B + T_{Rb} + T_{Rg} + T_{CI} + T_{CC} + T_{Geo} + T_{Mrp} + T_{Brp} + T_{Log} + T_{Oth}} \quad (74)$$

In the UT model, empirical relationships for estimating tunneling activity times, as presented in Eq. 74, are listed in Table 20. These equations are derived from statistical analysis of tunneling activities in Iranian projects constructed with shielded machines. As shown, these times, organized into 6 groups, relate to the 10 main tunneling activities (with zero or negligible overlap) and the delays that occur in them.

Total activity times per 1000 m of tunneling—such as net boring, segment erection, re-gripping, disc cutter replacement and inspection—can be estimated using parameters like penetration rate (PR), complete ring erection time, machine gripping time, number of required disc cutters, disc cutter replacement time, and rock abrasivity

Table 20 Empirical equations developed in the UT model for estimating routine tunneling activity times and unforeseen delay times due to geological issues in a tunnel constructed by shielded machine

Group	Tunneling activities/delays	Empirical equations for estimating times (h/km)	Eq. no.
(1) Machine advance	Net boring	$T_B = 1000/PR$	(75)
	Ring building for shielded machines	$T_{Rb} = (t_{Rb} \times 1000)/(60 \times L_{seg})$	(76)
	Re-gripping	$T_{Rg} = (t_{Rg} \times 1000)/(60 \times L_{st})$	(77)
(2) Geological problems	Cutters Inspection	$T_{CI} = \begin{cases} 0 & ABI < 200 \\ 0.091 \times ABI + 17.13 & ABI > 200 \end{cases}$	(78)
	Cutter replacement	$T_{CC} = (1000 \times t_{CC}/60)/H_m$	(79)
	Activities or delays related to adverse geological conditions (Water inflow, collapse, squeezing, stickiness, etc.) such as cleaning of machine, pumping, probe drilling, pre-grouting, foam injection, excavating behind the shield, crushing the boulders, etc	$T_{geo} \begin{cases} 0.2 \times T_B & \text{Non-problematic zones} \\ 0.5 \times T_B & \text{Slighty problematic zones} \\ (2.0 - 5.0) \times T_B & \text{Mod. problematic zones} \\ \text{Critical cond.;} & \text{High problematic zones} \\ \text{long delays;} \gg 5 \times T_B & \text{Very high problematic zones} \end{cases}$	(80)
(3) Machine	Routine machine repairments and maintenance	$T_{Mrp} = 150$	(81)
(4) Back-up	Routine back-up repairments and maintenance	$T_{Brp} = 100$	(82)

(continued)

Table 20 (continued)

Group	Tunneling activities/delays	Empirical equations for estimating times (h/km)	Eq. no.
(5) Logistics	Utilities installation, conveyor apparat, construction materials, etc	$T_{Log} = 100$	(83)
(6) Others	Other unforeseen activities/delays	$T_{oth} = \begin{cases} 100 \text{ Non-problematic zones} \\ 120 \text{ Slighty problematic zones} \\ 150 \text{ Mod.problematic zones} \\ 200 \text{ High problematic zones} \\ 500 \text{ Very high problematic zones} \end{cases}$	(84)

Legend $\mathbf{T_B}$**:** Total boring time, **PR:** Penetration rate (m/h), $\mathbf{T_{Rb}}$**:** Total time for erecting all segmental rings, $\mathbf{t_{Rb}}$**:** Time for erecting a full ring of segments (=30–60 min), $\mathbf{L_{seg}}$**:** Segment width (m), $\mathbf{T_{Rg}}$**:** Total time for regripping of the machine, $\mathbf{t_{Rg}}$**:** Time for one regrip of machine (=3–5 min); $\mathbf{L_{st}}$**:** Strock length (m),$\mathbf{T_{CI}}$**:** Total time for all cutter inspections, **ABI:** Abrasivity Index (= UCS × VHNR/100), $\mathbf{T_{CC}}$**:** Total time for all cutter changes, $\mathbf{t_{CC}}$: Time for one cutter replacement (= 40–60 min), $\mathbf{T_{Geo}}$**:** Total times related to adverse geological conditions, $\mathbf{T_{Mrp}}$**:** Total time for routine machine repairments and maintenance, $\mathbf{T_{Brp}}$**:** Total time for routine back-up repairments and maintenance, $\mathbf{T_{Log}}$**:** Total time for logistic

index (ABI). Geological conditions and the presence of problematic zones, such as fault zones with varying characteristics (Shayan et al. 2023, 2025a, 2025b), also influence geological delay estimates. Other tunneling times are based on experience from various projects in Iran.

12 Validation of the Model

The best way to evaluate the reliability of a newly developed model is to use data from the observed performance of the machines in the field. This involves comparing the results from the model with actual datasets. In this section, recorded data from recent projects will be utilized to achieve this goal.

It's important to mention that data collection in the studied tunnels continued after the UT model was established, and this data was used to validate the model. This study focused on two projects, KWCT and ZWCT-2, which were only partially completed at the time the initial model was developed. Additionally, many researchers in recent years have examined the validity of the UT empirical relationships. Some of their most significant findings will be presented in this section.

12.1 KWCT-2 Project (Hassanpour et al. 2015)

The second lot of the Karaj Water Conveyance Tunnel (KWCT-2), which is approximately 14 km long, was excavated in geological conditions similar to those of the first lot (KWCT-1), utilizing the same double-shield TBM (Fig. 5b). Due to the consistent geological conditions and the use of the same machine, insights gained from the first lot were applied to predict the machine's performance and schedule the project for the second lot. In 2009, the initial UT model was employed to forecast machine performance for this new section (Hassanpour and Rostami 2009).

The geological cross-section of this tunnel is illustrated in Fig. 47. The identified rock units along this section of the tunnel are listed in Table 21. By precisely locating the rock units along the tunnel route and evaluating their geomechanical properties, it was possible to predict and estimate the machine's performance using Eq. 52 and Fig. 22. It is important to note that in this study, Eq. 52, which was proposed for LT-I (see Table 4), is deemed appropriate since the tunnel is being excavated through the Formations, mainly pyroclastic and volcanic rocks.

The input data and analysis results are shown in Tables 22 and 23 and Fig. 48. As shown in Fig. 48 and Table 23, it was predicted that the total time required for excavating the second lot of the project is approximately 1049 working days, equivalent to around 35 months.

Figure 48 also displays the monthly and cumulative machine advances. It is observed that the maximum monthly advance was predicted to be 637 m, with the minimum at 124 m. The lowest machine advance was expected in months 18 and 19 at a known fault zone (chainage 6 + 700 m), where a lower daily advance rate was anticipated due to reduced machine utilization in that section.

The tunnel excavation began in the autumn of 2009 and was completed in late summer of 2012, lasting a total of 30 months. Figure 49a provides a visual representation of the predicted versus the actual monthly and cumulative machine advances. Notably, there is a significant difference between the predictions and the machine's actual performance.

Upon closer examination of the results, it became clear that the primary reason for the discrepancy was the assumption of a lower utilization factor in the initial predictions. Due to skill enhancements, increased crew experience, and the resolution of technical issues related to the machine and tunnel, a higher utilization factor was achieved in the KWCT-2 project compared to the KWCT-1.

When considering the actual utilization factor in the predictions, as shown in Fig. 49b, there is a strong correlation between the newly forecasted values and the actual advance. Although there are slight differences in the monthly advance values, a significant alignment can be seen in the cumulative advance, indicating accurate predictions of penetration values. The variations in monthly values can be attributed to the assumptions made in the initial prediction, which were used to estimate geological features across various engineering geological units.

Figure 50 presents a comparison between the actual values and the calculated values of FPI for the identified geological units along the KWCT-2 project, using

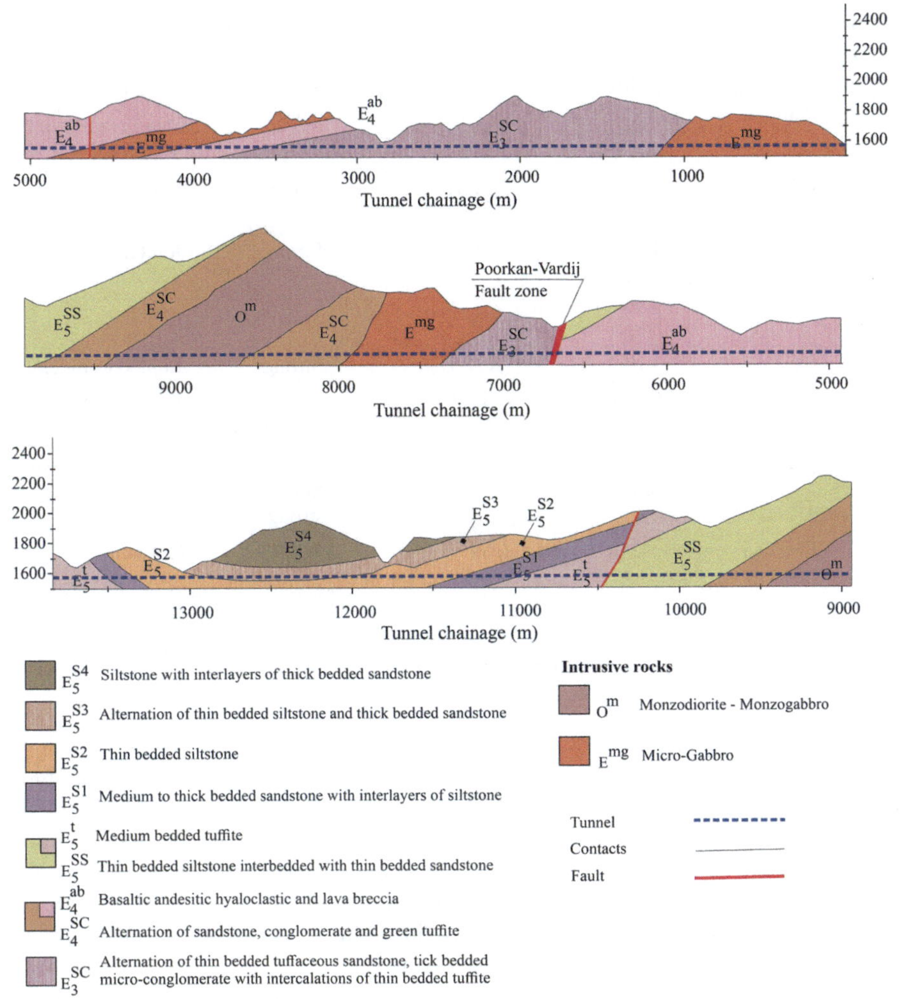

Fig. 47 Geological cross-section along the KWCT-2 project (Hassanpour et al. 2023a)

Eq. 52. The results show a strong agreement in both the values and the trends of the variations between the actual and calculated FPI. It is important to highlight that the engineering characteristics of the identified geological units, obtained during construction, were used to calculate the FPI values through the new developed empirical relationships.

In summary, these findings indicate that, despite minor discrepancies between the predicted and actual values (which are considered negligible), the prediction of the machine penetration rate using these new relationships is sufficiently accurate for practical applications in projects.

Table 21 Identified engineering geological units along the KWCT-2 project (Hassanpour and Rostami 2009)

No.	Stratigraphic unit	Lithology	Eng. Geo. unit	Strength degree	Structure
1	E^{mg}	Gabbro	MG	Very strong	Massive
2	O^{m}	Monzonite, Monzogabbro	MO	Very strong	Massive
3	$E_{3,4}{}^{SC}$, $E_4{}^{ab}$	Tuffite, sandstone, conglomerate, basalt	SC	Strong	Jointed to blocky
4	$E_5{}^{SS}$	Tuff, tuffite	GT-1	Strong	Jointed to blocky
		Sandstone, Siltstone	SS-1	Strong to very strong	Jointed to blocky
5	$E_5{}^{S2,3}$	Siltstone, sandstone	GS	Weak to moderately strong	Jointed/fractured
6	Crushed zones close to the main faults		Cz	Weak	Crushed

12.2 ZWCT-2 Project (Hassanpour et al. 2016)

Data from the second lot of the Zagros Water Conveyance Tunnel (ZWCT-2), spanning Chainage 5 + 300 to 15 + 000 m and excavated with a 7.53 m double-shield machine (Fig. 5a), was used to evaluate the accuracy of the proposed UT model equations. Notably, Eq. 53 was based on data from 35 sections of the first 5.3 km of the tunnel, specifically for the LT-II lithological type (Hassanpour et al. 2010), which closely matches the geological conditions of the next 9 km. By July 2013, about 15 km of this tunnel had been excavated, providing field geological and performance data for checking the validity of proposed equations.

Geologically, the ZWCT-2 project area lies within the Zagros zone, one of Iran's main geological and structural zones, specifically in the "simply folded" subzone. The main outcropping formations consisting of various sedimentary carbonate and argillaceous rocks. The geological cross-section in Fig. 51 illustrates the distribution of these stratigraphic units along the tunnel.

In this study, based on data obtained from geological as-built maps of construction phase, a total of 41 different rock units along the tunnel were identified and characterized (Fig. 51). These units mainly consist of shale, calcareous shale, and limestone, resembling the geological units in the first 5.3 km of the tunnel, used to develop empirical relationships in this chapter. Figure 52 displays the average UCS and RQD values for these units.

In the next step, the averages of actual values of operating and performance parameters of the machine were determined for all 41 units using statistical methods. Then, the performance parameters of the machine were calculated using Eqs. 53 and

Table 22 A part of the spreadsheet used for defining the location and specifications of rock units and calculating machine performance parameters for each rock unit (Hassanpour and Rostami 2009)

Rock unit	Chainage (m)		Length (m)	UCS (MPa)	RQD (%)	FPI (kN/c/mm/rev)	P (mm/rev)	PR (m/h)
	Start	Finish						
MG-1	0	1109	1109	160	95	22.6	7.8	2.35
SC	1109	1591	482	75	50	13.5	13.1	3.94
CZ-1	1591	1611	20	75	25	12.9	13.8	4.14
SC	1611	4015	2404	75	50	13.5	13.1	3.94
MG-2	4015	4612	597	160	90	22.4	7.9	2.38
CZ-2	4612	4632	20	100	25	14.6	12.2	3.65
SC	4632	5119	487	75	50	13.5	13.1	3.94
CZ-3	5119	5139	20	75	25	12.9	13.8	4.14
SC	5139	6516	1377	75	50	13.5	13.1	3.94
CZ -P	6516	6854	338	75	25	12.9	13.8	4.14
SC	6854	7287	433	75	50	13.5	13.1	3.94
MG-3	7287	7898	611	160	75	21.7	8.2	2.45
SC2	7898	8503	605	100	60	15.6	11.4	3.41
MO	8503	9336	833	160	75	21.7	8.2	2.45
SC2	9336	9704	368	100	60	15.6	11.4	3.41
GT-1	9704	10,402	698	75	75	14.2	12.5	3.75
CZ-4	10,402	10,422	20	75	25	12.9	13.8	4.14
GT-2	10,422	10,953	531	75	75	14.2	12.5	3.75
SS-1	10,953	11,284	331	125	50	17.4	10.2	3.07
GS	11,284	13,335	2051	25	25	10	17.7	5.32
SS-2	13,335	13,501	166	125	50	17.4	10.2	3.07

56 with the available data. The calculated results were compared to actual machine performance to quantify discrepancies. Figure 53 presents the comparison of actual and estimated PR (using Eqs. 53 and 56) across different geological units. Overall, the estimates show low error across various tunnel sections.

Additionally, Fig. 54 shows the comparison of actual and calculated FPI (using Eq. 53) across various geological units along the tunnel. The results demonstrate good agreement between the predicted and actual FPI values.

This study emphasizes the importance of geological similarity between the project site and the database used for model development in achieving accurate predictions of TBM performance. Therefore, local and site-specific models generally offer more precise results.

Table 23 Another part of the spreadsheet used for calculation of the advance rate and required time for excavating each rock unit along the tunnel route (Hassanpour and Rostami 2009)

Rock unit	Chainage (m)		Section length (m)	PR (m/h)	U (%)	AR (m/day)	Number of days (N_{days})	Cum. number of days	Cum. number of months
	Start	Finish							
MG-1	0	1109	1109	2 0.35	20	9.4	118	118	3.9
SC	1109	1591	482	3.94	20	15.7	31	148	4.9
CZ-1	1591	1611	20	4.14	5	4.1	5	153	5.1
SC	1611	4015	2404	3.94	20	15.7	153	306	10.2
MG-2	4015	4612	597	2.38	20	9.5	63	369	12.3
CZ-2	4612	4632	20	3.65	5	3.7	5	374	12.5
SC	4632	5119	487	3.94	20	15.7	31	405	13.5
CZ-3	5119	5139	20	4.14	5	4.1	5	410	13.7
SC	5139	6516	1377	3.94	20	15.7	87	497	16.6
CZ -P	6516	6854	338	4.14	5	4.1	82	579	19.3
SC	6854	7287	433	3.94	20	15.7	27	607	20.2
MG-3	7287	7898	611	2.45	20	9.8	62	669	22.3
SC2	7898	8503	605	3.41	20	13.6	44	713	23.8
MO	8503	9336	833	2.45	20	9.8	85	798	26.6
SC2	9336	9704	368	3.41	20	13.6	27	825	27.5
GT-1	9704	10,402	698	3.75	20	15.0	47	872	29.1
CZ-4	10,402	10,422	20	4.14	5	4.1	5	877	29.2
GT-2	10,422	10,953	531	3.75	20	15.0	35	912	30.4
SS-1	10,953	11,284	331	3.07	20	12.3	27	939	31.3
GS	11,284	13,335	2051	5.32	20	21.3	96	1036	34.5
SS-2	13,335	13,501	166	3.07	20	12.3	14	1049	35.0

12.3 *Koralm Tunnel (Brino et al. 2015)*

Brino (2015) and Brino et al. (2015) undertook a statistical analysis of actual data from 37 sections within the Koralm tunnel in Austria to investigate the accuracy and precision of various models. The Koralm Tunnel, which spans 32.9 km and consists of twin railway tunnels, is part of the railway network that connects the Austrian cities of Graz and Klagenfurt. The study focused on a segment approximately 1000 m long in the southern section of the tunnel (KAT2), situated within high-grade metamorphic rocks, including gneiss, mica schist, and marble. Each tunnel line (D = 9.93 m) was constructed using a double-shield machine, equipped with 80 17-inch disc cutters, manufactured by the Aker-Wirth company.

The researchers evaluated the validity of common models, including relationships presented in this chapter. Equations 52–56 have been investigated separately in this

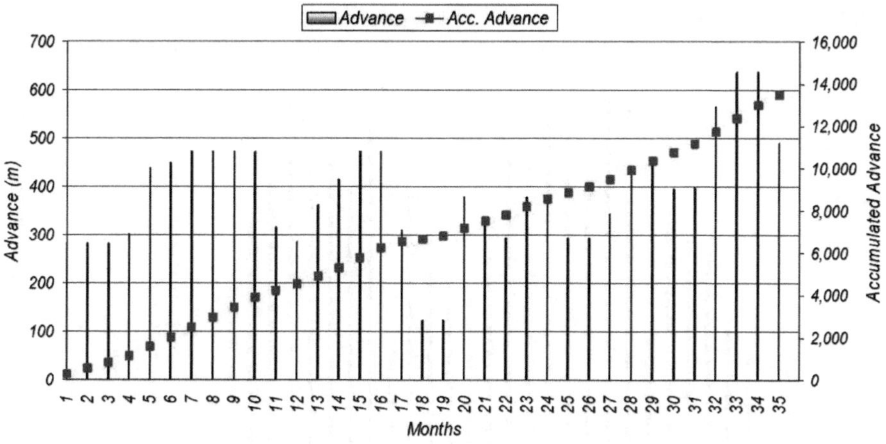

Fig. 48 Prediction of monthly boring machine advance (Hassanpour and Rostami 2009)

study. A subset of the researchers' findings is displayed in Fig. 55, comparing calculated penetration values, using the average values and 25th, 50th, and 75th percentiles of the cutter load in 37 selected tunnel sections, with the actual penetration values.

This study showed that the equations developed in the UT model align well with the actual results obtained from 37 selected sections of the Koralm Tunnel. In particular, the two equations related to the lithological types LT-I and LT-IV demonstrated significantly better results due to their closer resemblance to the geological conditions of the Koralm Tunnel, making them more accurate than many commonly used models.

12.4 The Follo Tunnel (Hansen 2018)

Hansen (2018) conducted a study as part of his MSc thesis at the Norwegian University of Science and Technology (NTNU), where he aimed to assess the validity of various models, including the empirical relationships presented in the UT model, by analyzing actual data from 18 sections of the Follo tunnel in Norway.

The Follo railway, which stretches a total length of 22 km between Oslo and Ski, features a 20-km twin tunnel, resulting in a total excavation of 40 km. Except for areas near urban centers where tunnel excavation employed drilling and blasting methods, most of the tunnel was excavated using four similar double-shield TBMs, manufactured by Herrenknecht. Each TBM with a diameter of 9.96 m was equipped with 71 19-inch disc cutters.

A significant portion of the tunnel route consists of various geological formations, including gneiss, granite gneiss, and amphibolite. These formations are generally classified as very hard to extremely hard rocks in terms of strength. The Rock Quality (RQD index) varies from very low to medium, while the Cutter Life Index (CLI)

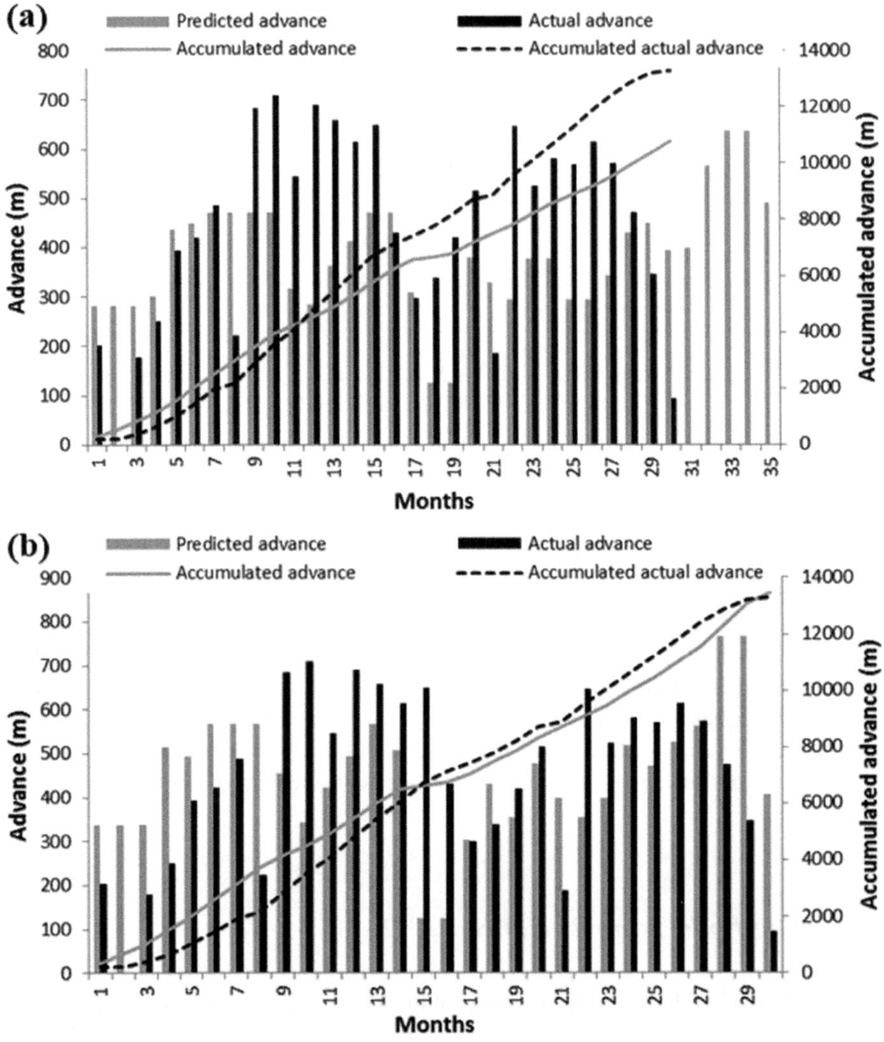

Fig. 49 Comparison between the actual and the predicted monthly advance of the TBM in the KWCT-2 project, **a** predicted based on the utilization factor experienced in the 1st lot, **b** predicted based on the actual utilization factor of the machine in the 2nd lot (Hassanpour et al. 2015)

ranges from extremely low to medium. Additionally, the quartz content in these rocks is between 10 and 50%.

The results of Hansen's (2018) study, which verify the empirical relationship established by Hassanpour et al. (2011), are summarized in Fig. 56. This figure illustrates the calculated average estimated penetration rate using Eq. 56, compared with the actual average penetration rates in the respective sections. The comparison shows a close correlation between the predicted and actual average penetration rates

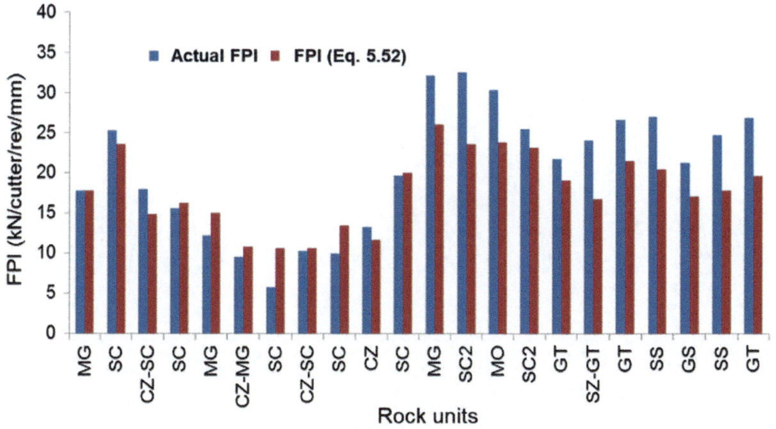

Fig. 50 Predicted (Eq. 52) versus actual FPI values across rock units in the KWCT-2 project (Hassanpour et al. 2015)

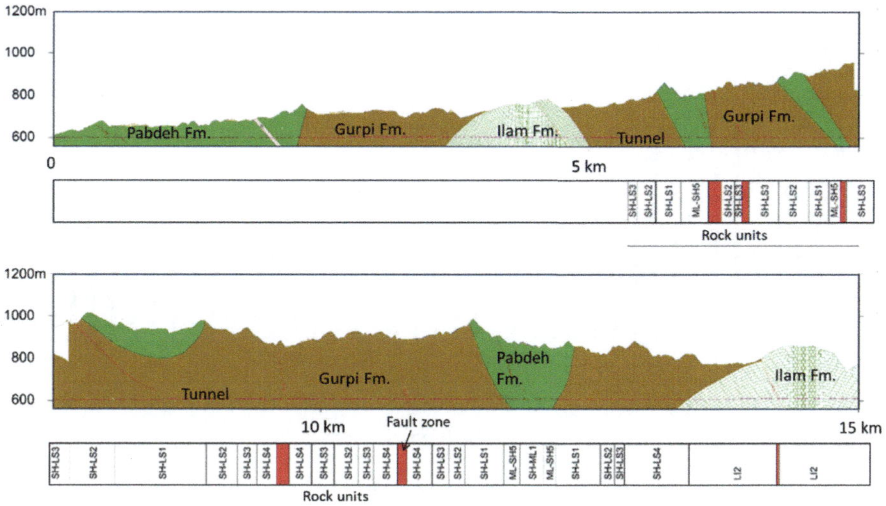

Fig. 51 Geological cross-section along the first 15 km of ZWCT-2 project (Hassanpour et al. 2016)

in some sections, although there are disparities in others. Additionally, the trend of variations generally exhibits reasonable conformity, except for the middle sections.

This study did not consider the limitations of the UT models, as the cutter diameter, tip width, and disc cutter spacing differ from the model assumptions. Additionally, the general model (Eq. 56) was used for estimations, though, based on the Follo tunnel's geological conditions, Eq. 55 would be more appropriate. The tunnel's geological features align more with lithological types LT-IV and LT-VI. Despite these limitations

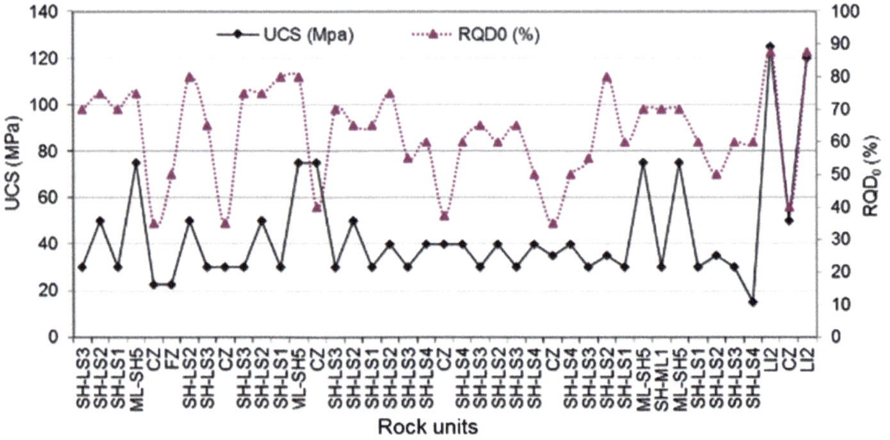

Fig. 52 The average values of UCS and RQD determined for rock units identified along the ZWCT-2 project (Hassanpour et al. 2016)

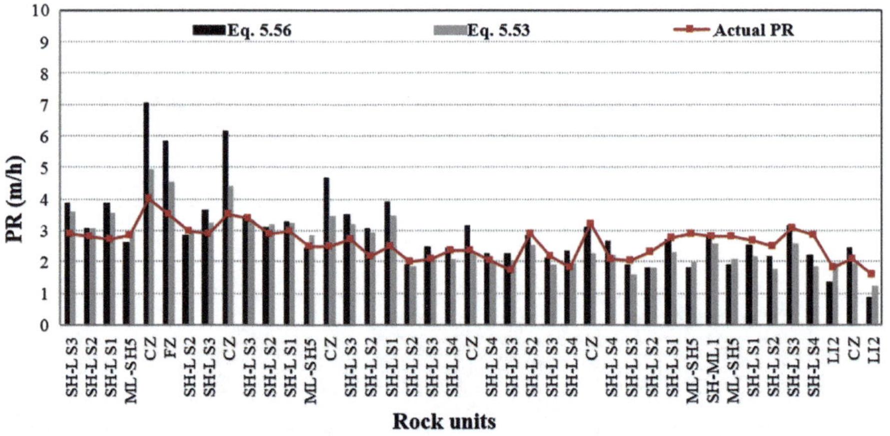

Fig. 53 Comparison of actual PR values with those calculated using Eqs. 53 and 56 across rock units in the ZWCT-2 (Hassanpour et al. 2016)

and the potential errors from averaging data over long tunnel segments, the overall performance of the UT model in Hansen's study can be considered acceptable.

12.5 Landivisiau Tunnel (Lehmann et al. 2022)

In a recent study, Lehmann et al. (2022) examined the performance of a small-diameter machine ($D_{TBM} = 2.2$ m) in the Landivisiau tunnel in Brittany, France.

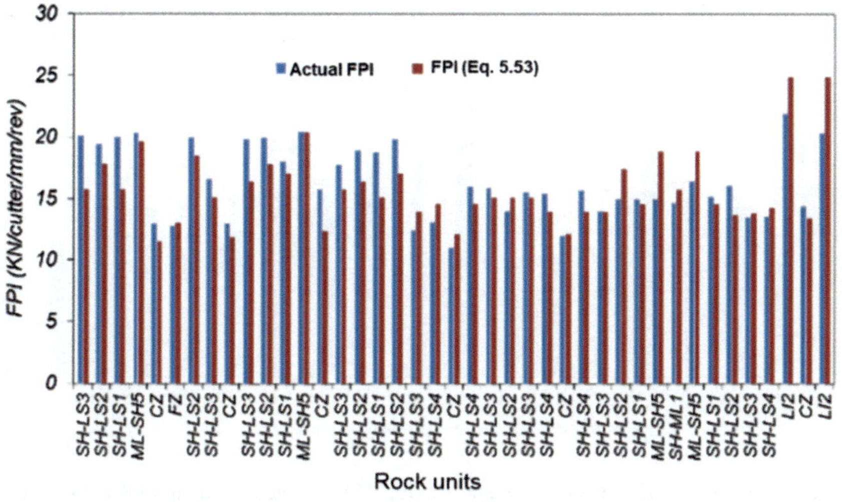

Fig. 54 Predicted (Eq. 53) versus actual FPI values across rock units in the ZWCT-2 project (Hassanpour et al. 2015)

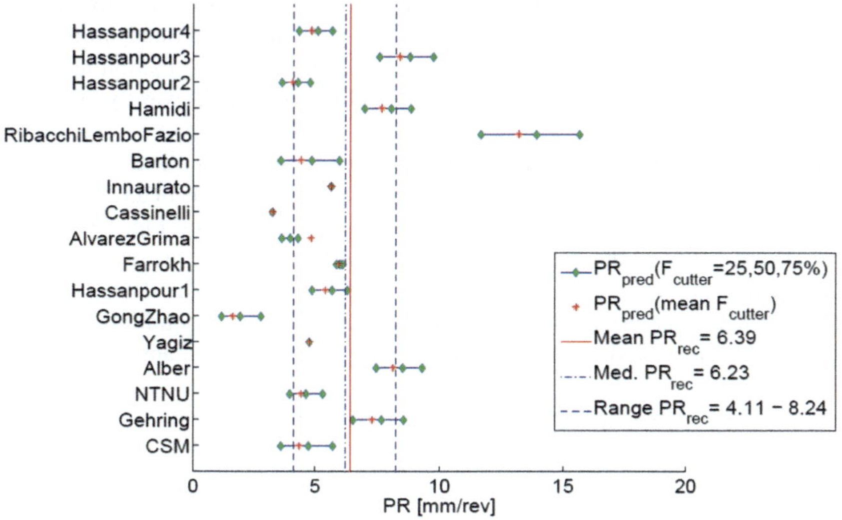

Fig. 55 The summary of the results of predicting the cutterhead penetration in the entire dataset (37 sections) using different models and comparing them with the recorded values; the green diamonds represent the calculated penetration values at 25%, 50%, and 75% cutter loads, and the red stars represent the calculated penetration at the average cutter load in 37 sections (Brino 2015; Brino et al. 2015)

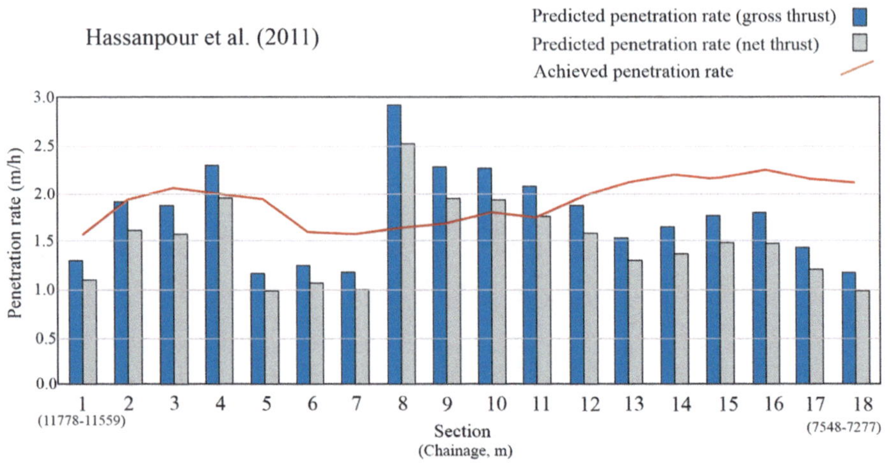

Fig. 56 Comparison of estimated average PR (Eq. 55) with actual average PR across eighteen tunnel sections (Hansen 2018)

They compared the actual performance parameters of this machine with the results obtained from various models. Geologically, the tunnel was excavated through highly heterogeneous rocks, including gneiss, granite, and schist, each exhibiting different levels of strength. The maximum strength (UCS) of these rocks was found to be 185 MPa.

The average penetration and the advance rate of the machine used in this project were 3.4 mm/rev and 17.3 mm/min, respectively. The variations in these two parameters along the tunnel are illustrated in Fig. 57. Additionally, Fig. 58 shows the relationship between the cutter load and the total penetration of the machine. As shown, these parameters display a wide range of variation. The data also indicate that, despite an increase in the cutter load up to 200 kN, the penetration do not reach a significant value, indicating incomplete chip formation in many sections of the tunnel.

Figure 59 presents a column chart that compares the actual cutterhead penetration with predictions made by four different models: Gehring (1995), CSM (Rostami 1997), Hassanpour et al. (2011), and Farrokh et al. (2012). As shown, the highest level of agreement between the actual data and the predicted values was observed with the models developed by Hassanpour et al. (2011) (Eq. 56) and Farrokh et al. (2012).

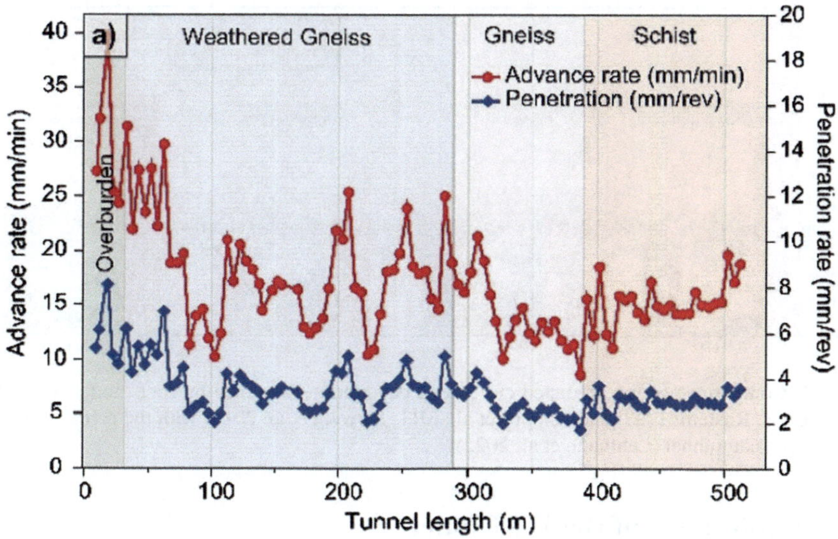

Fig. 57 The variations in two key factors-the cutterhead penetration and the advance rate-during excavation of Landivisiau tunnel (Lehmann et al. 2022)

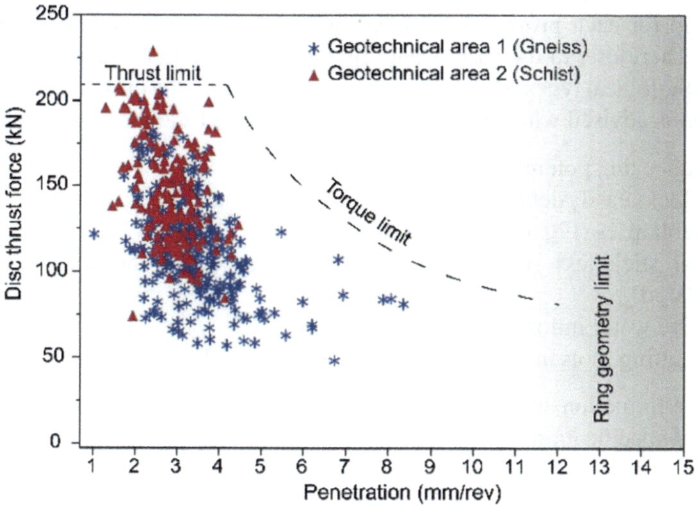

Fig. 58 Examining the relation between cutterhead penetration and cutter load in the Landivisiau tunnel (Lehmann et al. 2022)

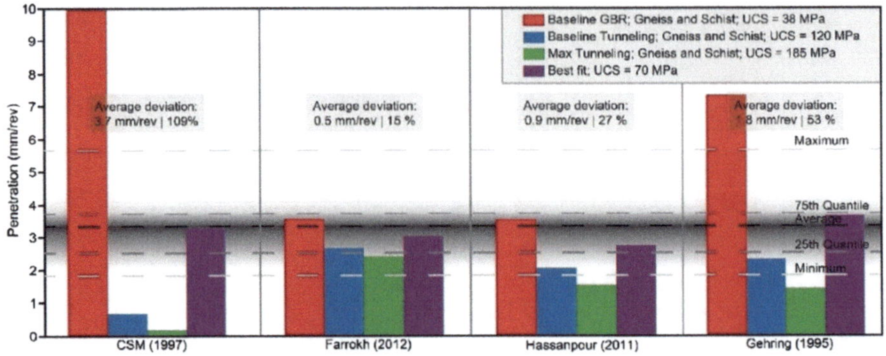

Fig. 59 Comparison of the predicted cutterhead penetrations obtained by four models (Gehring 1995; CSM, Rostami 1997; Hassanpour et al. 2011; Farrokh et al. 2012) with the actual values in the Landivisiau tunnel (Lehmann et al. 2022)

13 Limitations of the UT Model

The UT model and its related empirical equations have limitations that must be carefully considered to avoid errors in their application. Selecting the appropriate relationship for each project or geologic unit, based on its specific conditions, is essential. Therefore, in the early stages, geological conditions should be compared with the lithological types introduced in Sect. 4 to identify the most suitable equation.

Caution is advised when applying this model in the following situations:

- High squeezing potential;
- Rock blocks easily detaching due to unfavorable discontinuity orientation, causing minor collapses (Fig. 60a, b);
- Material stickiness risking clogging of cutting tools and transport systems (Fig. 60c, d);
- Excessive water inflow; and
- When cutting tools are heavily worn (Fig. 60e) or of poor quality.

Another limitation of this model pertains to the geometry, configuration, and mechanical specifications of the cutting tools, including the diameter and tip width of the cutter rings, as well as the spacing of disc cutters, all of which are constrained by the characteristics of the machines used in the projects examined in this study. The equations developed in the UT model are valid for disc cutters with a diameter of 432 mm (17 inches) and a tip width of 18–20 mm. The spacing between cutters should be 70–85 mm. Beyond these ranges, future correction coefficients should be developed.

It is important to highlight that the empirical equations in the UT model were primarily developed from experiences gained in projects constructed in Iran. Although the databases used for model development are considered appropriate in terms of both quantitative and qualitative criteria, users are encouraged to explore

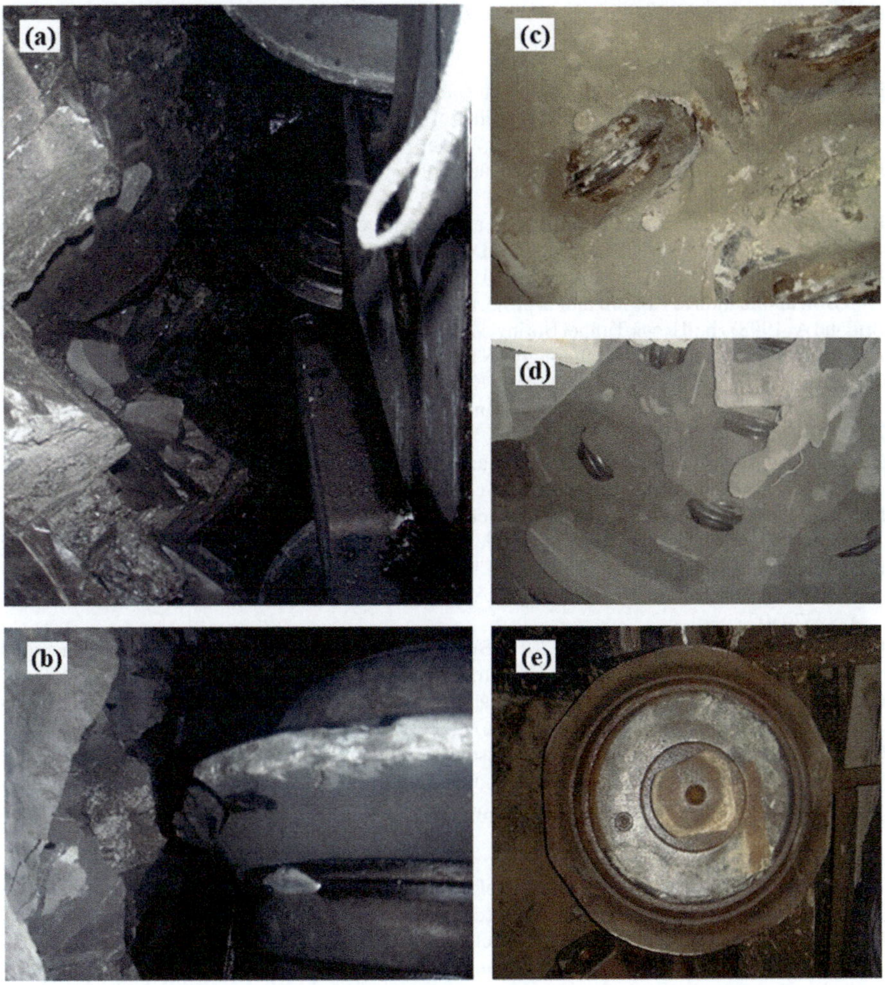

Fig. 60 Examples of limitations that are considered for the application of the model include: **a** and **b** examples of grounds where, due to discontinuities, rock blocks easily detach from the tunnel face, creating a rough surface. Chipping of the cutter ring is one of the consequences of this phenomenon; **c** and **d** clogging and jamming of the machine; **e** abnormal wear resulting from the jamming of disc cutter bearings in grounds susceptible to stickiness

other available models to evaluate machine performance and compare results. Furthermore, decisions regarding the appropriate penetration rate and the utilization factor for each tunnel section should be based on project-specific information.

References

Barton N (1996) Rock mass characterization and seismic measurements to assist in the design and execution of TBM projects. In: Proceeding of Taiwan rock engineering symposium. National Taiwan University Center for Education, Taipei

Bieniawski ZT (1973) Engineering classification of jointed rock masses. Trans s Afr Inst Civil Eng 15:335–344

Brino G (2015) Prediction of performance and cutter wear in rock TBM: application to Koralm tunnel project. MSc Thesis. Politecnico Di Torino

Brino G, Peila D, Steidl A, Fasching F (2015) Prediction of performance and cutter wear in rock TBM: application to Koralm tunnel project. Geoingegneria Ambientale e Mineraria 52(2):41–54

Bruland A (1998) Hard Rock Tunnel Boring, vol 1–10. PhD thesis. Trondheim (Norway): Norwegian university of Science and Technology (NTNU)

Copur H, Bilgin N, Tuncdemir H, Balci C (2003) A set of indices based on indentation tests for assessment of rock cutting performance and rock properties. J S Afr Inst Min Metall 103(9):583–599

Dardashti A, Ajalloeian R, Rostami J, Hassanpour J, Salimi A (2024) Performance predictions of hard rock TBM in subcritical cutter load conditions. Rock Mech Rock Eng Prev 57(1):739–755

Deere DU, Miller RP (1966) Engineering classification and index properties of rock. Technical Report No. AFNL-TR-65–116. US Air Force Weapons Laboratory, Kirtland AFB, New Mexico

Deere DU, Hendron AJ, Patton FD, Cording EJ (1967) Design of surface and near surface construction in rock. In: Proceeding of 8th U.S. symposium Rock Mech. AIME, New York, pp 237–302

Farrokh E, Rostami J, Laughton C (2012) Study of various models for estimation of penetration rate of hard rock TBMs. Tunnel Underground Space Technol 30:110–123

Fawcett DF (1993) The effects of rock properties on the economics of full face TBMs. In: Hudson et al. (eds) Comprehensive rock engineering, Pergamon, UK, pp 293–311

Gehring K (1995) Leistungs-und Verschleissprognosen in maschinellen Tunnelbau. Felsbau 13(6)

Goodarzi S, Hassanpour J, Yagiz S, Rostami J (2021) Predicting TBM performance in soft sedimentary rocks, case study of Zagros Mountains Water Tunnel Projects. Tunnel Underground Space Technol 109:103705

Hansen JN (2018) Comparison of existing performance prediction models for hard rock tunnel boring based on data collected at the Follo Line Project. Master thesis. Trondheim (Norway): Norwegian University of Science and Technology (NTNU)

Hassanpour J (2009) Investigation of the effect of engineering geological parameters on TBM performance and modifications to existing prediction models. Ph.D. Thesis. Tarbiat Modares University, Tehran, Iran

Hassanpour J, Rostami J (2009) Predicting TBM performance in second lot of Karaj Water Conveyance Tunnel. Proc Eurock 09:707–713

Hassanpour J, Rostami J, Khamehchiyan M. Bruland A (2009) Development new equations for performance prediction. Geomech Geoeng Int J 4(4):287–297

Hassanpour J, Rostami J, Khamehchiyan M, Bruland A, Tavakoli HR (2010) TBM performance analysis in pyroclastic rocks, a case history of Karaj Water Conveyance Tunnel (KWCT). J Rock Mech Rock Eng 4:427–445

Hassanpour J, Rostami J, Zhao J (2011) A new hard rock TBM performance prediction model for project planning. Tunn Undergr Space Technol 26:595–603

Hassanpour J, Rostami J, Zhao J, Tarigh Azali S (2014) Introduction of an empirical TBM cutter wear prediction model for pyroclastic and mafic igneous rocks; a case history of Karaj water conveyance tunnel Iran. Tunnell Underground Space Technol 43:222–231

Hassanpour J, Rostami J, Zhao J, Tarigh Azali S (2015) TBM performance and disc cutter wear prediction based on ten years experience of TBM tunnelling in Iran. Geomech Tunnell 8(3):239–247

Hassanpour J, Ghaedi Vanani AA, Rostami J, Cheshomi A (2016) Evaluation of common TBM performance prediction models based on data from the second lot of Zagros water conveyance tunnel (ZWCT2). Tunn Undergr Space Technol 52:147–156

Hassanpour J (2018) Development of an empirical model to estimate disc cutter wear for sedimentary and low to medium grade metamorphic rocks. Tunn Undergr Space Technol 75:90–99

Hassanpour J, Esmaeili Vardanjani S, Rostami J, Cheshomi A (2019a) Engineering geological studies used for redesigning and employing a hard rock TBM in soft rock formations of Chamshir water conveyance tunnel. Geopersia 9(1):1–20

Hassanpour J, Rostami J. Tavakoli HR, Firouzei Y (2019b) Optimizing TBM cutterhead design for application in very strong and abrasive rocks, case study of Kerman water tunnel. In: Proceedings of the WTC 2019 ITA-AITES world tunnel congress, pp 3803–3813

Hassanpour J, Firouzei Y, Hajipour G (2021) Actual performance analysis of a double shield TBM through sedimentary and low to medium grade metamorphic rocks of Ghomrood water conveyance tunnel project (lots 3 and 4). Bull Eng Geol Env 80:1419–1432

Hassanpour J, Saljooghi Khoshkar A, Ghasemi Farasani M, Hashemnejad A (2022) Investigating the relationships between rock mass classification systems based on data from mechanized tunneling projects in Iran. Bull Eng Geol Environ 81(147). https://doi.org/10.1007/s10064-022-02641-y

Hassanpour J, Rostami J, Khamehchiyan M (2023a) Engineering geology and tunneling: evaluation of TBM performance in rock tunnels. University of Tehran press, (in Persian)

Hassanpour J, Goodarzi S, Yagiz S, Rostami J, Davari Sarem M (2023b) Soft sedimentary rocks and their boreability characteristics in mechanised tunnelling projects. Geomechanik Und Tunnelbau 16(1):28–37

Hassanpour J, Kazemi C, Rostam J (2024) Introduction of a modified Q_{TBM} model for predicting TBM penetration rate in rock, based on data from mechanized tunneling projects in Iran. Bull Eng Geol Environ 83:165

International Association of Engineering Geology (IAEG), Commission on Engineering Geological Maps (1976) Engineering geological maps; a guide to their preparation. The Unesco Press, Paris

ISRM (1979) Suggested methods for determining the uniaxial compressive strength and deformability of rock materials. Int J Rock Mech Min Sci 16(2):135–140

Kanji MA (2014) Critical issues in soft rocks. J Rock Mech Geotech Eng 6:186–195

Karimi H, Khamehchiyan M, Hassanpour J (2024) Disc cutter wear/life analysis in a long tunnel excavated in strong to very strong, abrasive rocks. Bull Eng Geol Env 83(10):405

Lehman G, kasling H, Cambier A, Praetorius S, Thuro K (2022) Performance analysis of utility tunneling data: A case study of pipe jacking in hard rock in Brittany, France. Tunnell Underground Space Technol 127:104574

Maidl B, Schmid L, Ritz W, Herrenknecht M (2008) Hard rock tunnel boring machines. Ernst & Sohn, A Wiley company, Berlin

Nelson PP, O'Rouke TD, Kulhawy FH (1983) Factors affecting TBM penetration rates in sedimentary rocks. In: Proceeding of 24th U.S. symposium on rock mechanics, USA

Palmström A (1995) RMi- a rock mass characterization system for rock engineering purposes. PhD thesis. Oslo University, Oslo (Norway)

Pourhashemi SM, Ahangari K, Hassanpour J, Eftrekhari SM (2021a) TBM performance analysis in very strong and massive rocks; Case study: Kerman water conveyance tunnel project Iran. Geomech Geoeng Int J 17(4):1110–1122

Pourhashemi SM, Ahangari K, Hassanpour J, Eftrekhari SM (2021b) Evaluating the influence of engineering geological parameters on TBM performance during grinding process in Limestone strata. Bull Eng Geol Env 80:3023–3040

Pourhashemi SM, Ahangari K, Hassanpour J, Eftrekhari SM (2021c) Analysis of grinding and chipping processes beneath disc cutters of hard rock TBMs (Case study: The Uma-Oya water conveyance tunnel, Sri-Lanka). J Mini Environ (JME) 12(1):281–297

Rostami J (1997) Development of a force estimation model for rock fragmentation with disc cutters through theoretical modeling and physical measurement of crushed zone pressure. PhD thesis. Colorado School of Mines, Golden, Co. (USA)

Rostami J (2013) Study of pressure distribution within the crushed zone in the contact area between rock and disc cutters. Int J Rock Mech Min Sci 57:172–186

Samadi H, Hassanpour J (2025) Application of the stacking-ensemble machine learning algo-rithm to investigate the rock mass cuttability index in different lithological units using tunnel boring machine performance data. Rock Mech Rock Eng 58(7):8445–8475

Shayan F, Hassanpour J, and Uromeihy A (2023) Classification of fault zones in mechanized tunneling projects. In: Anagnostou, Benardos & Marinos (eds) Expanding Underground. Knowledge and Passion to Make a Positive Impact on the World, World Tunnel Congress, Athene, Greece, Taylor and Francis, pp 2245–2252

Shayan F, Samadi H, Hassanpour J, and Uromeihy A (2025a) Geological lessons learned during two decades' hard rock mechanized tunneling in fault zones of Iran. In: Johansson et al (eds) Tunnelling into a sustainable future- methods and technologies, World Tunnel Congress, Stockholm, Sweden, Taylor and Francis, pp 2809–2816

Shayan F, Uromeihy A, Hassanpour J (2025b) Classification framework for engineering geological challenges in fault zones during TBM tunneling: insights from mechanized projects in Iran. Accepted for publication, Engineering Geology

SOI Consulting Engineers (2016) Engineering geological reports of KrWCT-S project. Unpublished

Sutcliffe H (1996) Tunnel boring machines. In: Bickel et al. (eds) Tunnel engineering handbook, Chapter 11. Springer, pp 203–219

Terzaghi K, Peck R (1967) Soil mechanics in engineering practice. John Wiley and Sons Ltd., New York

The Performance of Three Hard Rock TBMs in Earthquake Zones and the Behavior of the Bored Tunnels Thereafter

Nuh Bilgin

1 Introduction

This chapter discusses and offers an overview of the performance of TBMs in adverse conditions of passing through some active fault zones as well as the impact of earthquakes on the TBM performance. The topic fits in the context of applying TBM in various conditions, specially as more TBM tunnel constructions are in extreme conditions that calls for special planning and operational settings. This includes the application of TBMs in tunnel construction at very high depths and in cases where the tunnel alignment has to go through active fault zones or places with frequent fault zones are on the tunnel alignment.

Earth's lithosphere, the planet's rigid outer shell including the crust and upper mantle, is fractured into seven or major plates. The relative movement of these plates on the magma ranges from zero to 10 cm annually. Faults created due to these movements tend to be geologically active, experiencing earthquakes, volcanic activity, uplift and subduction zones, and mountain-building, creating tectonic stresses on Earth's crust. These stresses are categorized into three main types: compressional, tensional, and shear, and they play a critical role in creating geological features such as mountains and rift valleys. Understanding tectonic stress is essential for predicting seismic activity and studying the dynamic nature of Earth's lithosphere (Read and Janet 1975).

Turkey is in a tectonically active region that experiences frequent destructive earthquakes. On a large scale, the region's tectonics are controlled by the collision of the Arabian Plate and the Eurasian Plate. A large piece of continental crust, almost the size of Turkey, called the Anatolian block, is being squeezed to the west. The block is bounded north by the North Anatolian Fault and southeast by the East Anatolian Fault.

N. Bilgin (✉)
Istanbul Technical University (ITU), Istanbul, Turkey
e-mail: bilgin@itu.edu.tr

J. Hassanpour and J. Rostami (eds.), *Performance of Tunnel Boring Machines (TBM) in Rock*, Springer Series in Geomechanics and Geoengineering,
https://doi.org/10.1007/978-3-032-20424-0_6

The East Anatolian Fault (EAF) is eastern Turkey's major strike-slip fault zone. It forms the transform-type tectonic boundary between the Anatolian and northward-moving Arabian Plates. Faults will slip when the ratio of shear to normal stress on the fault exceeds its frictional strength. The North Anatolian Fault (NAF) is an active right-lateral strike-slip fault in northern Anatolia that runs along the transform boundary between the Eurasian and Anatolian Plates. GPS data give a lateral sleep of 24 $\pm$ 4 mm/year along the North Anatolian fault, and also a left-lateral slip of 18 $\pm$ 6 mm/year on the East Anatolian fault and ~ 5 mm/year on the Northeast Anatolian fault generating an average shear stress building at 0.15 bar/year along most of these two major fault zones (Stein et al. 1997). Seven tunneling projects affected directly by the EAF and NAF faults are the Kargı, Doğançay, Gerede, Nurdağı, Ayaş, Uluabat, and Bolu tunnels, as located in Fig. 1 (Bilgin and Balcı 2025). The effect of active faults on tunnel boring emerges as face collapses, high deformations, excessive floor heave, squeezing of TBM, etc.

The performance of hard rock TBMs used in Kargı, Nurdağı, and Doğançay is one of the main topics of this chapter. Although the author of this chapter has published several articles in this respect, the current chapter covers some unpublished data relating to the tectonic influence of two major fault zones, the North Anatolian fault, and the East Anatolian fault. Doğançay and Nurdağı tunnels, which are influenced by tectonic activities of the East Anatolian Fault, are located in Fig. 2. The location of the Kargı Tunnel is seen in Fig. 1

The faults that affected the February 6th, 2023 February, Pazarcık and Kahramanmaraş earthquakes are extensions of the regional-scale strike-slip faults known from Anatolia and its surroundings. These include the East Anatolian Transform Fault, the Dead Sea Transform Fault, the Antakya Transform Fault, the Sarız-Saimbeyli Mega Shear Zone, the faults of the Foreland Fold-Thrust Belt, and the Karasu Graben

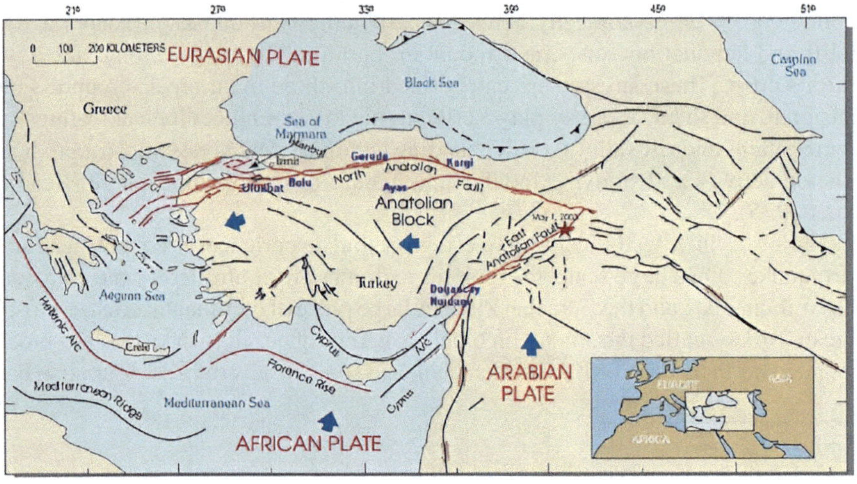

Fig. 1 North and east Anatolian faults in Turkey affecting tunnelling activities (Bilgin 2016)

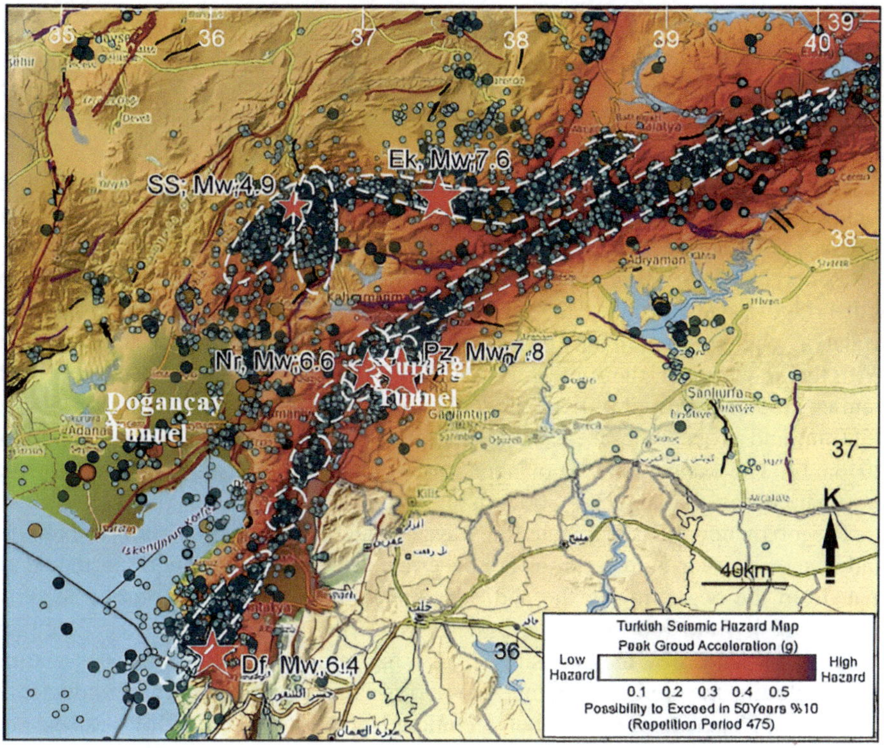

Fig. 2 Map showing the main shock epicenters of the earthquake region and the distribution of aftershocks with the location of Nurdağı and Doğançay Tunnels, (Modified after Yılmaz et al. 2025)

boundary faults. The interactions among these faults appear to have increased the magnitude of the earthquakes up to Mw 7.8 (Yılmaz et al. 2025). This powerful and shallow event caused widespread damage and casualties in southeastern Turkey and northern Syria, affecting an area of about 350,000 square kilometers, or the size of Germany. This earthquake was followed by an Mw 7.5 event approximately nine hours after the first. The total number of affected people in the Turkish and Syrian regions was about 14 million and 9 million, respectively (Galasso and Opabola 2024). This event allowed us to see the effect of earthquakes on the tunnels bored by TBMs.

2 Nurdağı High Speed Railway Tunnel

The Nurdağı tunnel is in the active East Anatolian Fault Zone, one of Turkey's most seismically active zones because it represents a plate boundary extending over 500 km between the Arabian and Anatolian plates. However, this zone was relatively quiescent in the last century when compared to historical records, and it was reported by Nalbant et al. (2002) that it accumulated significant stresses along its length. In particular, the segment tunnel reach near the project area had a high seismic risk, and the possible magnitude of future earthquakes was predicted to be approximately 7.3 Mw. As reported in Sect. 1 and various publications, an Earthquake of 7.8 Mw occurred in this area on February 6th, 2023.

Southeastern Turkey's Gaziantep province is characterized by a complex fractured rock within the Eastern Anatolian Fault. With a population of nearly 1.7 million, the province is overhauling its public transportation with a rail line between the towns of Bahçe and Nurdağı, being excavated by both NATM (850 m) and TBM. The railway will bridge a gap in one of the busiest railway sections between Toprakkale and Malatya, where trains can only proceed with the support of an additional locomotive. With the new project, the route between Bahçe and Nurdağı is 15 km shorter, with gentler gradients dropping from 27 to 15%, wider curves increasing from 500 to 1500 m, and proceeding at faster speeds from 40 to 160 km/h (Jordan and Bilgin 2018a, b). The excavation was planned to start from the chainage 13 + 450 km and terminate in the chainage 3 + 700 km.

The job owner is Turkish State Railways (TCDD), and the contractors are Intekar and Biscon. As far as the author is concerned, this project is unique in the sense that, from the beginning to the end, it is realized with the collaborative work of the University, the Job Owner, and the construction Company in a very complex geology-influenced by tectonic stresses of a big fault zone. Full-scale rock-cutting tests, including the prediction of machine performance parameters, were conducted in the Mining Department of Istanbul Technical University (ITU). TBM was selected and manufactured based on the results of rock-cutting tests. A risk analysis was carried out based on the geo-characteristics of the tunnel route, and the reality of the East Anatolian Fault (Bilgin and Balci 2016; Bilgin et al. 2017). As the result of the risk analysis, 1050 m from the inlet of the tunnel was recommended to open with NATM, and the use of the TBM was advised thereafter.

2.1 The Geology and Geomechanical Properties of the Rock Formations

The excavation was planned to start from the chainage 13 + 450 km and terminate in the chainage 3 + 700 km. The chainage from 13 + 450 km to 12 + 400 km includes the Karadag Limestone of the Mesozoic age, which is affected by the East Anatolian Fault (EAF), fracturing the rock formation to a great extent. High water ingress was expected in this area. Several springs are present along the EAF, a sizeable strike-slip fault about 550 km long extending from the Gulf of Iskenderun to the North Anatolian Fault with a strike of 60°E. This fault takes up most of the motion between the Turkish and Arabian plates. Risk analysis showed that the first 1050 m in the Nurdağı tunnel was risky for TBM excavation. Also, due to the technical difficulties and time necessary to set up the TBM, this part of the tunnel was planned to be excavated using the New Austrian Tunnelling Method (NATM) (Bilgin et al. 2016). However, TBM was ready to excavate sooner than expected, and the tunnel was excavated by NATM only 600 m of this tunnel. The geological formation from chainage 12 + 400 m to 4 + 850 m consists of middle-Ordovician-aged Kızlaç Formation of interbedded Metasandstone, metaquartzite, and metamudstone with very high strength and abrasive characteristics. This section was planned to be excavated with a single shield hard rock TBM. The geological longitudinal cross-section of the tunnel is shown in Fig. 3a, b. In some parts of the tunnel, severe groundwater and 50–300 l/s water inflows were measured, but some parts were occasionally dry. The overburden thickness reaches up to 640 m.

Compressive strength and tensile strength of the Metasandstone, metamudstone, and interbedded metasandstone-metamudstone are summarized in Tables 1 and 2.

Geological and geotechnical reports prepared by Fugro Sial showed that 1050 m of the tunnel from the inlet was affected deeply by the East Anatolian fault, making the rock formations fractured to a great extent. However, when boring the tunnel,

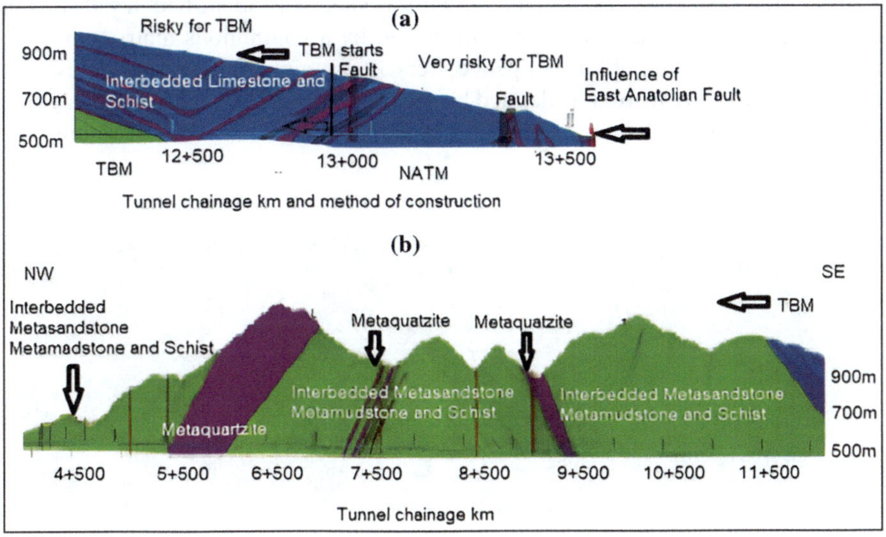

Fig. 3 **a**, **b** Geologic cross section of Nurdağı Tunnel (Fugro Sial 2014)

Table 1 Compressive strength of the rocks in the tunnel route

Rock	Maximum UCS* (MPa)	Minimum UCS (MPa)	Mean UCS ± st.dev (MPa)
Metasandstsone	301.3	177.9	213.0 ± 55.4
Metamudstone	327.4	89.4	136.1 ± 61.7
Interbedded metamudstone-sandstone	136.4	70.6	107.1 ± 25
Mean			151.2 ± 67.2

Table 2 Tensile strength of the rocks along the tunnel route

Rock	Maximum UTS* (MPa)	Minimum UTS (MPa)	Mean UTS ± st.dev (MPa)
Metasandstone	27.2	11.63	17.8 ± 2
Metamudstone	19.2	17.6	18.8 ± 1.1
Mean			18.1 ± 1.5

(*) UTS: Uniaxial tensile strength

the length of this fracture and disturbed zone was almost double than predicted. A petrographic analysis of the Metasandstone and metamudstone was done for the project by the late Prof Dr. Aral Okay. He defined the metasandstone sample as sandstone dominated by equant quartz grains with lesser grains of Feldspar, opaque rock fragments, and chlorite. A thin chlorite cement holds the grains together. The sandstone can be classified as quartz arenite. A thin section of the sample is given in Fig. 4.

The Metasandstone, as defined in Fugro Sial, was identified as a siltstone composed of angular quartz grains surrounded by a voluminous matrix composed of sericitic muscovite, chlorite, opaque, very fine-grained quartz, and Feldspar. A distinct cleavage in the rock is defined by the parallel orientation of the mica grains; the quartz grains also show subparallel orientation. The thin section examined is given in Fig. 5.

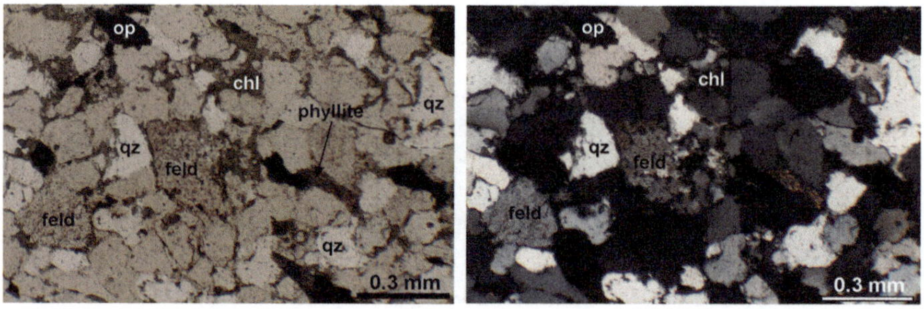

Fig. 4 Photomicrograph of the sandstone in plane-polarized light (left) and under crossed polars (right). Quartz (qz), Feldspar (feld), green chlorite (chl), opaque (op), and rock fragments (phyllite) are the main constituents. Note the green chlorite cement between the grains (private archive of the author)

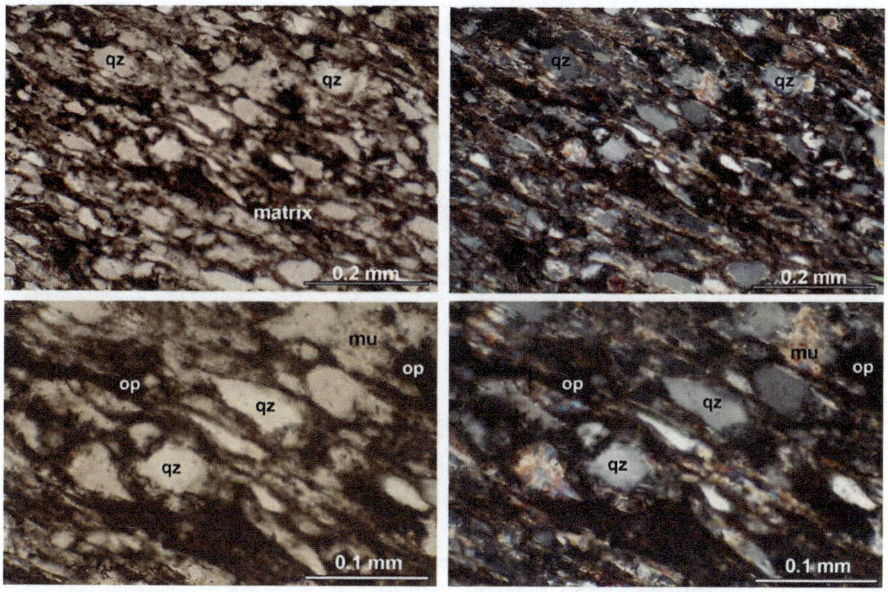

Fig. 5 Photomicrograph of the siltstone in plane-polarized light (left) and under crossed polars (right). The microphotographs in the lower panel show a more magnified view. Angular quartz (qz) grains are set on a voluminous matrix composed of very fine-grained sericitic muscovite (mu, bright colours under crossed polars), opaque (op), feldspar and quartz. Note the cleavage, which trends from upper left to lower right (from the archive of the author)

2.2 *Laboratory Rock-Cutting Tests to Identify Design Parameters of TBM and Performance Prediction*

Two block samples of metasandstone and metamudstone were subjected to full-scale laboratory rock-cutting experiments in the Mining Engineering Department of ITU. A constant cross-section disc cutter with a tip width of 12 mm and ring diameter of 330 mm was used in the experiments. A testing program with a constant cutter (line) spacing of 80 mm, including relieved (adjacent cutters) and unrelieved (single cutter) cutting patterns, were set, and the depth of cut was varied (3,5,7 mm). The metamudstone sample was cut parallel to the bedding planes. Based on the experimental results, the optimum line spacing to penetration ratio was determined for the TBM to be used in the tunnel. Cutter spacing is a constant value of a TBM cutterhead, which dictates that the predetermined cutting depth in the laboratory determine the optimum thrust values of the excavating machine for a given rock formation. Given these main rules, an intensive laboratory full-scale cutting test program was planned to obtain the two samples' relationships between cutting depth and s/d ratio. An optimum penetration value of 5 mm/rev was found for high-strength rocks. Recommended values for single-shield TBM were obtained and taken based on the laboratory cutting experiments. Recommended and actual parameters for a single shield 8 m diameter TBM are given in Table 3. The cutterhead with the shield and backup of the single shield TBM can be seen in Figs. 6 and 7.

Table 3 Recommended and actual parameters for single shield TBM of 8 m diameter

Parameter	Recommended value	TBM (Robbins) manufacturer value
Disc number and diameter	53, 19″ (483 mm)	53, 19″ (483 mm)
Maximum load per disc	311 kN	311 kN
Disc spacing	80 mm	77.6 mm
Total power	3250 kW	10 × 330 kW
Cutterhead torque	4500 kNm	4588 kNm at 6.9 rpm
Exceptional torque	N/A	14,453 kNm at 0–3.3 rpm
Recommended thrust based on disc bearing capacity	16,483 kN	16,503 kN
Maximum thrust	N/A	88,543 kN

Fig. 6 Cutterhead with the shield of the TBM (from the archive of the author)

Fig. 7 The backup of the TBM (from the archive of the author)

2.3 The Performance of TBM in the Zone Affected by the North Anatolian Fault, also in Non-disturbed Areas

The performance of TBM was recorded throughout the tunnel length. However, for this chapter, in view of showing the effect of the East Anatolian Fault on the performance of TBM, the boring data is analyzed in five different zones, as shown in Table 4. However, if the readers are interested in the performance of TBM covering the full length of the tunnel, they should refer to the paper by Gökçeoğlu (2021). The first three zones in Table 4 are examples of the areas where the East Anatolian fault disturbed the rock, mainly slate, within 2000 km of the tunnel route. TBM was trapped three times within these zones. Trapping of the TBM mainly occured by accumulating the very weak rock formation in front of the cutterhead of the TBM. The top core sample box in Fig. 8 represents the area where the TBM is trapped. The bottom core sample box is an example of the area where TBM is unaffected by the fault zone.

The important points emerging from Table 4 are summarized below:

- In zones 1,2 and 3, the rotational speed of the TBM changed between 3.8 and 4.5 rpm, resulting in penetration from 15 to18 mm/rev, since higher rpm would increase the penetration with the risk of trapping/jamming the TBM by a higher amount of muck accumulated in front of the cutter head. In these zones, Q values varied between 0.2 and 0.5.
- TBM was trapped three times within zones 1, 2 and 3 in the areas with Q values between 0.1 and 0.2. This supports the findings of Barton and Bilgin (2016), as shown in Fig. 9. A window was cut in the shield (Fig. 10) to release the TBM, and later, a gallery is opened with wooden supports (Fig. 11). Almost 60 days have been spent in this operation.
- Daily advance rates in zones 1,2 and 3 remained between 6 and 8 m/day (4–5.5 ring) since high water ingress going up to 50–300 l/s was encountered in these zones (Fig. 12), making the time spent for cleaning the cutterhead and the tunnel

Table 4 The performance of TBM in different selected zones where the rock formations are highly fractured due to the effect of the East Anatolian Fault (zones 1, 2, and 3) and in unaffected areas (zones 4 and 5)

Ring	rpm	mm/rev	Thrust kN	Torque kNm	Thrust/ Torque	m/day	Geological formation
845/911-951/956 **Zone 1**	3.8	18	14,479	925	15.6	6	Highly fractured and weathered slate, Q = 0.20–0.5; GSI= 10–20 Before and after the first squeezing zone, TBM trapped at the 913 ring where thrust force reached 65,000 kN and Q changed between 0.1 and 02
1018/1068 **Zone 2**	4	16	12,317	708	17.3	7	Highly fractured and weathered slate, Q = 0.20–05; GSI= 10–20 Before the second squeezing zone, TBM trapped at ring 1071 where thrust force reached 61,500 kN and Q changed between 0.1 and 02
1291/1324 **Zone 3**	4.5	15	12,684	1381	9.5	8	Highly fractured and weathered slate, Q = 0.20–05; GSI= 10–20 Before the third squeezing zone,TBM trapped at ring 1328 and Q changed between 0.1 and 0.2
2094/2122 **Zone 4**	6.7	4	20,241	1847	10.7	14	Metasandsone, metaquartzite GSI= 50–70
2247/3108 **Zone 5**	6.5	6	18,119	1926	9.8	15	Interbedded metamudstone, meta-sandstone, shale, GSI= 40–70

as high as 26% in a working pie chart (Fig. 13), 17% of all the shift time is spent for breakdown and maintenance, and 24% is spent for boring the tunnel.

- For the non-disturbed area by the East Anatolian Fault, in hard Metasandstone, metaquartzite, and interbedded metasandstone-metamudstone slate, the penetration was between 5–6 mm/rev as predicted in the laboratory full-scale cutting tests (Bilgin 2016). However, as noticed in the muck upon the belt conveyor, TBM cut a 3 m thick quartzite vein in the rings 2018–2019 (rings number 25–26 in Figs. 14,

Fig. 8 Representative samples of the areas where TBMs were (top picture) and the disturbed area by the East Anatolian Fault (bottom area)

Fig. 9 The variation of machine utilization with Q values in the Uluabat Tunnel (Barton and Bilgin 2016)

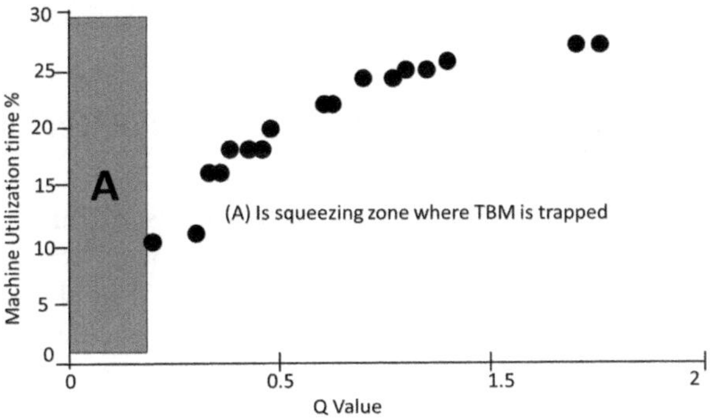

15 and 16. Although the penetration in quartzite was lower than that of the other hard rocks recorded as 2.2 mm/rev in ring 2018 and 2.7 mm/rev in rig 2019, a significant difference is not found in thrust and torque values as seen in Figs. 14, 15 and 16. The rotational speed was around 6.6 rpm, and the daily advance rate was between 14–15 m/day (9.3–10 rings) with a mean value of 40% for machine utilization time as shown in Fig. 17.

Fig. 10 Cutting shield of TBM for rescuing the trapped TBM in squeezing zone 1 (from the archive of the author)

3 Doğançay Tunnel, the Effects of Tectonic Stresses

Doğançay tunnel is located in the north-east of Turkey, affected by the East Anatolian Fault, and it is a part of a hydroelectric project. The tunnel length is 6655 m. The first 223 m of the headrace tunnel is excavated by the conventional tunneling method, and the remaining 6232 m is bored with a double shield TBM. During the boring of the tunnel, TBM was trapped 10 times, and around 160 days were spent on rescue operations. During field studies of the tunnel by the author, it was concluded that the main reason for the squeezing of the tunnel was high overburden reaching 1000 m and tectonic stresses exerted by the complex East Anatolian Fault. The tunnel is around 30–35 km away from the fault zone. According to Yönlü et al. (2017), this complex fault zone moves to the West at a rate of 2.5–10 mm/year.

The excavation started in September 2012 and ended in July 2015, making a daily advance rate of 6.2 m/day with all the delays included. Maximum advance per month was in December of 2012 at 450 m/month. Except for significant downtimes like TBM squeezing, the total excavation machine utilization time changed between 16 and 45%. The tunnel was excavated by a double-shield Herrenknecht M-1665, having a diameter of 4.1 m. During field studies by the author, it was observed that the

Fig. 11 Gallery opened to rescue the trapped TBM (from the archive of the author)

performance of the TBM was completely different than that of the Nurdağı Tunnel, one being very close to the East Anatolian Fault, the other 30–35 km away from this fault zone.

3.1 The Geology and Some Mechanical Properties of the Rock Formations

As seen in Fig. 18, the geological profile of the tunnel is composed of three zones: in the first zone, laminated shale and interbedded-sandstone -limestone; in the second zone, mainly sandstone, quartzite, and in the third, primarily limestone, sandstone quartzite, and shale are found. Limestone and shale are Carboniferous age, and sandstone and quartzite are upper and lower Permian age. RMR varies between 10 and 70. Only two boreholes could be opened in the tunnel route. Figures 19 and 20 show two shale samples from where TBM was trapped in chainage 5 + 004 km on 22.01.2013 and chainage 4 + 700 km. Traces of shear stresses in the fault zone

Fig. 12 High water ingress in a fractured and disturbed zone by fault zone (from the archive of the author)

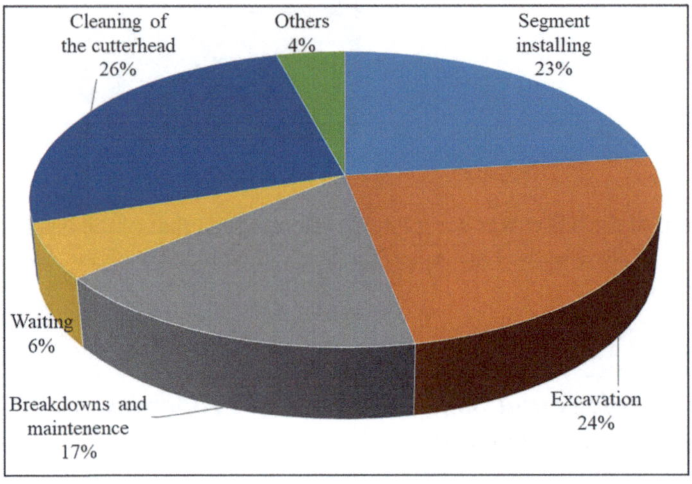

Fig. 13 Pie chart of the works done in disturbed zone by East Anatolian Fault (from the archive of the author)

were apparent. The overburden within 3 km of the tunnel route was around 1000 m and only two boreholes could be opened in the tunnel route before the excavation started. Tectonic stresses squeezed the TBM several times, causing considerable delays in tunnel drivage. Figure 18 shows the general layout of the tunnel route with the main faults (F1, F2, F3, F4 and F5). The squeezing of the TBM occurred at the

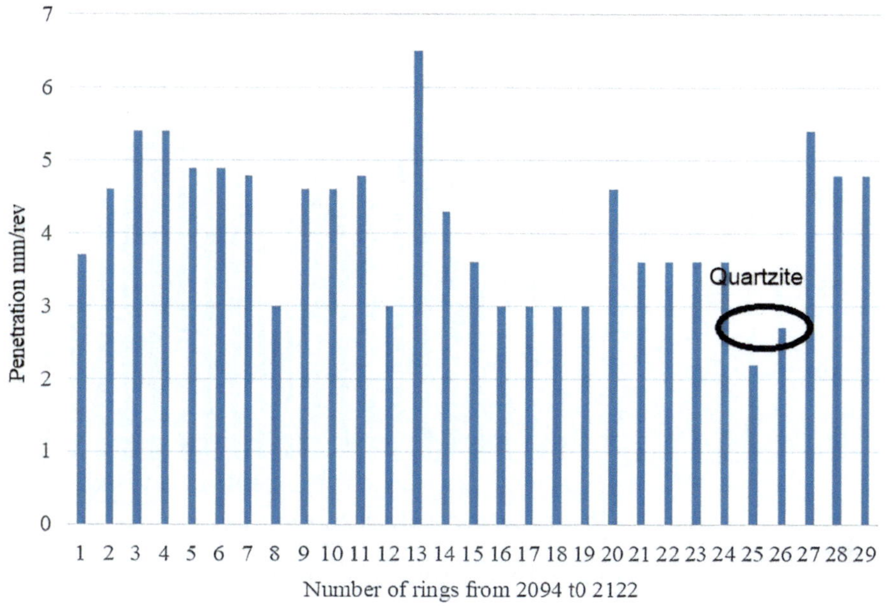

Fig. 14 The variation of the penetration in mm/rev with ring number in Metasandstone and meta quartzite (from the archive of the author)

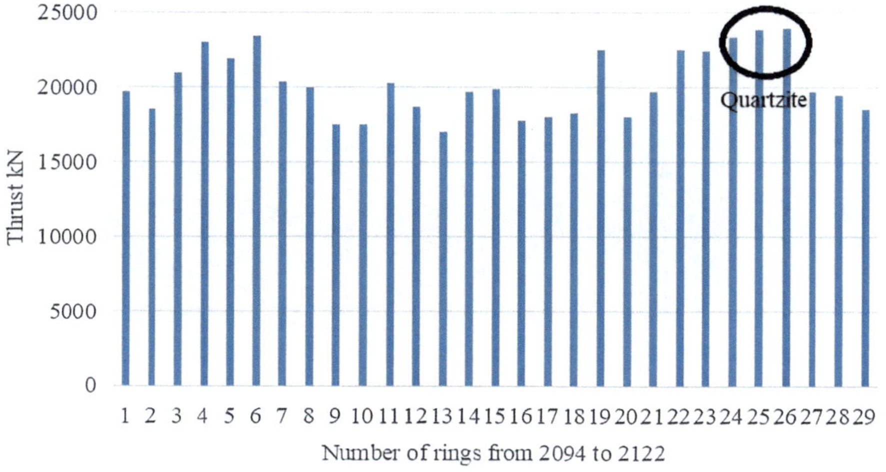

Fig. 15 The variation of the Thrust in kN with ring number in the zone Metasandstone and metaquartzite (from the archive of the author)

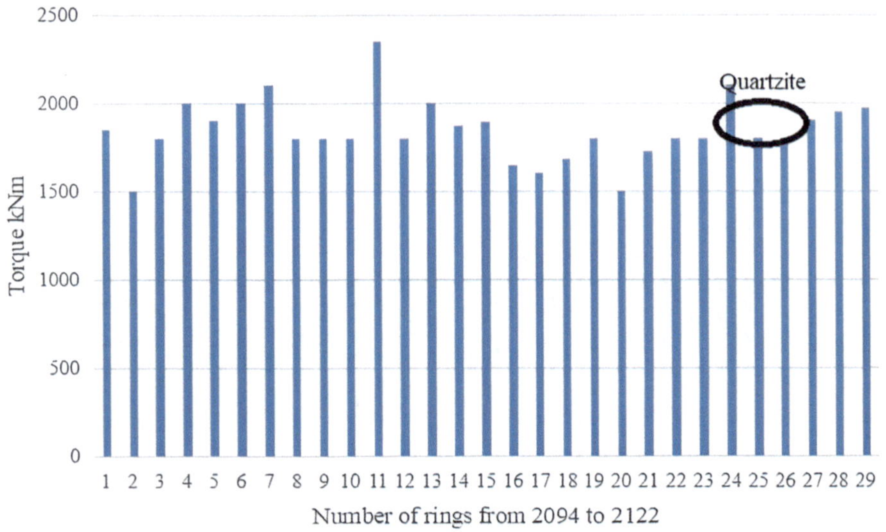

Fig. 16 The variation of the torque in kNm with ring number in Metasandstone and metaquartzite (from the archive of the author)

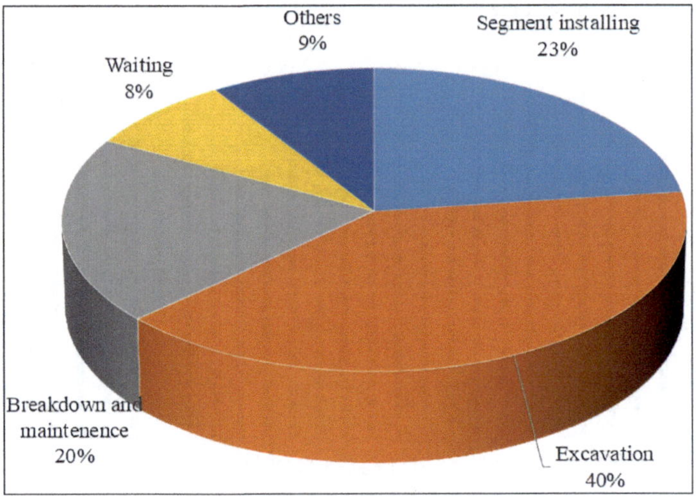

Fig. 17 Pie chart of the works done in massive rock (Metasandstone and Metaquartzite) (from the archive of the author)

contact areas of two faults, F1 and F2. Figures 19 and 20 clearly shows that rock samples bear traces of tectonic stresses, suggesting that the main reason for the TBM jamming was tectonic stresses. In the site investigations carried out by the author, it was also estimated that the same type of squeezing would also occur at the contact

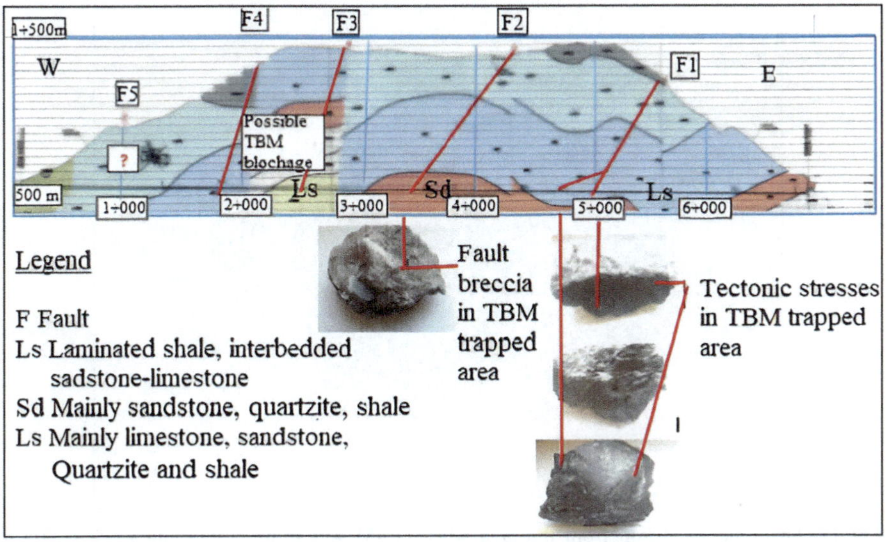

Fig. 18 The geologic profile of Doğançay tunnel route (from the archive of the author)

points of the faults with the tunnel route. The TBM passed fault F3 with some minor problems, but it jammed around Fault F4 on March 14, 2015. However, the TBM was rescued in 21 days by opening two rescue galleries. It is highly recommended that the orientation and magnitude of the tectonic stresses should be measured in tunnels planned to be opened around big fault zones. Stress field in this area is explained in detail by Aydan et al. (2024a, b) referring to the 2023 massive earthquakes of Turkey.

During field studies, some samples were collected from the muck and subjected to point load tests. The compressive strength of laminated shale, interbedded sandstone-limestone (Ls) was found to change from 60.2 to 70 MPa, sandstone, quartzite, shale (Sd) from 180.1 to 61.7 MPa, and limestone, sandstone, shale from 77.2 to 70.2 (Ls). Two samples subjected to point load tests from chainage 3 + 075 km and 2 + 875 km are shown in Fig. 21.

3.2 TBM Used in the Project

TBM used in Doğançay Tunnel is seen in Fig. 22, and the general specifications of double shield Herrenknecht M-1665 TBM are given in Table 5.

Fig. 19 Shale sample taken from the area where TBM was trapped in chainage 5 + 004 on 22.01.2013. Traces of shear stresses in the fault zone are apparent

Fig. 20 Shale sample taken from the area where TBM was trapped in chainage 4 + 700 m on 30.06.2013. Traces of shear stresses in the fault zone are apparent

3.3 Performance Jamming of the TBM

The performance of TBM was analyzed in detail between chainages 3 + 020 and 1 + 520 since this area is the representative of all the tunnel routes, explaining to the readers the effect of the East Anatolian Fault squeezing the tunnel with a moving

Fig. 21 Two samples are subjected to a point load test; please note the quartz vein in the second sample (from the archive of the author)

Fig. 22 A view of the cutterhead used in Doğançay tunnel (from the archive of the author)

rate of 2.5–10 mm/year. In the studied route, there are F3, F4 and F5 faults effecting directly squeezing and the performance of TBM. The variation of Thrust, Torque, Torque/Thrust ratio, and the penetration with chainage between February 2 and April 14, 2015, is seen in Fig. 23, and critical performance values of TBM are tabulated in Table 6. The performance of TBM within 35 m of the squeezing point is also given in Fig. 24.

Table 5 Design parameters of TBM, double shield TBM Herrenknecht M-1665

Parameter	Value
Bore diameter	4.11 m
Weight of the shield	79.7 t
Weight of the cutter head	37.3 t
Maximum Thrust force	19,704 kN
Nominal torque	1749 kNm
Intermittent torque	2274
Installed power	1400 kW
Rotational speed	0–12 rpm
Disc diameter	432 mm (17 inch)
Number of single discs	22
Number of double center discs	4

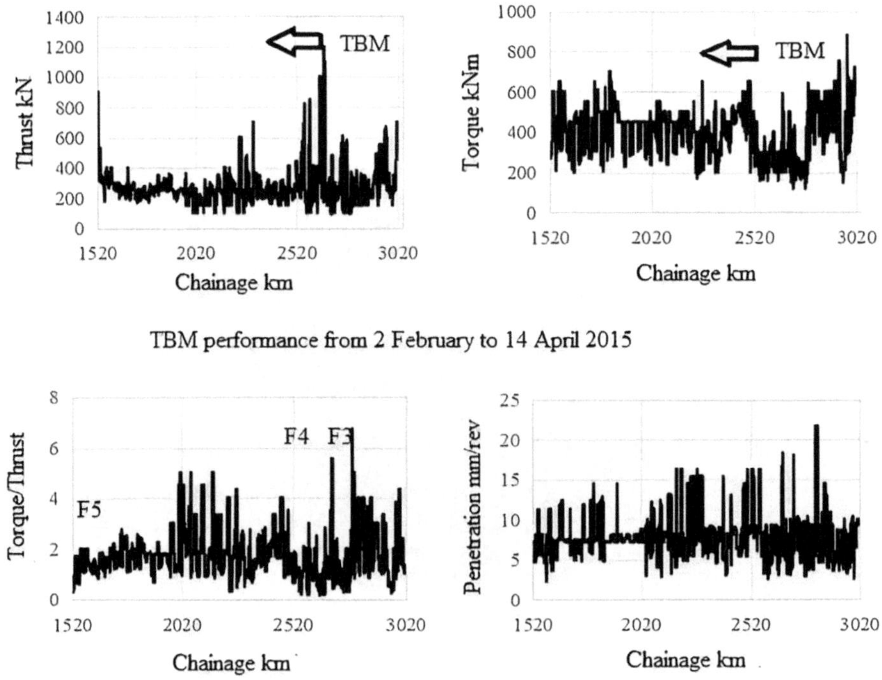

Fig. 23 The variation of thrust, torque, torque/thrust ratio, and the penetration with chainage between February 2 and April 14, 2015, in the Doğançay tunnel

Table 6 TBM mean performance values in different zones

TBM performance parameters	Between chainages 3 + 020 and 1 + 520 km	Between chainages 2 + 755 and 2528, this zone is effected with faults F3 and F4	Chainages between 1 + 558 and 1 + 1523 TBM tapped within the fault F5
Mean torque kNm	397.1	278.2	410.8 TBM trapped at 270 kNm
Mean thrust kN	266.2	310.0	370.0 TBM trapped at 900 kN
Mean torque/thrust	1.5	0.9	1.3 TBM trapped at 0.3 torque/thrust
Rotational speed, rpm	8.4	7.7	9.4 TBM trapped at 9 rpm
Penetration mm/rev	7.7	7.7	6.6 TBM trapped at 4.8 mm/rev

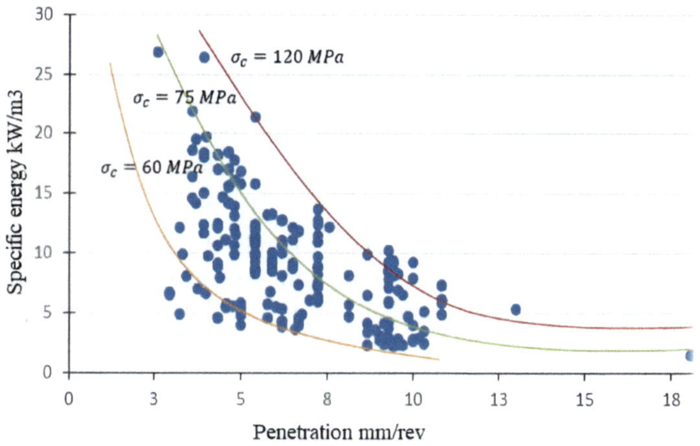

Fig. 24 The performance of TBM within 35 m of the squeezing point

From these figures and Table 6 the following conclusions may be drawn:

(a) Within 1500 m of tunnel driving, mostly on average, Torque (in kNm) was greater than thrust in absolute values (kN), which is unusual. The author believes that the high overburden and squeezing effect of the East Anatolian Fault to the West, with a rate of 2.5–10 mm per year, is the main reason for this anomaly, emphasizing that this needs detailed investigation. However, there are some more cases where torque is higher than thrust force; one of these cases is the Melen Istanbul strait tunnel, (Balci et al. 2014). Melen tunnel is bored underneath Istanbul strait Bosphorus. It has a bored diameter of 6.11 m and is 5550 m long. A 3400 m long part, which includes the section under the sea, was constructed using a shielded TBM. The TBM was designed to sustain a maximum hydrostatic pressure of 13.5 bar during standstills and operate in EPB mode at up to 4 bar pressures. At the deepest section of the alignment, the TBM operated 135 m below the sea level and 70 m below the seabed. The bedrock has similarities with the rock in the Doğançay tunnel, consisting of alternating calcareous shales, clayey or sandy limestones, and sandstones. Due to thick granular alluvial deposits, however, the minimum rock cover was as little as 35 m in the marine section of the tunnel (Anagnostou 2014).

It is worth mentioning in this section that in high overburden, the initial stress of the surrounding rock mass affected the tunneling operations. High overburden may induce brittle failure of the surrounding hard rock mass, significantly impacting TBM performance. A case is reported by Lu et al. (2021) in the southern section of the Yinhan Water Conveyance Tunnel in China, which is mainly composed of massive to intact granite, with a maximum overburden of nearly 2000 m. The paper discussed the relationship between various brittle failure degrees at the typical tunnel faces and TBM performance. The typical patterns, such as undamaged, slabbing, and rockburst at the tunnel face, and the corresponding characteristics of the TBM performance parameters were revealed. TBM operating parameters, in a typical undamaged tunnel face were reported as mean values of Thrust 18,000 N, Torque 2150 kNm, penetration 4.5 mm/rev, rotational speed 5 rpm, in typical slabbing tunnel face Thrust was 12,000, Torque 8000 kNm, penetration 12.6 mm/rev and rotational speed 2 rpm. However, in the rock mass where the rockburst occurred, the Thrust was 15,000 kN, Torque 8000 kNm, penetration 8 mm/rev, and rotational speed was 3 rpm.

(b) Decreasing the ratio of Thrust/Torque is a good indicator of the TBM squeezing. Specific energy is the energy spent per unit volume of the excavated material, which indicates the boring efficiency. Figure 25 shows the variation of specific energy with penetration for different rock strengths, data representing the date between January 2 and January 29, 2015, the graph levels off around 8 mm/rev. Table 6 shows a mean penetration value of 7.7 mm/rev within the chainages studied. It may be concluded that, although TBM worked in difficult ground conditions, the contractor achieved success regarding the operation of the TBM.

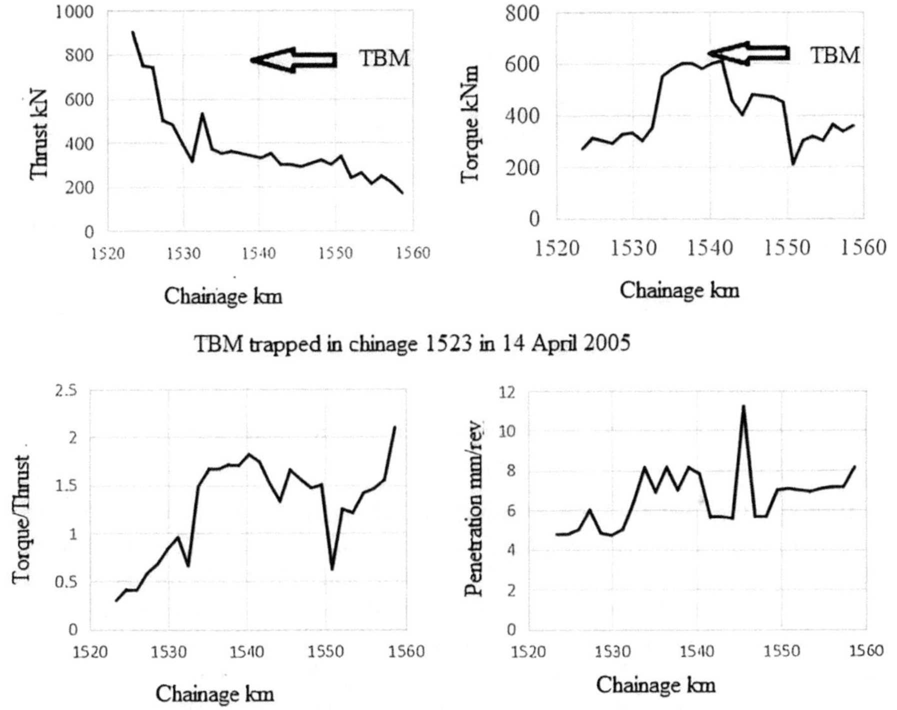

Fig. 25 The variation of specific energy with penetration for different rock strengths, data representing the date between January 2 and January 29, 2015

4 Kargı Hydropower Tunnel

4.1 Project Description, the Geology, and Some Geotechnical Properties

The Kargi Hydropower Project is situated on the Kızılırmak River, downstream of Osmancık Town in the Çorum Province of Turkey. The project area is located within the Northern Anatolian Fault Zone, which is primarily responsible for the majority of earthquakes in Turkey. Between 1 March 1990 and 1 March 2025, 7261 earthquakes of magnitudes on the Richter scale up to 5.8 occurred within 100 km of Kargı. Nine of the earthquakes were above magnitude 5, 68 earthquakes were between magnitude 4 and 5, and 609 earthquakes were between magnitude 3 and 4.[1]

This large fault zone is approximately 1500 km long and extends from the junction with the East Anatolian Fault in eastern Turkey, across northern Turkey, and into the Aegean Sea. It fractured the rock mass within its influence zone, extending from east

[1] https://allquakes.com/place/313384/earthquakes/karg/archive/1900-03-01-2025-03-01.html.

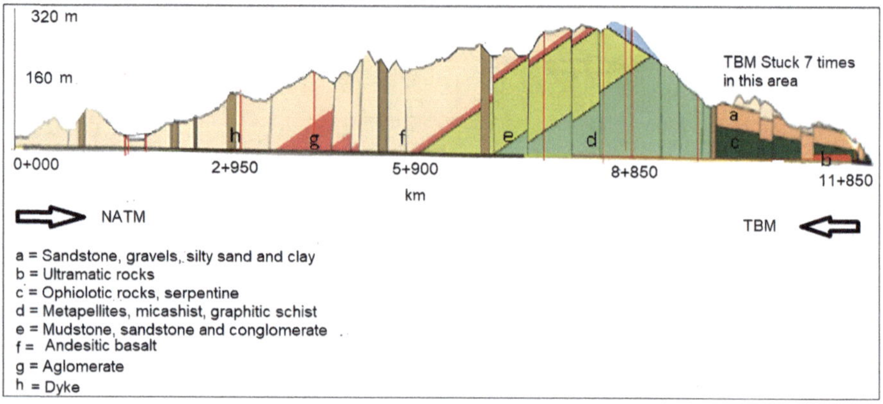

Fig. 26 The geologic cross section of Kargı hydropower tunnel (Bilgin 2016)

of Turkey to the west, creating several thousand inactive fault segments with gauge fillings, such as smectite, which have swelling characteristics. The Kargı Tunnel is situated within the influence of this large zone and is a good example of how it affected the performance of a double shield, 9.84 m diameter, hard rock TBM.

It is worth noting that the tunnel was driven through a mountainside with up to 600 m of overburden. The expected geology along the tunnel alignment consisted of Kirazbaşı complex Kargı ophiolites (including sandstone, siltstone, and marl and graphitic shist) for the initial 2300 m, followed by 1000 m of Kundaz metamorphites (including marble and metapelite), and the remaining 8500 m consisted of Eocene aged Beynamaz Volcanites (including basalt, agglomerates and andesite). The anticipated strength of the rock was up to 130 MPa. Multiple fault zones and transition zones added to the complexity of the geological conditions (Yurt et al. 2014). The author of this Chapter had the opportunity to follow the project from start to finish, and this Chapter is a result of analyzing the data collected during his field observations. The geologic cross-section of the tunnel is shown in Fig. 26. Some geotechnical properties of the rock formations are described in Table 7.

4.2 *TBM Specifications in the Project*

A Robbins crossover double shield TBM with a 9.84 m diameter, as seen in Fig. 27, was used in the project. The general characteristics of the machine are as given by Clark (2015): 19-inch back-loading hard rock disc cutters, 13 × 370 kW variable frequency drive motors, 0–8.05 rpmcutterhead speed, 20,904 kN maximum operating cutterhead thrust, 9864 kNm cutterhead torque at 3.66 rpm, and 1900 mm thrust cylinder stroke. As Clark mentioned in his article published in 2015, "Custom-made gear reducers were ordered and retrofitted to the cutterhead motors to mitigate the effects of squeezing ground or collapse. They were installed between the drive motor

Table 7 Some technological properties of the rock formations along the tunnel route; the values are taken from the baseline report of Hidrokon (2010)

Chainage	Rock formation	σ_c (kg/cm^2)	RQD (%)	Rock mass description
11 + 873/10 + 475	Grey, green metapelites Graphitic schist Phyllite Marble	25–45 30–40 120	50	Very weak rock
10 + 475/8 + 900	Magmatic and ophyolitic Graphitic schist Marble Metabasites	15–40 70–100 70–120	40–50	Very poor rock
8 + 900/7 + 400	Marble with metapelites Grapitic schists Marble Metabasits	15–40 70–100 70–120	50–90	Good rock
7 + 900/5 + 900	Mudstone, sandstone, conglomerate	65–130	60–90	Good rock
5 + 900/0 + 000	Volcanites Agglomerate Andesite and basalt Tuff	40–100 70–120 30–80	75–100 Rarely 50–75	Good rock

and the primary two-stage planetary gearboxes. During standard boring operations, the gear reducers operate at a 1:1 ratio, providing no additional reduction and allowing the cutterhead to achieve the design speeds required for hard rock boring. When the machine encounters loose or uneven ground, the reducers are engaged, which reduces the cutterhead speed but increases the available Torque. After installing the canopy drill and increasing available cutterhead torque, the TBM traversed several sections of adverse geology, including stretches of severe convergence, without becoming trapped. The typical thrust force for standard boring operations using the main thrust cylinders on the Kargı machine is approximately 21,000 kN. Several times, a thrust force of up to 136,000 kN was applied through the auxiliary thrust system before the machine could be freed from the squeezing ground (Clark 2015).

4.3 Overall TBM Performance

The overall performance of the TBM is summarized in Table 8. The field geologist described the rock samples collected from the tunnel face. The table contains the selected tunnel chainage's Thrust, Torque, and Penetration values. As summarized from this table, the Thrust of the TBM changed between 10,180 kN (188.5 kN per disc) and 18,233 kN (337.6 kN), the Torque changed between 2063 and 4100

Fig. 27 Robbins double shield XRD TBM used in Kargı project (from the archive of the author)

kNm, and the Penetration changed between 5 and 12 mm/rev. Penetration is one of the most fundamental parameters for predicting daily advance rates, provided it is used in conjunction with a machine utilization factor that incorporates a thorough examination of the geology across the tunnel line, including the precise determination of faults, water ingress into the tunnel, squeezing, and the geotechnical characteristics of the rock formations with transition zones.

Despite the slow progress initially, TBM made some remarkable advances once modifications were made to the TBM. An advance rate of 600 m was achieved in one month in March 2013, including a daily best of 39.6 m in April 2014. In so doing, the TBM significantly outperformed a drill-and-blast heading progressing from the opposite end of the tunnel (Clark 2015; Home 2015, 2016). The tunnel, comprising both TBM and drill-and-blast portions, is 11.8 km long. Mean advance rates in different ground conditions are summarized in Table 9. It is clear from this table that the mean daily advance rate of TBM after modification was doubled in different geological conditions. This point is critical in mechanized tunneling, so the contractor and the TBM manufacturer should jointly control the TBM performance and make necessary modifications to improve tunneling efficiency. The comparison in non-self-supporting rock again shows a double increase in the production rates of the TBM after the modifications were carried out. This increase can be attributed to the custom-built pipe roof drill, increase in cutterhead Torque, and the mucking system that enabled faster clearance of the buildup of loose material in the telescopic shield area of the TBM.

Table 8 The overall performance of the TBM, the description of samples was done by a field geologist (from the archive of the author)

Chainage (km)	Rock	Thrust (kN) ± sd	Torque (kNm) ± sd	Penetration (mm/rev) ± sd
9 + 000/9 + 830	Schist, calcite, clay	10,415 ± 1423	2063 ± 481	12 ± 1.0
9 + 000/9 + 830	Schist, calcite, clay metasiltstone	12,410 ± 1985	3055 ± 932	11 ± 0.9
8 + 340/8 + 220	Micritic limestone	15,500 ± 1320	4072 ± 601	7 ± 0.5
8 + 340/8 + 220	Schist, calcite, clay metasiltstone	10,294 ± 2515	2681 ± 704	10 ± 0.8
7 + 420/7 + 230	Basaltic dyke, serpentine	11,762 ± 845	3438 ± 334	11 ± 0.7
7 + 420/7 + 230	Limestone, marble	16,638 ± 954	3428 ± 763	5 ± 0.1
	Conglomerate	13,534 ± 2278	3388 ± 539	7 ± 0.6
7 + 110/6 + 990	Limestone, marble	18,233 ± 700	4100 ± 567	6 ± 0.4
7 + 110/6 + 990	Andesitic dyke	10,290 ± 4094	3448 ± 836	10 ± 0.8
7 + 110/6 + 990	Schist	10,180 ± 4014	3437 ± 728	10 ± 0.8
6 + 910/6 + 725	Graphitic schist	10,454 ± 2028	2857 ± 517	11 ± 0.4
6 + 910/6 + 725	Limestone, marble	16,754 ± 1948	2857 ± 516	7 ± 0.3
6 + 200/5 + 900	Conglomerate, sandstone	14,296 ± 2649	3361 ± 564	11 ± 0.9
6 + 200/5 + 900	Dyke	17,357 ± 1997	3361 ± 564	9 ± 0.4
5 + 880/5 + 750	Conglomerate, sandstone	16,094 ± 901	3050 ± 482	10 ± 0.6
5 + 880/5 + 750	Dyke fractured	12,115 ± 1446	2165 ± 404	12 ± 0.8
5 + 880/5 + 750	Dyke fractured, sandstone	13,529 ± 765	3024 ± 313	11 ± 0.7

Table 9 Advance rates in different ground conditions (Clark 2015; Home 2015)

Ground condition	TBM before modification	TBM after modification
Good stable rock	N.A	22.5 m/day
Fairly table rock	9.7 m/day	19 m/day
Non self-supporting rock	4.7 m/day	8.5 m/day
Running/squeezing rock	1.5 m/day	1.8 m/day

4.4 A Close Look at the Squeezing of the TBM

TBM was launched in April 2012. After boring only 80 m, the TBM became stuck in a section of the collapsed mixed ground face, which comprised hard rock and running ground. Within the first 2 km of the tunnel, the TBM jammed seven times, necessitating the construction of bypass tunnels to release the TBM. For remedial works, 285 days were spent on rescue galleries, setting the umbrella arch, replacing the seals of the main bearing, cleaning the cutterhead and buckets, and installing the belt conveyor. The distribution of the TBM stoppages is seen in Fig. 28. After a lengthy discussion between the contractor and project owner, it was agreed that the delays in TBM drives necessitated hybrid tunneling as an alternative to decrease the job completion time. Thereafter, it was decided to excavate 7.8 km of the total tunnel length of 11.8 km using the XRD TBM and 4 km from the inlet part using the drill and blast tunneling method. A recent book by Bilgin and Balcı (2025) compares the TBM and drill-and-blast tunneling efficiencies.

One of the most critical factors in TBM squeezing is the presence of swelling minerals, such as smectite (including montmorillonite and vermiculite), anhydrite, and some pyrrhotites, in calcareous shale. The type of cation present has a significant impact on the degree of swelling. Na+, for example, will cause a high degree of swelling, while CA+ could cause a lower rate of swelling. However, this was a rare case. During tunnel advancement in the Kargı Tunnel, rock samples were collected

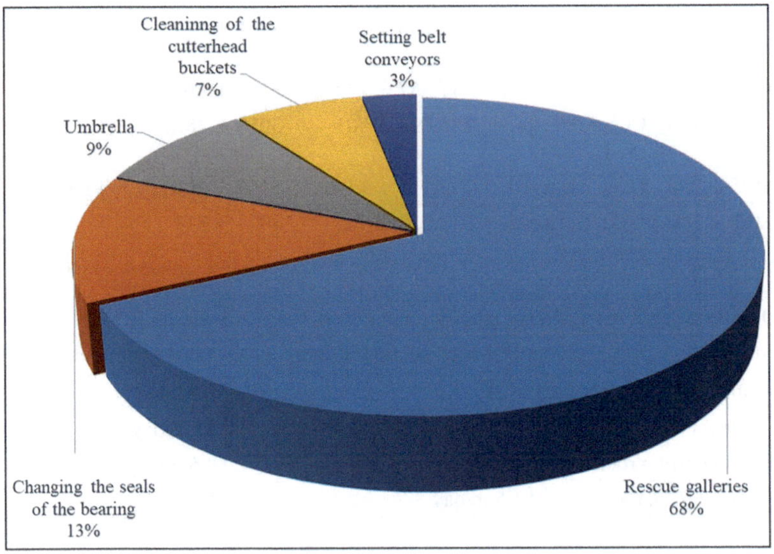

Fig. 28 Distribution of the TBM stoppages of 285 days due to face collapses and squeezing of TBM by the date of 5 August 2013; from the launch of the TBM in March 2012 to the end of July 2013, the TBM encountered 2228 m of extreme geological conditions, which required seven rescue galleries, and nine umbrella arches

in 12 different tunnel excavation areas for detailed geotechnical, mineralogical, and petrographic analyses. Most of the analysis indicated that smectite was present in the rock samples. One of the primary reasons for the TBM jamming in the Kargı Tunnel was the presence of swelling clay minerals along the tunnel route. The second reason for TBM squeezing was the weak rock and high overburden conditions. Three similar methods include empirical, semi-empirical, and theoretical-analytical models to assess the squeezing potential in rock masses. Empirical methods are mainly based on two parameters, including rock mass classification systems and tunnel depth (Bilgin and Algan 2012). Semi-empirical methods focus on estimating potential by utilizing the expected deformation of the rock mass surrounding the tunnel in a hydrostatic stress field. The typical starting point for all these methods for quantifying a rock's squeezing potential is the "NC, Competency Factor," defined as the ratio of the uniaxial compressive strength of the intact rock mass to the overburden or in situ stress. This competency factor was proposed by Aydan et al. (1993). Theoretical and analytical approaches involve closed-form solutions and numerical methods. Based on experiences with tunnels in Japan, Aydan et al. (1993) proposed the relationship in Eq. 1.

$$NC = \frac{\sigma_c}{\sigma_v} = \frac{\sigma_c}{\gamma H} \tag{6.1}$$

where σ_c is rock uniaxial compressive strength, γ is rock mass unit weight, and H is tunnel depth below the surface. According to Aydan et al. (1993), squeezing conditions will occur if the NC is less than 2.0.

Table 10 summarizes the causes and effects of various factors on TBM squeezing in the Kargı tunnel.

In most locations, for example, cases 3, 4, and 5, where the TBM jamming occurred, this ratio is less than 2, suggesting that the TBM jamming is the result of the swelling clay and weak ground characteristics. In cases 2,6,7, the broken rock accumulated in front of the cutterhead, and the TBM was trapped, in other words, face collapse had stopped the advance. In these cases, where excessive broken rock was detected through probe drilling injection was used to consolidate the ground. This was done by drilling and installation of perforated pipes in the ground ahead of the cutterhead to form an umbrella arch, which helped to penetrate the tock and implement injection into the ground. The injection was performed at pressures of up to 120 bar, depending on the material's properties of the rock and the grout. Fast-setting micro-fine Portland cement (Rheocem 650) was used for injection. As well as microcement, two-component polyurethane injection resin (MasterRoc MP 355 and Geofoam) and for filling huge voids, two-component urea-silicate (MasterRoc MP 367) injection were used (Bilgin et al. 2016b). Figure 29 shows a case where loose rock was bound with resin injection.

The ratio of Thrust force to torque, along with the time dependency of these parameters, is a good indicator for estimating TBM jamming cases, as explained by Bilgin and Algan (2012). To understand the mechanism of TBM jamming, it is

Table 10 The causes and effects of different factors in the TBM squeezing in Kargı tunnel

Remarks	Jamming of TBM and Rescue Galleries						
Gallerie no.	1	2	3	4	5	6	7
Ring no.	70	224	709	792	942	1041	1429
Chainage	11 + 756	10 + 796	10 + 671	10 + 478	10 + 445	10 + 295	9 + 971
Geotechnical values from İTÜ report*	σ_c = 20 MPa γ = 2.39 g/cm^3	σ_c = 15 MPa γ = 2.09 g/cm^3	UCS = 5 MPa γ = 1.99 g/cm^3	It is interesting to note that between chainage 10 + 879/10 + 147 in Kirazbaşı Complex UCS = 5–23 MPa; 2.3 g/cm^2 the rock is weaker than predicted before			
NC swelling index	16.7	4.8	1.3	1–1.5	1–1.5	4.6	2.2–10
Peak value of thrust/torqueBefore jamming	6.5	**2.7**	20.2	**0.2**	17.4	8.6	**1.8**
Mean value thrust/torque before jamming	5.4	**2.5**	12.8	11.3	**2.8**	4.6	4.6
Mean value of thrust/torque during jamming	3.6	6.5	8.2	6.4	3.6	12.3	2.6
Mean value thrust/torque after jamming	**2.5**	6.14	13.9	**3.2**	4.3	**2.2**	**1.2**
Reason of jamming	Fault	Swelling clay, very weak rock	Swelling clay, very weak rock	Swelling clay, weak rock	Fault, Swelling clay, weak rock	Fault, Swelling clay, weak weat rock	Weathered dyke

In the geology report by the contractor, the rock is of the Kirazbaşı complex- mélange with various ophiolitic rocks, faults are well determined, different samples tested in ITU showed that they contain smectite which is a group of clays having swelling characteristics.
Tests are carried out in İTÜ, Geology Deprment, under the supervision of Prof.Dr.Mustafa Kumral.

also essential to follow the theoretical concepts given by Ramoni and Anagnostou (2010a, 2010b).

The variation of Thrust and Torque values in seven different cases of jamming is illustrated in Fig. 30. As seen from this Figure, the Torque was higher than the thrust force in cases 6 and 7, where the trapping of the TBM was due to the accumulation of the broken rock in front of the cutterhead.

Fig. 29 Polymer is binding loose rocks

5 Behavior of Tunnels During Earthquakes

Dowding and Rozen (1978) presented one of the first surveys of case studies on the effect of earthquakes on rock tunnels. They collected information on 71 tunnels and compared their behavior with estimated peak ground accelerations and peak ground velocities. Their conclusions can be summarized as tunnel collapse from shaking occurring only under extreme conditions: Tunnel collapse mainly occurs associated with the movement of an intersected fault. Kontogianni and Stiros (2003) share the findings of Dowding and Rozen (1978), emphasizing that the Wrights tunnel affected by the San Francisco earthquake in 1906, the Urban Kern Country railway tunnel affected by the Kern Country railway earthquake, and the Twin Bolu tunnels affected heavily by Düzce Earthquake in 1999. All these tunnels were constructed using conventional tunneling methods or NATM. It is usually accepted that bored tunnels by TBMs are more resistant to failure and deformation than those constructed by conventional tunneling methods. However, readers of this book should remember that even the tunnels mined with TBMs in the earthquake zones are cautiously designed against its effect. A typical example is the Avrasya Tunnel, which is located in a seismically active region about 16 km from the North Anatolian Fault Zone in Turkey. To decrease the seismic stresses/strains below the permissible levels, two flexible seismic joints with displacement limits of $\pm$ 50 mm for shear and $\pm$ 75 mm for extraction/contraction were established. The design earthquake magnitude was specified as a moment magnitude of 7.25 (Arıoğlu et al. 2023).

The tunnel engineering committee of the Japan Society of Civil Engineers concluded that there were a few recorded structural damages to shield tunneling

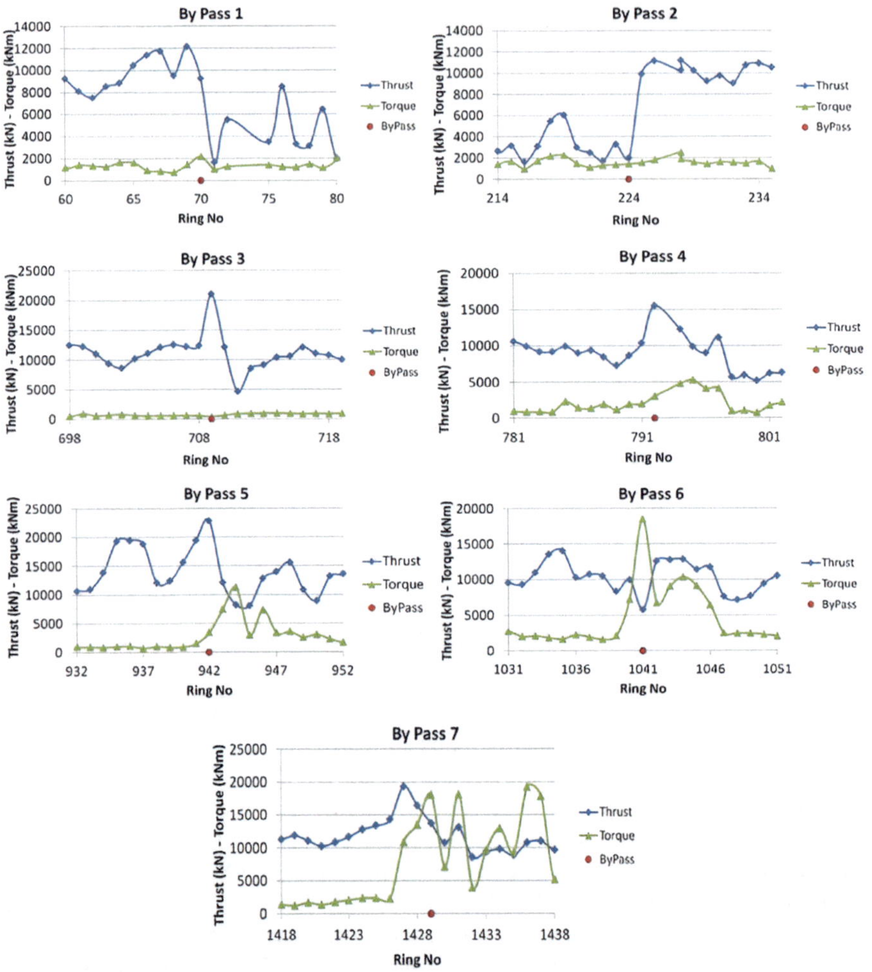

Fig. 30 The variation of thrust and torque in seven cases where TBM was jammed

during the Hyogoken-Nanbu Earthquake and no obvious failures of major structural members (JSCE 2007). However, they pointed out that in the following conditions, careful consideration should be given as the tunnel would be subject to large earthquake effects.

- Where the lining structure changes quickly, for example, at tunnel connections and the connections with shafts.
- When boring soft ground.
- Where the ground conditions change quickly, such as geology and overburden change suddenly.
- When the alignment includes sharp curves.

When the tunnel is in loose, saturated sand, there is a possibility of liquefication.

Gökçeoğlu (2023) investigated the effect of big earthquakes on around 100 tunnels located in the influence area of these earthquakes inTurkey. He concluded that most of the tunnels stayed in a good state; however. The T3 railway tunnel collapsed due to the displacement of the Erkenek Faul segment. Although the Malatya-Erkenek twin tunnels are closer to the Erkenek segment, only some failures in the final linings were observed in these tunnels, having dimensions of 10.30 × 7.47 m. This tunnel has been in service since 2017. The final lining has a thickness of 40 cm and is made of C25/C30 concrete (Gökçeoğlu 2023). The total length of the railway tunnels in this area is 30,723 m, and the total length of the highway tunnels is 38,598 m (Gökçeoğlu and Karahan 2024). It is also important to note that although Nurdağı Town is completely destroyed, Nurdağı tunnel, which is only a few km away from the town, stayed intact; in Nurdağı tunnel, only some precast segments between the chainage 11 + 696 and 13 + 143 km are affected by big Earthquake happened on February 6th, 2023. The rest of the segments stayed intact, and private communication with Prof. Dr. C. Gökçeoğlu.

Aydan et al. (2024, b) made some observations in the same area and concluded that "The damage by faulting was quite severe at the railway tunnel near Ozan village, and the relative slip was more than 200 cm. The damage to the concrete lining of the new Erkenek tunnels occurred in several places. The damage in the new Erkenek tunnels may be related to mass movements caused by heavy ground shaking. Sinkholes occurred in several locations, including urbanized areas. This study presents the damage to underground structures by the massive earthquakes with some aftershocks, emphasizing on Erkenek tunnels. The causes of damages are described, and their implications on seismic design of underground structures are discussed."

Interestingly, Apostolaki et al. (2025) concluded that the field survey conducted following the big Earthquake in 2023 in Turkey revealed limited damages to the tunnels in the affected region despite the very high recorded ground motions. Therefore, they examined the reasons for these limited damages and then investigated whether the available state-of-the-art methods and tools for the seismic risk assessment of tunnels could have predicted this behavior. Pitilakis (2023) also pointed out the importance of seismic design in protecting tunnels from the severe effects of earthquakes.

6 Concluding Remarks

This chapter summarizes the effect of high stress ground as well as presence of fault zones and squeezing ground on TBM tunneling, using examples of East Anatolian Fault and North Anatolian Fault zones. It examined the impacts of such conditions on the performance of TBMs with reference to some of the recent projects including the Nurdağı high-speed railway tunnel, Doğançay, and Kargı hydropower tunnels.

The Nurdağı tunnel is in the active zone of the East Anatolian Fault, one of Turkey's most seismically active zones because it represents a plate boundary extending over 500 km between the Arabian and Anatolian plates. An Earthquake of 7.8 Mw occurred in this area on February 6th, 2023, affecting millions living in this area. The town of Nurdağı is completely destroyed during this event. However, in the Nurdağı Tunnel, some of the rings between the chainage 11 + 696 and 13 + 143 km were affected. This part of the tunnel was evaluated as risky for TBM, and a conventional tunneling method was recommended for this part of the tunnel. The other part of the tunnel stayed in a good state, and the performance of the TBM was as predicted from the laboratory cutting experiments. The hardest rocks of Turkey are found in this area. A single shield TBM was used in this tunnel consisting of two parallel 9.75 km tunnels, T1 and T2. In this tunnel, TBM showed different performances in the affected and non-affect zones by the East Anatolian fault. The rotational speed of the TBM changed between 3.8 and 4.5 rpm, resulting in a high penetration from 15 to18 mm/rev, since higher rpm could increase the penetration with a risk of trapping/ jamming the TBM by the accumulation of a higher amount of fractured rock in front of the cutter head. Almost 60 days were spent rescuing the TBM. Daily advance rates in this zone remained between 6 and 8 m/day since high water ingress of up to 50–300 l/s was encountered, making the time spent cleaning the cutterhead and the tunnel as high as 26% and 24% of all shift time spent for boring the tunnel. The rotational speed and penetration in hard and non-disturbed rocks are found to be 6.6 rpm and 5–6 mm/rev, as predicted in the laboratory full-scale cutting tests. The advance rate changed between 14–15 m/day with a mean value of 40% machine utilization rate. In the Nurdağı tunnel, only some precast segments between the chainage 11 + 696 and 13 + 143 km were affected by a big earthquake that happened on February 6th, 2023. The rest of the segments stayed intact (based on a private communication with Prof. Dr. C. gökçeoğlu).

Doğançay tunnel is located in the north-east of Turkey, affected by the East Anatolian Fault, and it is a part of a hydroelectric project. The tunnel length bored with a 4.5 m diameter double shield TBM is 6232 m. During the boring of the tunnel, TBM was trapped 10 times, and around 160 days were spent on rescue operations. It was concluded that the main reason for the squeezing of the tunnel was high overburden reaching 1000 m and tectonic stresses exerted by the complex East Anatolian Fault. A 6.2 m/ day daily advance rate was obtained with all the waiting included. The recorded maximum advance per month was recorded at 450 m. Except for significant downtimes like TBM entrapment in squeezing ground, the total excavation machine utilization time changed between 16 and 45%. On average, Torque was found to be greater than Thrust (kN) in absolute values, which is unusual needing further studies. The author believes that the high overburden and squeezing effect of the East Anatolian Fault to the West, with a rate of 2.5 mm- 10 mm per year, is the main reason for this anomaly, emphasizing that this needs detailed investigation. However, there are more cases where Torque is higher than Thrust force in absolute values; one is the Melen Istanbul strait tunnel (Balci et al. 2014). In his project, TBM operated 135 m below the sea level and 70 m below the seabed under a hydrostatic pressure of 13.5 bar (Anagnostou 2014). The bedrock has similarities with the rock in the Doğançay tunnel, consisting of alternating calcareous shales, clayey or sandy limestones, and sandstones. The other important point emerging from the study of the TBM performance is that decreasing the ratio of Thrust/Torque is a good indicator of the TBM squeezing.

Kargı hydropower tunnel is located within the Northern Anatolian Fault Zone, primarily responsible for most earthquakes in Turkey. This big fault fractured the rock mass within its influence zone, extending from the east of Turkey to the West. It created several thousand inactive fault segments with gauge fillings, such as smectite, which have swelling characteristics. The Kargı Tunnel is situated within the influence of this large zone and is a good example of how it affected the performance of a 9.84 m diameter double shield hard rock TBM, within 2000 km of the tunnel route. After being bored for only 80 m, the TBM got jammed 7 times, necessitating 285 days for rescue operations. One of the primary reasons for the TBM jamming in the Kargı Tunnel was the presence of swelling clay minerals along the tunnel route. The second reason for TBM squeezing was the weak rock and high overburden conditions. Despite the slow progress initially, TBM made some remarkable advances once modifications were made to the TBM. An advance rate of 600 m was achieved in one month in March 2013, including a daily best of 39.6 m in April 2014. In so doing, the TBM significantly outperformed a drill-and-blast heading progressing from the opposite end of the tunnel. Thrust of the TBM changed between 10,180 kN (188.5 kN per disc) and 18,233 kN (337.6 kN), the Torque changed between 2063 and 4100 kNm, and the penetration changed between 5 and 12 mm/rev. Penetration was one of the most fundamental parameters for predicting daily advance rates of the planned tunneling projects, provided it is used in conjunction with a machine utilization factor that incorporates a thorough examination of the geology across the tunnel line, including the precise determination of faults, water ingress into the tunnel, squeezing, and the geotechnical characteristics of the rock formations with transition zones. The observed ground behavior reinforces that theory that if the machine can maintain a reasonably high daily advance rate, the chances of jamming and entrapment can be reduced. This is due to the fact that the ground does not get a chance to deform and converge on the shield when a steady and reasonably good daily advance is maintained. Alternatively, if there are long delays in the tunneling in grounds with potential for squeezing, the time dependent behavior and creep allows the ground to converge and impose heavy load against the shield, hence leading it to entrapment.

It is also concluded that bored tunnels are more resistant to earthquake than the tunnels opened with conventional tunneling methods.

References

Anagnostou G (2014) Some critical aspects of subaqueous tunnelling, muir wood lecture. WTC 2014, Iguacum, Brazil

Apostolaki S, Karahan S, Riga E, Tsinidis G, Gokceoglu C, Pitilakis K (2025) Seismic performance of tunnels and verification of available seismic risk models for the 2023 Kahramanmaras¸ earthquakes. Tunnelling Undergr Space Technol 156:106185.https://doi.org/10.1016/j.tust.2024.106185

Arıoğlu B, Arıoğlu E, Gökçe B (2023) Eurasia tunnel, emphasis on seismic design. In: Tunnel 2023, an international symposium on the effect of natural hazards on tunnels, organized by the Turkish Tunneling Society in Istanbul, was held in 23–24 Nov. The presentation is open access on the Turkish Tunneling Society web page

Aydan Ö, Akagi T, Kawamoto T (1993) The squeezing potential of rock around tunnels: theory and prediction. Rock Mec. Rock Eng. 26(2):137-163

Aydan Ö, Ulusay R, Kumsar (2024a) Faulting by the 2023 great earthquakes of Türkiye and associated stress field and its effects on built environment. Geoenvironmental Disasters 11(38):25

Aydan O, Malistani N, Ulusay R, Kumsar H (2024b) The damage to underground structures by the 2023 February 6 great Turkish earthquakes with a special emphasis on Erkenek tunnels. In: ISRM European rock mechanics symposium, Alicante, Spain, 15–19 Julu 2024, pp 710–715

Balci B, Bilgin N, Copur C, Comakli R, Polat C, Sarbangholi FS (2014) Analysis of the tunnel boring machines (TBM) performance parameters using boreability Index in Turkey. In: Proceedings of the world tunnel congress 2014—tunnels for a better Life. Foz do Iguaçu, Brazil

Barton N, Bilgin N (2016) Fast or slow progress with TBM in ideal or faulted conditions. In: Ulusay et al. (eds) Rock mechanics and rock engineering: from the past to the future. Taylor & Francis Group, London, ISBN 978-1-138-03265-1

Bilgin N, Algan M (2012) The performance of a TBM in a squeezing ground at Uluabat, Turkey. Tunnel Undergr Space Technol 32:58–65

Bilgin N (2016) An appraisal of TBM performances in Turkey in difficult ground conditions and some recommendations. Tunn Undergr Space Technol 57:265–276

Bilgin N, Copur H, Balci C (2016) TBM Excavation in difficult ground conditions, case studies from Turkey. Wiley, Germany, p 346

Bilgin N, Balci N (2016) Tunnels and tunneling in Turkey. In: Feng X-T (ed) Rock mechanics and engineering, volume 5: surface and underground projects. CRC Press, a Francis and Taylor Balkema book

Bilgin N, Copur H, Balci C, Tumac D (2017) TBM performance prediction using laboratory cutting tests in very hard and abrasive rock formations. In: International conference on tunnel boring machines in difficult grounds (TBM DiGs), Wuhan, 16–18 Nov 2017

Bilgin N, Balcı C (2025) Critical issues in selecting conventional and mechanized tunneling methods. Spinger and Nature

Clark J (2015) Extreme excavation in fault zones and squeezing ground at the Kargı HEPP in Turkey. In: "SEE Tunnel: promoting tunneling in SEE Region" ITAWTC 2015 congress and 41st general assembly May 22–28, 2015, Lacroma Valamar Congress Center, Dubrovnik, Croatia

Dowding CH, Rozen A (1978) Damage to rock tunnels for earthquake shaking. J Geotech Eng Div Am Soc Civil Eng 104:p175-191

Fugro Sial Inc. (2014) Nahce–Nurdağı, high speed railway tunnel, geological and geotechnical project report, vol 1, P2013.0015, 91p (in Turkish)

Galasso C and Opabola EA (2024) The 2023 Kahramanmaraş Earthquake sequence: finding a path to a more resilient, sustainable, and equitable society. Commun Eng 3:24. https://doi.org/10.1038/s44172-024-00170-y. www.nature.com/commseng

Gökçeoğlu C (2021) Assessment of rate of penetration of a tunnel boring machine in the longest railway tunnel of Turkey, SN Appl Sci 14

Gökçeoğlu C (2023) Latest big earthquakes in Turkey and their effects on tunnels. In: Tunnel Turkey 2023 international tunneling symposium organized by Turkih Tunneling Society, Istanbul, basic concepts of Eurocode 8

Gökçeoğlu C, Karahan S (2023) Seismic performance of transportation tunnels in the region affected by the 6 February 2023 Türkiye earthquake sequence: a general assessment, ARMA 23–0960, p 8

Hidrokon (2010) Geological and geotechnical report for Kargı Project, Sept 2010

Home L (2015) Hard rock TBM tunnelling in challenging ground: Developments and lessons learned from the past. In: International conference on tunnel boring machines in difficult grounds (TBM DiGs) Singapore, 18–20 November 2015

Home L (2016) The Kargı challenge, a study of TBM boring vs conventional tunneling. Tunnel Expo organized by The Turkish Tunneling Society, Istanbul

Jordan D, Bilgin, N (2018a) Excavating Turkey's hardest rock at the Bahce-Nurdag railway tunnel, the world tunnel congress (WTC 2018) Dubai, Apr 20–26

Jordan D, Bilgin, N (2018b) Turkey's hardest rock, the Bahce-Nurdag high speed railway tunnel Tunnel. Off J Stuva 22–28

Kontogianni VA, Stiros SC (2003) Earthquakes and seismic faulting: effects on tunnels. Turkish J Earth Sci (Turkish J Earth Sci) 12:1–27

Lu J, Gong Q, Yin L (2021) Zhou X (2021) Study on the tunneling response of TBM in stressed granite rock mass in Yinhan Water conveyance tunnel. Tunn Undergr Space Technol 118:104197

Nalbant SS, McCloskey J, Steacy S, Barka AA (2002) Stress accumulation and increased seismic risk in eastern Turkey. Earth Planet Sci Lett 195:291–298. https://doi.org/10.1016/S0012-821X(01)00592-1

Pitilakis K (2023) Seismic design and risk assessment of tunnels and underground structures basic concepts of Eurocode 8, Tunnel Turkey 2023. In: International tunneling symposium organized by Turksih Tunneling Society, Istanbul

Ramoni M, Anagnostou G (2010a) Tunnel boring machines under squeezing conditions. Tunn Undergr Space Technol 25:139–157

Ramoni M, Anagnostou G (2010b) Thrust force requirements in squeezing ground. Tunn Undergr Space Technol 25:433–456

Read HH and Janet W (1975) Introduction to geology, V2, earth history. Wiley, pp 13–15. ISBN 978-0-470-71165-1. OCLC 317775677

JSCE, Tunnel Engineering Committee (2007) Standard specifications for tunneling—2006: shield tunnels, published by Japan Society of Civil Engineers, p 269. IBN 978-4-8106-0568-6

Stein RS, Barka AA, Dieterich JH (1997) Progressive failure on the North Anatolian fault since 1939 by earthquake stress triggering. Geophys J Int 128:594–604

Yılmaz Y, Gürer ÖF, Yiğitbaş E (2025) Field data and some thoughts on the 2023 Kahramanmaraş Earthquakes faults. Geol Bull Turk 68(2025):113–146. https://doi.org/10.25288/tjb.1577026

Yönlü Ö, Altunel E, Karabacak V (2017) Geological and geomorphological evidence for the southwestern extension of the East Anatolian Fault Zone, Turkey, Earth and Planetary

Yurt Y, Özturk A, Arslan Z, Nuhoğlu C, Oystein L, Erdoğan E, Atlar B, Palakçı, Y, Bilgin N (2014) Factors affecting the performance of a double shield TBM in a very complex geology in Kargi Turkey. In: Proceedings of the world tunnel congress 2014—tunnels for a better Life. Foz do Iguaçu, Brazil

Feature Optimization: Identifying Critical Parameters for TBM Performance Assessment

Marlene C. Villeneuve and Arsham Moayedi Far

1 Introduction

The assessment of Tunnel Boring Machine (TBM) performance is influenced by a multitude of factors ranging from geological conditions to machine-specific parameters (Rostami 2016). Each of these elements has a distinct impact, particularly when they occur under variable conditions. For example, the efficacy of TBM chipping is significantly affected by the rock's strength (Villeneuve 2017), while strength and hardness are known to correlate with wear on disc cutters (Hassanpour et al. 2015; Macias et al. 2016), thereby potentially affecting advance rate due to poor penetration and utilization arising from more frequent cutter replacements. The complexity escalates when multiple intact and rock mass characteristics, such as the strength of the intact rock, the presence, orientation, and condition of discontinuities, and the composition of the rock interact to affect TBM performance.

Given the growing application of machine learning (ML) and artificial intelligence (AI) across diverse fields, including civil engineering (Alishvandi et al. 2024; Moayedi Far and Zare 2025), rock engineering (Saadati et al. 2024), and tunneling (Hansen et al. 2024; Morgenroth et al. 2022), more sophisticated analytical tools for the analysis and prediction of TBM performance are increasingly being used. These advanced tools enable the integration and analysis of complex datasets, and provide the opportunity to gain deeper insights into how various parameters collectively influence TBM performance.

Choosing which are the influential parameters in order to optimize ML predictions in terms of computational cost by reducing unnecessary parameters can be challenging, as in tunneling projects the number of descriptive variables may be large,

M. C. Villeneuve (✉) · A. Moayedi Far
Subsurface Engineering, Technical University of Leoben, Leoben, Austria
e-mail: marlene.villeneuve@unileoben.ac.at

J. Hassanpour and J. Rostami (eds.), *Performance of Tunnel Boring Machines (TBM) in Rock*, Springer Series in Geomechanics and Geoengineering,
https://doi.org/10.1007/978-3-032-20424-0_7

connections between these variables and decisions may be complicated, measurements are subject to noise, and it is not known beforehand how many variables influence the decision. Feature importance (FI), feature engineering (FEng), and feature selection (FS) are among the methods that help explore the relation between various parameters, and are combined in this chapter under the general term "feature optimization". Feature importance provides insights into which features contribute most to the model's predictive power, guiding further refinement or validation of the feature set used in modeling. Feature engineering combines features from raw data into new features, thereby helping to manage high-dimensional data. Feature selection aids in choosing the most relevant subset of features to improve model efficiency. In addition, these tools provide the opportunity to better understand the features that affect or reveal tool-rock interaction in order to improve conceptual and predictive models of the TBM excavation process.

This chapter summarizes a selection of methods appropriate for feature optimization prior to machine learning model development. It provides insights into the relationships between different features to promote the selection of the most meaningful features and the reduction of complexity in models. The methods were selected for the applicability to the type of data frequently encountered in TBM performance analysis and prediction and are illustrated using a case study.

In data processing and analysis, having a defined workflow ensures consistency and streamlines processes, enabling predictable and reliable outcomes. However, in complicated cases that require frequent double-checks and iterative updates based on new insights, adhering to a rigid workflow can hinder performance. This limitation arises because such a defined approach often demands that reality conform to the structure rather than allowing the structure to flexibly adapt to the evolving complexities of the task. Therefore, proposing a sequential process or a fixed workflow for using FI, FEng, and FS can be misleading, as it risks jeopardizing both the flexibility and efficiency of the process. Here, we will illustrate this through examples.

Feature importance aids in understanding the initial correlations between variables and targets, guiding subsequent refinements while feature engineering benefits from applying new inputs derived from the understanding gained during feature importance analysis. Incorporating the knowledge from feature importance enables the development of features that might not emerge solely through mathematical or statistical processes. At any stage in this process, feature importance offers insights into how these newly engineered features influence the target variable compared to others. Thus, to fully capitalize on the potential of this process, it is best to use an iterative, collaborative approach where feature engineering and feature importance work together, providing in continuous feedback and refinement.

In fields such as tunneling and rock engineering, the role of expert judgment is particularly important due to the inherent uncertainties associated with geological conditions and the often-limited ability to quantify events or properties reliably. For example, the strength data even in the same rock can easily be $\pm$ 20%, whereas the distribution of RQD could highly variable, the Alpha angle of the joints relative to the tunnel axis could also vary within a wide range and the favorable or unfavorable attitude of the joint with the same angle can impact the results. Additionally, geological characterization is typically subjective, which can significantly impact quantification of geology in preparation for application of ML and AI. Unlike sectors like finance or IT, where data tend to be more structured, tunneling and rock engineering demand a higher reliance on expert judgment to address complex ambiguities (known and unknown). This might also lead to more connected feature engineering and feature selection in these fields, with both processes relying heavily on prior experimental knowledge. While feature importance can validate the effectiveness of proposed features, the emphasis here lies on the relationships between feature engineering, selection, and importance. This complexity demonstrates that these techniques cannot be confined to a fixed framework or flowchart. Instead, the focus should be on understanding their applications and adapting them to the specific challenges of fields like tunneling, rather than attempting to simplify the process into a direct A-to-B route.

As a result, in this chapter, relevant methods that enable us to derive a ranking of the effectiveness of independent parameters on dependent variables fall under the category of feature importance (FI) methods. Techniques for generating new features from existing features are classified as feature engineering (FEng), while methods that assist in choosing among the set of raw and engineered features are denoted as feature selection (FS). This will then be demonstrated using a case study of TBM data feature optimization. With this case study, we highlight how some of the presented methods are combined in an iterative process, and how the decision-making and insights are guided by domain expertise supported by conventional data analysis results.

2 Feature Importance, Feature Engineering and Feature Selection Techniques and Tools

2.1 Feature Importance (FI)

FI encompasses various methods and tools designed to rank the effectiveness of independent variables in influencing a target variable. FI provides the ability to explore the complex relationships underlying physical phenomena. In this context, it provides insights into the interplay between parameters related to the geology and those related to the TBM, particularly during excavation along routes with variable geological conditions.

Because TBM performance assessment involves continuous, binary and categorical variables, different methods for these different data types are needed. Consequently, we will explore these concepts using several widely utilized and practical approaches for targets with continuous variables, including Correlation Analysis, Decision Tree Regressors, Random Forest, XGBoost, and Permutation Importance, and for targets with binary or categorical variables, Logistic Regression, Random

Forest Classifier, and Mutual Information. These methodologies represent a selection of the most effective tools available for assessing feature importance in complex engineering and geological scenarios.

2.1.1 Feature Importance for Continuous Target Variable

Correlation analysis

Correlation analysis is a statistical method employed to evaluate the strength and direction of relationships between variables. One of the most common techniques across different fields (Monedero et al. 2012; Wang et al. 2018) is the Pearson Correlation Coefficient, which measures the linear relationship between two continuous variables (Cohen et al. 2009). This coefficient proposed by Karl Pearson (Deng et al. 2021) ranges from − 1 to + 1, where − 1 indicates a perfect negative linear relationship, + 1 indicates a perfect positive linear relationship, and 0 signifies no linear correlation.

In order to use the Pearson correlation coefficient (ρ) to reveal the intensity of correlation between parameters we consider m and n as two parameters to be investigated (Eq. 1):

$$\rho_{(m,n)} = \frac{\mathrm{cov}(m, n)}{\sigma_m \sigma_n} \tag{1}$$

In which cov, represents covariance, while σ_m and σ_n are standard deviation of m and n, respectively.

For modeling nonlinear but monotonic relationships between variables, nonlinear correlation coefficients (Deng et al. 2021) should be used. Spearman's rank correlation coefficient (ρ) and Kendall's rank correlation coefficient (τ) are two widely used methods (Alvo and Cabilio 1995; Kendall 1938; May and Looney 2020; Spearman 1904; van den Heuvel and Zhan 2022; Wang et al. 2024a; Zhang and Wang 2023) and are formulated as follows (Crysdian 2022):

$$\rho = 1 - \frac{6\sum d_i^2}{n\left(n^2 - 1\right)} \tag{2}$$

$$\tau = \frac{nc - nd}{\frac{1}{2}n(n - 1)} \tag{3}$$

In which, nc and nd represent the counts of concordant and discordant pairs, respectively, n signifies the total number of observations, and d indicates the difference between the ranks of a pair of observations. Both methods yield a correlation coefficient that spans from − 1 to 1. A coefficient of − 1 indicates a perfect negative correlation, meaning as one variable increases, the other decreases monotonically; a coefficient of 1 signifies a perfect positive correlation, where an increase in one variable corresponds to a monotonic increase in the other; while a score of 0 suggests no correlation, implying that the variables change independently of each other.

Regressors

Regressors such as Decision Trees (DT), Random Forests (RF), and eXtreme Gradient Boosting (XGBoost) are useful in the context of FI (Devi et al. 2019; Hwang et al. 2023; Wang et al. 2024b). These models, available in libraries like scikit-learn and XGboost, offer built-in methods to compute feature importance, which can be used to rank and visualize the relative influence of each feature on the model's predictions. In order words, by fitting these algorithms on the data set they provide the magnitude of each parameter's contribution in prediction of the output.

These algorithms are widely used in machine learning (ML) for predictive modeling and share foundational principles in their operation. Given that the primary focus of this chapter is on the application and interpretation of FI, rather than the mechanics of the regression models themselves, we provide only a brief introduction to the employed regressors. For readers unfamiliar with these algorithms, we recommend referring to foundational texts or resources on machine learning for a more in-depth understanding.

The Decision Tree Regressor, found in scikit-learn as "DecisionTreeRegressor", computes the feature importance of a feature as the (normalized) total reduction of the criterion (function to measure the quality of a split) brought by that feature, which is also known as the Gini importance (Pedregosa et al. 2011). The Random Forest Regressor, also implemented in scikit-learn as "RandomForestRegressor", also provides feature importances using the Gini importance. For datasets with a considerable number of unique values (high cardinality) impurity-based feature importances can be misleading and permutation feature importance could be an alternative option (Pedregosa et al. 2011).

XGBoost, through the xgboost library's XGBRegressor class, also provides a framework for conducting feature importance (FI) analysis. XGBoost offers several metrics to evaluate feature importance, depending on the importance type parameter (XGBoost: Scalable and Flexible Gradient Boosting 2024):

- Weight: Represents the frequency with which a feature is used to split the data across all trees.
- Gain: Reflects the average improvement in performance (gain) achieved by splits involving the feature.
- Cover: Measures the average number of data points (coverage) affected by splits involving the feature.
- Total Gain: Aggregates the gain across all splits where the feature is used, providing a cumulative measure of its impact.
- Total Cover: Sums the coverage across all splits where the feature is utilized, offering a comprehensive view of its reach.

When the model is trained on multi-class, multi-label, or multi-target datasets, the feature importance scores are averaged across all targets. For instance, if the importance_type is set to "total_gain," the score represents the sum of loss changes for each split from all trees, averaged across all targets (XGBoost: Scalable and Flexible Gradient Boosting 2024).

Permutation feature importance

Permutation feature importance is a model evaluation technique designed to assess the contribution of each feature to the statistical performance of a trained model on a given dataset. Particularly advantageous for non-linear or complex estimators, this method involves randomly shuffling the values of a specific feature and observing the resulting changes in the model's performance metrics. A key advantage of permutation feature importance is its model-agnostic nature, allowing its application to any trained estimator regardless of its underlying structure. Additionally, the method can be repeated with multiple permutations of the feature, providing a more robust and comprehensive assessment of the variability in feature importance estimates for the trained model. This adaptability makes it a valuable tool for unbiased evaluation in diverse scientific contexts (Pedregosa et al. 2011). It can be applied by utilizing the KNeighborsRegressor as the fitted estimator for feature importance analysis.

2.1.2 Feature Importance for Binary or Categorical Target Variable

In the assessment of TBM (Tunnel Boring Machine) performance, it is important to recognize that the target variables are not limited to just continuous parameters, such as ROP (Rate of Penetration) and torque. In some cases, prediction tasks may involve classification or binary outcomes. For example, when estimating TBM jamming (stuck TBM), which can be treated as a binary target (e.g., 1: occurrence, 0: non-occurrence), appropriate methods should be employed. This sub-section explores suitable tools for such tasks, including Logistic Regression, Random Forest Classifier, and Mutual Information.

Logistic Regression

Logistic regression is appropriate in contexts where the target variable is binary (e.g., yes/no, good/bad, 0/1) (Goh et al. 2017; Hosmer et al. 2013). The predicted outcome from logistic regression models is the probability of an event occurring, with values constrained between 0 and 1. Furthermore, this method is widely utilized in feature importance analysis, as its coefficients can be interpreted to assess the relative influence of each predictor variable on the binary outcome, providing insights into which features most significantly drive the model's predictions.

Random Forest Classifier

Generally, classification-based techniques are more suitable when investigating the FI for binary or categorical target variables as they are designed the solve these problems, for the same reason regressors were previously suggested for conducting FI on continuous variables. In this regard, the random forest classifier is a suitable method as it selects, at each node, a random subset of variables and then determines the best split over these latter variables only (Louppe et al. 2013). Random forest classification has also been shown to offer advantages when handling categorical independent variables with many values (Breiman 2001).

Breiman (2001, 2002) proposed that to evaluate the importance of an independent variable for predicting another variable, the weighted impurity decreases could be added up for all nodes where the independent variable is used, then averaged over all trees in the forest (Louppe et al. 2013). When the Gini index is used as the impurity function, the resulting measure is referred to as Gini Importance or Mean Decrease Gini (Louppe et al. 2013). In this way random forest assesses how much each feature decreases the impurity (e.g., Gini impurity or entropy) across all trees to derive the feature importance.

Mutual Information

Mutual information is a measure of the amount of information that one random variable has about another variable (Cover 1999). It is equal to zero if and only if two random variables are independent, and higher values mean higher dependency. The function relies on nonparametric methods based on entropy estimation from k-nearest neighbors distances, as described in (Kraskov et al. 2004; Pedregosa et al. 2011; Ross 2014). For a categorical target variable the mutual information can be estimated using "sklearn.feature_selection.mutual_info_classif" provided by scikit learn (Pedregosa et al. 2011).

2.2 Feature Engineering (FEng)

Feature engineering is an important yet labor-intensive step in data preparation for machine learning (Bengio et al. 2013). It involves constructing new features from existing ones to improve predictive performance (Nargesian et al. 2017). This process encompasses various methodologies, ranging from leveraging domain knowledge to relying on intuition or patterns derived through data analysis (Verdonck et al. 2024). It should be noted that manual feature engineering is often challenging and time-consuming, requiring substantial effort and expertise (Verdonck et al. 2024).

The effectiveness of machine learning models depends on the representation of the feature vector and much of the effort in deploying these models focuses on designing preprocessing pipelines and applying data transformations (Bengio et al. 2013). Feature engineering typically expands the feature vector by introducing new features, which are derived through calculations based on existing features (Coates et al. 2011). These derived features can include products, differences, or other mathematical transformations that capture meaningful relationships in the data, similar to widely known features like body mass index (BMI), wind chill, or Triglyceride/HDL cholesterol ratio, which help to better understand and model complex interactions between variables (Heaton 2016).

Many feature engineering techniques exist, and it is not always clear which methods qualify as feature engineering and which do not (Verdonck et al. 2024). Additionally, there is a distinction between feature engineering, feature selection, and feature extraction, although these terms are sometimes used interchangeably (Verdonck et al. 2024). In the following sections, we will provide an overview of three of the most popular and often successful feature engineering techniques, as they could have more usage in tunneling projects. There are also other types of feature engineering including: feature engineering for multimedia data, network data, text data (Verdonck et al. 2024) that readers could consider for acquiring more in-depth knowledge in this field.

2.2.1 Decorrelation

In sensor data acquisition, it is common to encounter artifacts such as repetitions of past data, akin to echoes in acoustic signals (Duboue 2020). Decorrelation serves as a method to mitigate these artifacts. When domain knowledge suggests that the relationships of interest are inherently non-linear, linear dependencies between instances or features may be interpreted as artifacts resulting from the acquisition process (Duboue 2020). Decorrelation typically involves applying a linear filter to discount past data influences on the current data, effectively reducing these unwanted artifacts (Duboue 2020). In tunneling and rock engineering, for instance, geotechnical monitoring devices may be influenced by factors such as thermal expansion or gradual ground movement. These influences can cause the recorded displacement or stress distribution values to exhibit repeated or slowly changing patterns, which may introduce linear dependencies in the data. As a result, the true behavior of the ground may be obscured, leading to misleading interpretations if not properly addressed.

2.2.2 Smoothing

When a feature is affected by independent errors, these errors can significantly deviate the observed value from its true value, rendering it less effective for machine learning (Duboue 2020). To mitigate this, the feature value can be adjusted by comparing it to other feature values in the vicinity of the instance and moving it closer to those values. The rationale behind this approach is that independent errors tend to cancel each other out when aggregated, allowing the true signal to emerge. This technique, known as smoothing, is a foundational concept in statistics (Leek 2015). This method, which typically works best when dealing with features such as machine torque and thrust in a single project, will take care of the anomalies and instantaneous peaks and drops, and is highly recommended in cases of dealing with data recorded over time or space (Duboue 2020; Leek 2015).

2.2.3 Discretization

This approach focuses on identifying thresholds to segment continuous features into distinct intervals or categories (Duboue 2020). By doing so, it conveys to the machine learning model that differences within certain ranges, for example, between 15 and 45 degrees, are not significant for the specific task or domain. The goal of discretization is to define meaningful boundaries in the data such that the distribution of feature values within each interval is reasonably balanced (Duboue 2020). This method is particularly useful when small variations in densely populated regions of the data are considered more relevant than larger differences in sparse areas. Research has shown that this technique can improve the accuracy of machine learning predictions (Kotsiantis and Kanellopoulos 2006).

For instance, in tunneling and rock engineering, discretization can be applied to classify rock strength into categories such as weak, moderate, and strong based on uniaxial compressive strength values. This classification can help machine learning models focus on critical thresholds that influence tunnel stability, rather than treating minor variations in strength within these categories as equally important. This is achievable when combined with research in rock mechanic and rock engineering, which can provide meaningful ranges for a considerable number of influential indices. Alternatively, other methods were proposed to find these meaningful thresholds, for instance, the class-attribute interdependence maximization (CAIM) algorithm, which uses the mutual information between the feature intervals and the target class to provide the limits (Duboue 2020; Kurgan and Cios 2004). Simpler methods of discretization also include equal-width and equal-frequency binning, where the data are either encoded according to bins of equal size (e.g. strength bins of 5 MPa) or equal frequency (e.g. using the average of 5 consecutive datapoints).

2.2.4 Expansion

In some cases, significant value can be derived from generating new features based on existing ones, which can later be refined by applying feature selection techniques (Duboue 2020). This process might involve simple operations, such as multiplying or dividing two parameters, or more advanced methods like Kernel-Induced feature expansion. An example of the former is the Field Penetration Index (FPI; Farrokh et al. 2012, Hassanpour et al. 2011, Salimi et. al. 2022), or Penetration Index (PI; Villeneuve 2017) for TBMs, which are ratios of the net cutter thrust and the penetration rate. This index operates in an implicit higher-dimensional feature space, based on the concept that a problem appearing complex in a lower-dimensional space may be more easily solved with a simpler representation, such as a hyperplane, in a higher-dimensional space (Duboue 2020).

2.2.5 Univariate Feature Engineering

Univariate feature engineering for continuous variables serves as a valuable approach to enhance data characteristics such as symmetry, normality, or model fit. It is used to make the variable more suitable for modeling by addressing issues like skewness, non-linearity, or scale differences. The Box–Cox transformation, a widely recognized family of power transformations, is frequently employed for this purpose and is defined by the following formula (Osborne 2010; Verdonck et al. 2024):

$$f(x;\lambda) = \begin{cases} \frac{x^{\lambda}-1}{\lambda}, \lambda \neq 0 \\ \log(x), \lambda = 0 \end{cases} \tag{4}$$

In which x is the original dataset and λ is the transformation parameter. Considering different values would result in different scenarios; $\lambda = 1.00$: no transformation needed; produces results identical to original data; $\lambda = 0.50$: square root transformation; $\lambda = 0.33$: cube root transformation; $\lambda = 0.25$: fourth root transformation $\lambda = 0.00$: natural log transformation; $\lambda = -0.50$: reciprocal square root transformation; $\lambda = -1.00$: reciprocal (inverse) transformation, and so forth (Osborne 2010).

The Box–Cox transformation is monotonic and differentiable with respect to x, as well as monotonic and continuous with respect to its parameter. However, it can only be applied to strictly positive values. If the data contain non-positive observations, a positive constant can be added to shift all values into the positive range (Verdonck et al. 2024). Alternatively, the Yeo–Johnson power transformation can also be applied. The λ parameter for both the Box–Cox and Yeo–Johnson transformations is generally determined through maximum likelihood estimation but can also be set based on expert judgment or experimental testing (Verdonck et al. 2024).

2.2.6 Feature Extraction/Reduction

An equally important effort, complementing the previously discussed approaches, is to reduce the number of features, as uninformative features can burden machine learning models with unnecessary parameters. Excessive parameters may lead to suboptimal results by increasing the risk of overfitting or necessitating larger training datasets to achieve robust performance (Duboue 2020).

Dimensionality reduction is the process of finding a related representation for the input features into a lower dimensional space. The key is to find the meaningful subspace of the considered dataset and to operate inside of it. Finding this underlining structure within the data is also the primary objective of unsupervised learning and many dimensionality reduction techniques fall squarely within it (Brink et al. 2016). Moreover, feature reduction can be achieved through explicit feature selection, or by transforming sparse feature vectors into lower-dimensional, denser representations. This includes methods like Principal Component Analysis (PCA), which creates new, denser features, and explicit feature selection, which retains a subset of the original features. In contrast, feature selection specifically refers to choosing a subset of the original features without altering them. While feature selection preserves the interpretability of the original features, feature reduction via transformation may produce new features that capture essential patterns in the data but are not physical and thus are more difficult to interpret.

PCA

Principal Component Analysis (PCA) is a commonly used method for feature reduction, whereby it reduces data dimensionality by generating new features, known as principal components (PCs), which are linear combinations of the original variables (Verdonck et al. 2024). The first component captures the largest variance, with each subsequent component orthogonal to the previous ones and accounting for the next highest variance (Abdi and Williams 2010). These components are geometrically represented by the projections of observations onto them, known as factor scores, which allow for a clearer understanding of the data's intrinsic geometry and relationships (Abdi and Williams 2010). The PCs aim to capture the maximum variance in the original data while summarizing the information with fewer features, thereby simplifying the dataset while retaining essential variability (Verdonck et al. 2024). PCA serves to extract the most significant information from datasets, compress data size by focusing on key variances, simplify data representation, and analyze the structure of both observations and variables (Abdi and Williams 2010). In TBM performance prediction models and data analysis PCA is often used and it offers the rock strength (UCS), joint frequency, RQD, machine thrust (FN or TH), and cutterhead speed RPM as the most important input parameters in prediction of ROP, and hence the use of FPI as the objective function and UCS and RQD as the main input parameters is a very common theme in recent models (e.g. Hassanpour et al. 2015, Salimi et al. 2019).

Feature extraction

Feature extraction involves the application of mathematical or statistical transformations to compress or summarize raw data while preserving its most informative patterns, thereby facilitating its use in predictive modeling. It is particularly focused on automatically generating new representations of the data, often through dimensionality reduction techniques such as Principal Component Analysis (PCA), to yield a more compact and meaningful set of features. In image analysis, for instance, feature extraction may involve edge detection (using e.g., Sobel operator) to highlight object boundaries by identifying sharp changes in brightness, effectively reducing high-dimensional image data to a smaller set of informative variables (e.g., edges or outline). These extracted features can then serve as a foundation for further feature engineering, wherein new features are constructed to better characterize and distinguish elements within the data. Engineered features such as the area or perimeter of detected edges can offer more descriptive representations, which may enhance the detection and differentiation of objects such as voids and air bubbles in concrete structures, tumors in medical images, or chemical elements in different materials.

2.2.7 Domain Specific Feature Engineering

Feature engineering generally involves a data scientist drawing upon their domain expertise, engaging in iterative trial and error, and employing model evaluation to refine data features (Nargesian et al. 2017). However, in many instances, collaborating with individuals who possess substantial experience in the relevant field can significantly enrich the feature engineering process conducted by the data scientist. For instance, business experts might recognize that a peak in usage just before the expiry date of a service could serve as an early indicator that a customer is using up their remaining credit, signaling potential churn behavior (Verdonck et al. 2024). In human resources (HR) analytics, a binary feature might be developed indicating whether an employee has recently updated their LinkedIn profile. Such an update could be a precursor to the employee seeking new job opportunities, thus indicating a risk of churn (Verdonck et al. 2024).

The field of tunneling and rock engineering also benefits from feature engineering, with extensive work in recent decades aimed at introducing new indices and parameters to predict Tunnel Boring Machine (TBM) performance. For example, Yagiz (2008) considered the alpha angle to account for the effect of weakness planes on TBM efficiency:

$$\alpha = \arcsin\big(\sin\alpha_f.\sin(\alpha_t - \alpha_s)\big) \tag{5}$$

where α_f and α_s represent the dip and strike of the weakness planes, and α_t is the azimuth of the tunnel axis (in degrees). This is based on knowledge developed by earlier work (e.g. Bruland 2000, Macias et al 2014, Sanio 1985) showing the effect of anisotropy on TBM cutter performance. Hamidi et al. (2010) developed indices based on the Rock Mass Rating (RMR) and Rock Mass Quality Index (Q), while more complex indices, such as the Penetration Coefficient of (Bruland 2000) or the critical thrusts described in (Villeneuve 2017) and (Bruland 2000) were developed by investigating TBM performance data in minute detail. These parameters represent just a few of the options available in the literature on TBM tunneling, each composed of several sub-parameters designed with the expectation of improving predictions of TBM performance. Therefore, a rock engineering and tunneling expert could potentially identify and generate at least some significant features from the initial dataset of a case study, thereby enhancing the predictive capabilities for TBM performance.

Moreover, the inherent uncertainty within rock masses, coupled with the limited accuracy of empirically developed formulas applicable to slightly varied rock types, complicates the reliability of data from tunneling case studies. Qualitative indices such as the Geological Strength Index (GSI) can yield significantly different values when assessed by different experts, contributing further to this variability. These factors collectively impact the accuracy of the datasets collected from tunneling projects. This also underscores the necessity of integrating expert knowledge into feature selection and engineering in tunneling applications. Such an approach would likely yield outcomes more aligned with the actual performance of Tunnel Boring Machines (TBMs) than would relying solely on statistically or mathematically derived methods. Instead of merely applying thresholds or rigid models, incorporating domain expertise can adapt the feature engineering process to be more compatible with the nuanced realities encountered in tunneling operations.

2.3 *Feature Selection (FS)*

Feature selection (FS) and feature importance (FI) are highly connected processes and sometimes the border between them is blurry. In some cases, FS can be considered as a broader definition which uses FI as an internal step within its procedure. However, in some other cases, FI is considered as the main concept (Stańczyk and Jain 2014), as FS would be carried out to eventually find the finalized important features. But a possible differentiating point is that FI can be considered as an evaluation step, while, FS can be considered as an action-oriented step, which uses knowledge gained during evaluation to delete/select the unnecessary/necessary features.

The selection of influential features can often rely on domain knowledge. In the tunneling industry, domain expertise has provided insights into the importance of various variables, leading to the development of engineered indexes and proposed metrics. Leveraging this domain knowledge to interpret the results of feature importance (FI) analyses and to eliminate unnecessary features within the framework of feature selection (FS) is preferred.

A common source of confusion lies in the differing roles of expert knowledge in feature selection (FS) versus feature engineering (FE). To illustrate this distinction, consider a scenario where domain expertise leads to including the Brittleness Index (BI) in a dataset, based on the belief that it may significantly influence model predictions. If BI is not directly available, it might be derived using established formulas based on other available parameters, such as uniaxial compressive strength (UCS) and tensile strength, an approach characteristic of the FE phase, where new features are constructed from existing ones. In contrast, in a tunneling project situated within a massive rock mass, engineering judgment might lead to deprioritizing features such as Rock Quality Designation (RQD) or Discontinuity Spacing in the selection procedure, recognizing that they offer limited predictive value in such geological conditions. This decision could be strengthened by access to the results of FI, and which would be considered as the usage of expert knowledge in explicit FS.

However, when such knowledge is unavailable or insufficient, three primary approaches to feature selection, filters, wrappers, and embedded methods, are commonly utilized (Stańczyk and Jain 2014) and are described here.

2.3.1 Filter Methods

In the filter method, the model starts with all features and selects the best features subset based on statistical measures such as Pearson's correlation (Biesiada and Duch 2008), and mutual information (Kwak and Choi 2002). In other words, filter algorithms remove features that are unlikely to be helpful for later data analysis, no matter which prediction model is used. A basic filter method works by using a function that gives a relevance score J(S|D,C) (Biesiada and Duch 2007), showing how useful a group of features S is for a certain task C, based on the dataset D (Biesiada and Duch 2007). Since the dataset and task usually stay the same and only the group of features changes, this score is often written as J(S) (Biesiada and Duch 2007). The score can be calculated using simple methods, like a correlation value, or using other approaches such as entropy-based measures (Biesiada and Duch 2007).

2.3.2 Wrappers

Wrappers evaluate feature subsets by using a modeling algorithm as a black-box evaluator, often yielding better-performing subsets than filters because they rely on actual model performance (Jović et al. 2015). Feature selection using wrappers can proceed either incrementally or reductively. Forward feature selection begins with an empty set, adding features one at a time based on their contribution to improving a specific metric, a process also known as incremental construction (Duboue 2020). In contrast, backward feature selection starts with the full feature set, removing features iteratively by evaluating their impact on the target metric and eliminating those with the smallest negative effect, a technique referred to as an ablation study or Recursive Feature Elimination (RFE) (Duboue 2020; Guyon et al. 2002). Both methods can be streamlined using utility metrics when dealing with numerous features.

Wrapper methods for feature selection can follow either a greedy or non-greedy approach. Greedy methods select features based on the best immediate improvement in performance, which may lead to suboptimal results by settling at a local optimum (Kuhn and Johnson 2019). In contrast, non-greedy methods re-evaluate previous feature combinations and can explore initially unfavorable directions if they promise long-term benefits, allowing them to escape local optima and potentially achieve better overall outcomes (Kuhn and Johnson 2019).

Wrapper methods are computationally intensive, as they involve running the learning algorithm multiple times for various feature subsets, which becomes unfeasible with large datasets or complex training processes. Additionally, the subsets are biased toward the specific algorithm used for evaluation, making independent validation and alternative model testing essential for reliable generalization (Jović et al. 2015; Stańczyk and Jain 2014).

2.3.3 Embedded

Wrapper methods use learning algorithms to evaluate the quality of feature subsets and provide better results but at higher computational costs than filter methods (Jović et al. 2015; Venkatesh and Anuradha 2019). Embedded methods address this by integrating feature selection into the training process, reducing computational demands while ensuring that feature selection occurs during model learning (Venkatesh and Anuradha 2019). Common embedded approaches include various types of decision tree algorithms, such as random forests (Sandri and Zuccolotto 2006), as well as other methods like multinomial logistic regression and its variants (Cawley et al. 2006). Certain embedded methods utilize feature weighting through regularization models, employing objective functions designed to minimize fitting errors while simultaneously constraining feature coefficients to be small or precisely zero. Techniques such as Least Absolute Shrinkage and Selection Operator (LASSO) (Ma and Huang 2008) and Elastic Net (Zou and Hastie 2005) are commonly applied in this context, typically in conjunction with linear classifiers like Support Vector Machine and others, effectively penalizing features that do not contribute to the model.

The LASSSO is a regression analysis method that simultaneously performs parameter estimation and variable (model) selection (Muthukrishnan and Rohini 2016). LASSO is a specific form of penalized least squares regression that uses an L1-penalty function as a way of discouraging the model from relying on too many features. This technique acts as an embedded feature selection method by reducing the number of features used in the model, not by eliminating them from the training data, but by removing their impact on the model, for example by driving coefficients of unimportant features to zero (Duboue 2020).

It should be noted that Lasso and Recursive Feature Elimination (RFE) differ fundamentally in their approach to feature selection. Lasso integrates feature selection into the training process through L1 regularization, which can set some coefficients to zero, effectively selecting features during model optimization. This makes Lasso an embedded method where feature selection and model training occur simultaneously. In contrast, RFE employs a wrapper approach, iteratively training a model, evaluating feature importance, and removing the least significant features in separate steps, requiring multiple model fits. While Lasso provides a direct, computationally efficient way to select features within most commonly linear models, RFE offers flexibility by working with various algorithms but at the cost of additional computational effort due to its recursive nature.

3 Case Study: Feature Optimization of TBM Data from an Alpine Tunnel

To demonstrate the application of FI, FEng and FS, we conduct feature optimization on a dataset from a case study for a particular target variable. For this case study, feature optimization begins with domain specify feature engineering, whereby the features of a dataset are modified and expanded with both novel and commonly used TBM performance indices derived from many decades of conventional TBM performance analysis. Feature importance analysis is then conducted to identify relationships between features and the impact of different features on the target variable. Based on the results of this step, explicit feature selection is conducted to reduce the number of features after which automated feature selection is used to rank the remaining features. Another round of feature importance is then conducted, this time with linear regressors and permutation, complemented by a round of feature engineering to discretize the features. Feature engineering using feature extraction/ reduction is then applied to the raw dataset to determine which features should be carried through to the model development stage. Finally, domain knowledge is used to comment on the validity of the optimization results by relating them to the mechanics of TBM operation and performance and to make the final selection of features for modelling.

3.1 Dataset

3.1.1 Raw Data

The TBM excavation dataset containing the selection of TBM features listed in Table 1 was extracted from (Villeneuve 2008). While TBM data acquisition systems collect a great number of features which could be evaluated, the features for this case study were selected to provide indications of cutter-rock interaction. The TBM data were collected from a 9.58 m diameter gripper TBM with 62, 432 mm diameter cutters. The average cutter spacing was 77.25 mm, maximum thrust was 20 MN and the maximum torque was 8526 kNm. The TBM excavated through blocky to massive crystalline rock ranging from granite to schist and gneiss at an overburden of approximately 1500 m. Over a 385 m length of the tunnel, the TBM excavation data coincide with point load test data at roughly 1 m spacing, conducted on 50 mm diameter core drilled ahead of the TBM cutterhead. The point load tests were conducted both in the axial and diametral directions, allowing for calculation of strength anisotropy. Descriptions of the rock types along the 385 m length were grouped into categories labelled A-E and G-J, which are fully described in Villeneuve (2008).

To create the case study dataset on which to demonstrate feature optimization, the TBM data, collected at 10 s intervals, were pre-processed by removing data points

Table 1 Averaged TBM and geological features selected for feature optimization

Parameter	Units
Penetration rate (straight average)	mm/revolution
Thrust	kN
Torque	% of max
Torque	kNm
Mid-stroke tunnel meter	m
Stroke length	m
Speed	mm/min
Revolutions per minute	rev/min
Face stability class	Unstable, stable
Point load index, axial	MPa
Point load index, diametral	MPa
Point load index, anisotropy	–
Rock class	Categorical (A, B, C, D, E, G, H, I and J)

where the thrust was less than 4 MN to remove the effect of starting and stopping the TBM and then averaging for each stroke (approximately 2 m length), resulting in 192 datapoints for each feature. Face instability occurred over this 385 m interval and was thus recorded twice per shift by observation of the tunnel face through the cutterhead. Stable was assigned where depth of failure is 0, and unstable was assigned where depth of failure is > 0. Strokes without observations were assigned according to the category of the nearest observation, thus introducing some amount of error. The raw and pre-processed TBM data were also investigated in (Villeneuve 2008, 2017) to enable interpretation of TBM efficiency. The results therein will be discussed considering the results of the feature optimization presented with this case study.

The dataset is dominated by continuous features, however, there is one binary feature, face stability, and one categorical feature, rock class. Continuous methods for feature importance are applied to the full dataset, with the binary and categorical features pre-processed to encode them as integers to allow them to be utilized by the algorithms. The binary feature is assigned either a − 1 (unstable) or 1 (stable) so that each value of the features has the same weight in the algorithm.

In addressing the categorical variable "Rock Class," which comprises nine distinct categories within the dataset, a hybrid encoding approach was employed to mitigate challenges related to dimensionality, ordinal bias, and data leakage. Direct application of one-hot encoding would have expanded the feature space by nine additional dimensions, potentially compromising model performance due to the curse of dimensionality. Conversely, traditional label encoding risked introducing spurious ordinal relationships that certain algorithms might misinterpret as meaningful. Furthermore, target encoding, which assigns values based on the mean of the target variable for each category, posed a risk of data leakage by embedding target information into the features. To circumvent these issues, a hybrid encoding methodology was adopted. Initially, the dataset was partitioned into training and testing subsets to prevent leakage. Subsequently, target encoding was applied to the training set to compute the mean of the target variable for each "Rock Class" category, followed by label encoding of the categories on a scale from 0 to 8, ordered by the magnitude of these means (Table 2). This approach effectively minimized leakage while retaining the predictive utility of the encoded variable, balancing the trade-offs between encoding techniques.

3.1.2 Domain Specific Feature Engineering

The TBM features given in Table 1 were engineered by smoothing (as described in Sect. 3.1.1 for the averaging) and feature expansion to derive new features aimed at interpreting the cutter-rock interaction (Table 3). The active TBM time is the amount of time the cutterhead was rotating during the excavation of the stroke, and the number of revolutions is derived by multiplying the average RPM by the active TBM time (Villeneuve 2008). A second formulation for average penetration rate, termed "weighted" was calculated by dividing the stroke length by the total number

Table 2 Hybrid encoding of rock class: mean target variable and label assignments

Categorical values	Target encoding mean	Label encoding
G	23.66	0
A	24.91	1
B	29.36	2
C	30.51	3
H	31.01	4
D	31.21	5
J	32.83	6
E	32.96	7
I	34.85	8

of revolutions. The average thrust per stroke was univariately feature engineered first by removing the frictional effects, and second by dividing by the number of cutters. This results in a net cutter thrust friction removed value, which allows comparison between different TBMs operating with different shield forces and numbers of cutters. The shield friction was derived using a script to identify instances where the TBM head is moved freely (typically at the end of the stroke, when the TBM head is moved back slightly) and to determine an average value over several strokes (Villeneuve 2008). The net advance rate (NAR) was developed by (Villeneuve 2008) as a feature specifically to capture the impact of face instability on the TBM operation. It comprises a measure of the stroke length divided by the active TBM time, expressed in the same units as speed: mm/min. This was developed based on observations that when face instability is encountered the cutterhead will be rotated in the absence of thrust to clear fallen blocks from the tunnel face.

The final expanded features are the penetration index (PI; Villeneuve 2017), which is the average thrust minus shield friction divided by the penetration rate (straight average), and the drillability index (Villeneuve 2008), which is the inverse of PI, both of which and provide a measure of the ease or difficulty in excavating the rock (Villeneuve 2017). The PI is the same as the FPI (Farrokh et al. 2012) where both increase with rock and rock mass strength (Villeneuve 2017).

Table 3 Feature engineered TBM features

Parameter	Units
Active TBM time	min
Number of revolutions	–
Penetration rate (weighted average)	mm/revolution
Net thrust minus shield friction	kN/cutter
Net advance rate	mm/min
Penetration index	kN.rev/mm
Drillability index	mm/kN.rev

All data were standardized prior to conducting the feature importance analysis to ensure that the analysis was not influenced by the varying magnitudes or scales of the feature values, which could otherwise skew the results. For this purpose, the StandardScaler() function from the scikit-learn library was applied. This method transforms the data by subtracting the mean and dividing by the standard deviation for each feature, resulting in a standardized dataset with a mean of 0 and a standard deviation of 1.

3.1.3 Target Variable

To demonstrate the FI approach, the dataset is processed for the target variable net advance rate (NAR) in mm/min. This feature was selected as the target variable because it is a novel and complex feature for which the most important features to include in a predictive model is not clear. (Villeneuve 2008, 2017) highlighted that face instability will impact TBM operation by both improving the ability of the TBM to penetrate the rock (penetration index decreases), while also decreasing the efficiency of the excavation process due to large blocks at the cutterhead (net advance rate decreases). For example, in penetration tests at locations with stable and unstable face conditions, (Villeneuve 2017) showed in both strong (granite 1 and 4) and weak (schist 5 and 6) rocks (Fig. 1) that face instability allows the TBM to achieve optimal penetration rates with lower net cutter thrust than in rock masses with stable faces.

Villeneuve (2008) argued, however, that while face instability can increase penetration rate, its effects on utilization are negative, as captured by the comparison of net advance rate and speed. Speed is measured directly by the TBM data acquisition system as a function of the rate of extension of the thrust pistons during operation,

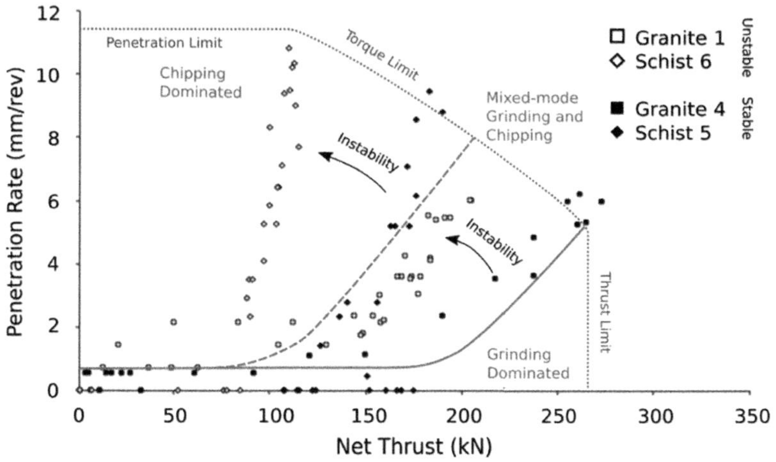

Fig. 1 Penetration curves for selected schist and granite samples under stable and unstable face conditions (from Villeneuve 2017)

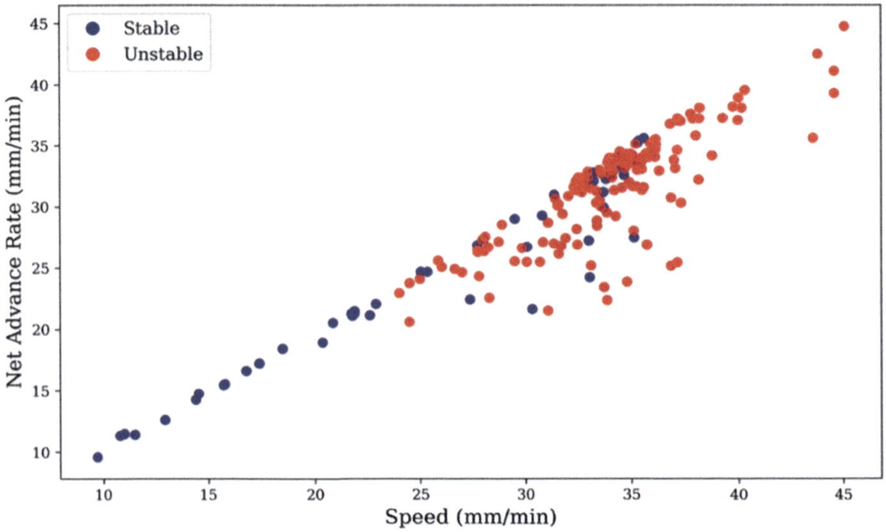

Fig. 2 Net advance rate compared to speed, color-coded according to face stability

and its average for the case study dataset was calculated only for the instances where thrust is greater than 4 kN, thus excluding instances where the TBM is not advancing. Conversely, the net advance rate is calculated as a function of stroke length and actual excavation time for instances where RPM is greater than 0, regardless of the thrust, thereby including instances where the TBM is not advancing (speed is 0 mm/min) but the cutterhead is rotating (the TBM is active and not merely stopped). This is to capture the pauses for cutterhead cleaning in unstable face conditions. Figure 2 highlights that, while the strokes with face instability tend to have higher NAR and speed in general, they also tend to have more occurrence of NAR lower than speed than those with stable faces. This means that utilization is being decreased, and the optimal speed is not achieved. Figures 1 and 2 highlight that there are complex relationships between face instability and TBM efficiency, providing an interesting case to explore how feature importance can be used to explore these features and identify those that could be most promising for developing a model.

3.2 Feature Importance by Correlation Matrix (Pearson and Spearman)

The correlations between each of the considered parameters are shown in Figs. 3 (Pearson) and 4 (comparison of Pearson and Spearman). The Pearson correlation heatmap highlights how strongly each feature relates linearly to the others. Many of these make mechanical sense, for example speed (mm/min) has a strong positive relationship with the straight average of penetration rate and NAR, and a strong negative relationship with number of revolutions, active TBM time and penetration index. Conversely, it has very weak relationships with torque, stroke length and point load index anisotropy. Indeed, stroke length is independent of all variables in the dataset except for weighted average penetration rate for which it is an input parameter. It also functions as an input parameter for the NAR at the feature engineering stage, but, along with tunnel meter, can otherwise be excluded for the dataset during feature selection. Interesting to highlight is the face stability, for which the two highest positive correlations are with thrust and penetration index, which are higher when the tunnel face is stable, and negative correlation with penetration rate, which is higher when the face is unstable. This matrix also highlights features which are duplicated, such as torque in % and torque in kNm, as well as average thrust and net cutter thrust. In this case, domain expertise can be used to determine which formulation of the features to keep during feature selection. For example, if only one TBM is being evaluated, average thrust may be sufficient.

A comparison of the Spearman and Pearson correlations for the target value, here NAR, reinforces the results of each. For example, in Fig. 4, the Spearman and Pearson coefficients for all features with respect to NAR are ranked from most negative to

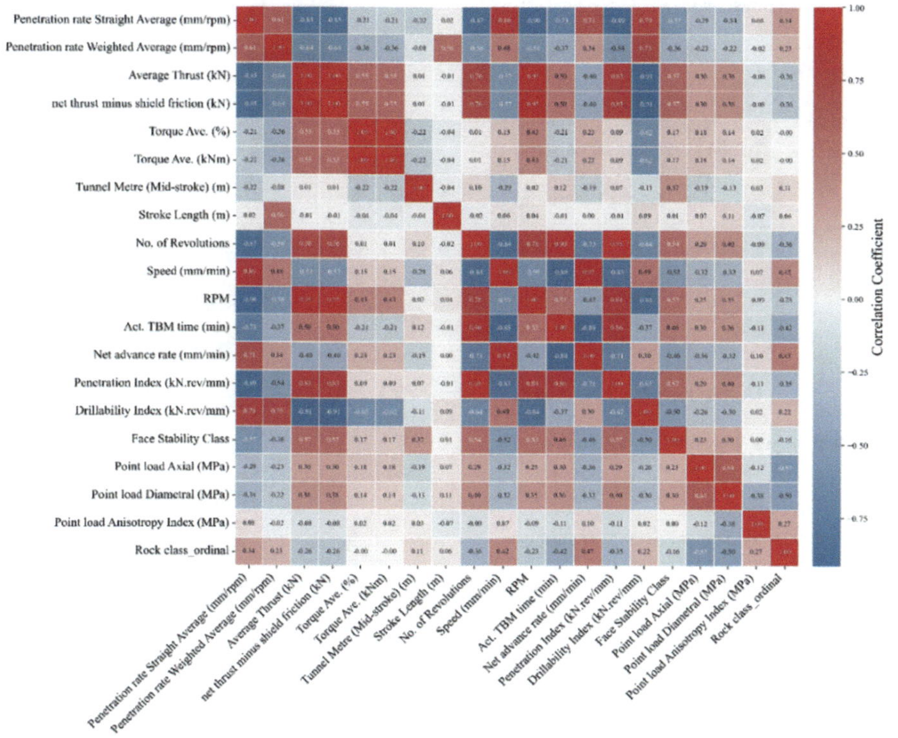

Fig. 3 Pearson correlation heatmap for the full case study dataset

most positive. Speed (mm/min) shows a strong positive linear (Pearson: 0.92) and monotonic (Spearman: 0.87) relationship, while penetration index demonstrates a high negative linear correlation (Pearson: – 0.71) but a much weaker negative monotonic trend (Spearman: – 0.22), suggesting that while the variables align linearly, the rank consistency might be disrupted by outliers or clusters (groups or subsets of data points within the dataset that exhibit a similar relationship among themselves but deviate from the overall trend when viewed as a whole).

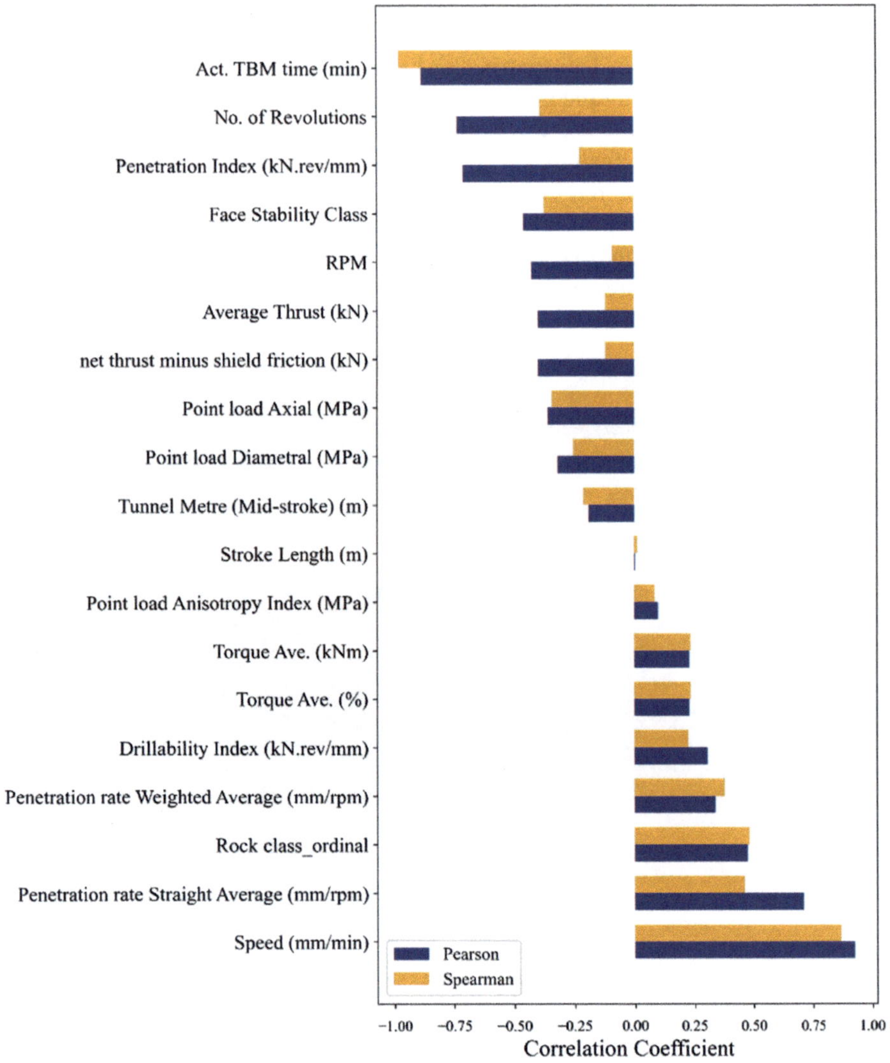

Fig. 4 Correlation comparison between Spearman and Pearson for the target variable net advance rate

For features like active TBM time, Pearson's (− 0.88) weaker correlation compared to Spearman's (− 0.98) indicates robust monotonicity even when deviations from linearity exist, which is interesting since active TBM time is an input parameter for calculating net advance rate. Face stability also shows a deviation between Spearman (− 0.37) and Pearson's (− 0.46). Also worth highlighting, are features with no or little importance, such as stroke length and tunnel meter (also referred to as chainage or face location). Neither of these should impact net advance rate, since stroke length is determined by the TBM design and on this machine was approximately 2 m, while tunnel meter is the location along the alignment where the TBM was located when the data were recorded. Indeed, stroke length does not impact net advance rate, however tunnel meter does have a very weak negative relation, suggesting that NAR decreases with increasing tunnel meter. This may be the result of the short distance (< 400 m) represented by the case study dataset whereby a monotonic variation in the geological conditions along the alignment is possible.

Over a longer distance, the geological conditions should change in a sufficiently random manner over the alignment as to reduce correlation between TBM performance features and tunnel meter. Conversely, if the overburden increases (or decreases) with tunnel meter, this could result in monotonic changes to TBM performance parameters and should be investigated via human expertise. Duplicated features, such as thrust (average and net) and torque (% and kNm), as well as highly similar features, such as penetration rate (straight and weighted average) have identical and similar correlation, respectively, helping to identify these features as candidates for reduction during feature selection. Ranking approaches such as this can help identify the important features more easily than using the confusion matrix, especially once the target feature has been identified.

3.3 Feature Selection

Selecting influential features often depends on domain knowledge, as shown for example with the explicit feature reduction. However, in cases where such knowledge is limited or unavailable, feature selection typically involves three main approaches: filters, wrappers, and embedded methods (Stańczyk and Jain 2014). For this case study, we use explicit feature selection to reduce the number of features, followed by wrappers and embedded methods to rank the remaining features according to how well they are related to or impact NAR.

3.3.1 Explicit Feature Selection

Explicit feature selection can be accomplished using the insights arising from the simplest forms of feature importance analysis. For the case study dataset, duplicated and highly similar features can be reduced by only keeping one of the features that relates to a particular aspect of TBM performance. For example, thrust and net thrust were already identified as duplicates using the correlation matrix. The Pearson and Spearman correlations also identified features that are not important for the target variable, such as stroke length, tunnel meter and point load anisotropy.

Prior to applying other methods for feature optimization, it is possible at this stage to select features that are most likely to result in a meaningful model. For the case study, duplicate parameters for thrust and torque, as well as stroke length, tunnel meter and point load anisotropy are excluded from further analysis. Drillability index is simply the inverse of penetration index and shows lower correlation with NAR and can thus be excluded. In addition, domain knowledge allows further exclusion of features used during feature engineering to calculate the feature of interest. In the case of NAR, these features are active TBM time and stroke length. Additionally, speed is similar to NAR, but unaffected by face instability (Fig. 2), as is penetration rate, and because these features risk obscuring the impact of other features they are also excluded.

3.3.2 Wrappers

Recursive Feature Elimination (RFE) is employed to systematically identify the most influential features for predictive modeling of the net advance rate. RFE operates by recursively training an external estimator, in this case, a linear regression model, on the dataset to assess feature importance (Pedregosa et al. 2011). The implementation involved simulating the dataset with 8 features, which were ranked based on their predictive contribution. Using RFE, the linear regression model iteratively eliminated less significant features to refine the selection. The rankings of all features are tabulated (Table 4), and the two top features could then be chosen for building a future regression model.

Table 4 Feature ranking and selection based on RFE

Feature	Ranking	Chosen
Average thrust	1	Yes
Penetration index	1	Yes
Rock class ordinal	2	No
Face stability class	3	No
Point load diametral	4	No
Point load axial	5	No
RPM	6	No
Torque average (%)	7	No

3.3.3 Embedded

We utilize LASSO regression, specifically LassoCV, as an embedded approach for feature selection within the broader context of machine learning model development. Lasso regression, characterized by its L1 regularization, inherently performs feature selection by driving some coefficients to zero, thereby removing the impact of less relevant features from the model. The application of LassoCV involves cross-validation to optimize the regularization parameter, alpha, ensuring that the model's performance is not only maximized but also generalized across different data subsets. This methodology is particularly beneficial for datasets where multicollinearity might exist, as it can discern which among correlated features contribute uniquely to the model's predictive power.

The Lasso coefficient magnitude for the case study dataset (Fig. 5) make it clear that only a subset of features is deemed significant for predicting NAR, where penetration index has a considerable coefficient, underscoring its linear relation to NAR under stable face conditions, as also observed with some of the other methods. The thrust also has a high coefficient, while features like rock class, face stability and point load index follow, suggesting they also have some likelihood of being related. However, torque and RPM are given low coefficients, implying they might not directly or linearly affect NAR in the Lasso context. This is similar to the ranking produced by RFE, and based on these findings, penetration index and average thrust would be selected for modelling.

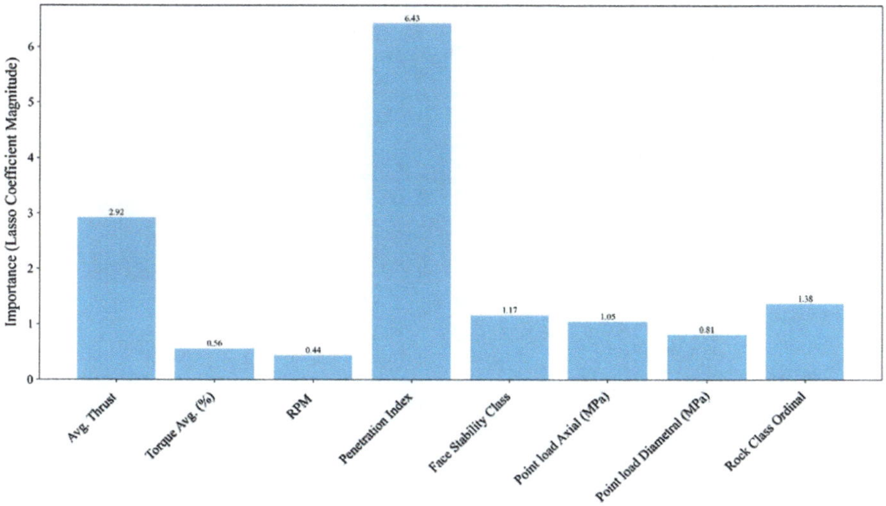

Fig. 5 Lasso coefficient magnitude for the 8 features

3.4 Feature Importance by Regressors and Permutation

The regressors XGBoost, random forest and decision tree along with permutation were applied to show the magnitude of each feature's contribution in prediction of the target variable NAR when fitting these algorithms on the data set. Similar to the Lasso coefficient magnitude, these algorithms produce a "scaled importance" measure, which can be listed or displayed in a heatmap as in Fig. 6.

The results show a strong consensus on the importance of penetration index. In all three regressor methods, other features have notably lower importance, implying that once penetration index is accounted for, other features have a minimal additional impact, or their effects are already captured by this top feature. The permutation method, in contrast, also highlights face stability, followed by rock class as being more important than the other features, and in general a broader range of importance, showing its ability to identify different relationships than methods founded on regression principles, and its potential to provide broader insight into the relationships between features. This is especially true for nonlinear or complex relationships, such as those involving NAR.

Important to note is that during the iterative process, when the regressors were applied to the dataset with all features, then a reduced set of 10, 9 and finally 8 features, they tended to converge on a single feature with extreme importance, while all other features had markedly less importance. When the feature with extreme dominance was removed from the dataset, another feature took its place as the feature of extreme

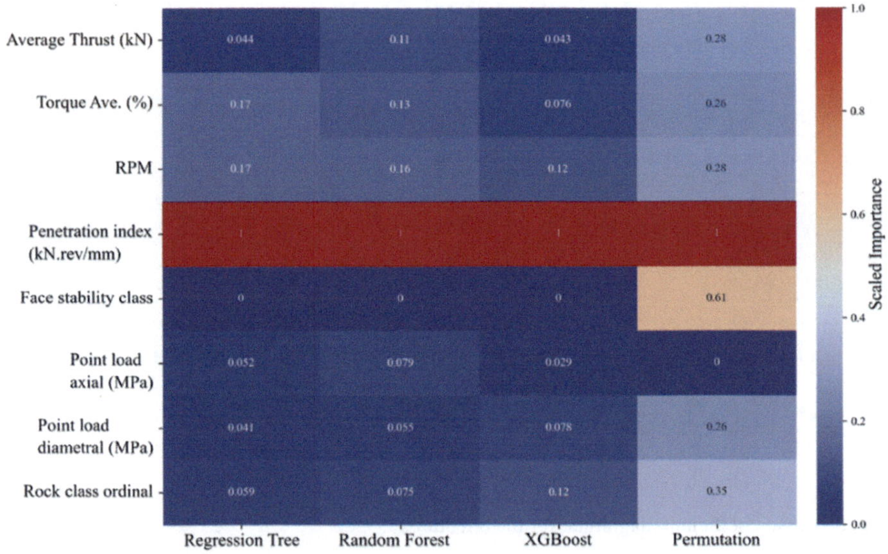

Fig. 6 Heatmap of each feature's contribution on the target variable net advance rate using four different regressor algorithms; note that each contribution type was scaled for comparison across all types

importance. This can be a sign of overfitting, high cardinality and redundancy. As explained at the explicit feature selection stage, some redundant features (e.g. the similar thrusts) and some features used to calculate the target feature (e.g. active TBM time) were excluded from the dataset. Nevertheless, as seen in Fig. 6 the regressors still converge on one feature with extreme dominance. This could be a result of the high cardinality of the numerical features in the dataset, which can lead to splits that perfectly separate the data at a node, resulting in a reduction in impurity. The model is then essentially overfitting to noise in the data. A way to mitigate this could be to transform high-cardinality features. To demonstrate this, we applied equal-width binning to the 6 numerical features in the dataset and show that, while the regressors still identify penetration index as the most important feature, there is now better differentiation between the other features (Fig. 7).

Another way to interpret feature importance from various methods is to derive an overall feature importance metric of all features. For the case study, dataset we derived a scaled average feature importance metric for NAR by normalizing feature importance scores from permutation, XGBoost, random forest, and decision tree models using MinMaxScaler to a 0–1 range and computing their mean across models for each feature, with the resulting ranking shown in Fig. 8. The outcome for the raw values shows that penetration index is ranked first, while all other features are ranked similarly and much lower (Fig. 8a). Here torque and RPM are ranked above the other features, which is not consistent with the ranking by the feature selection (wrapper and embedded) methods, which assigned a lower ranking to these features. For the

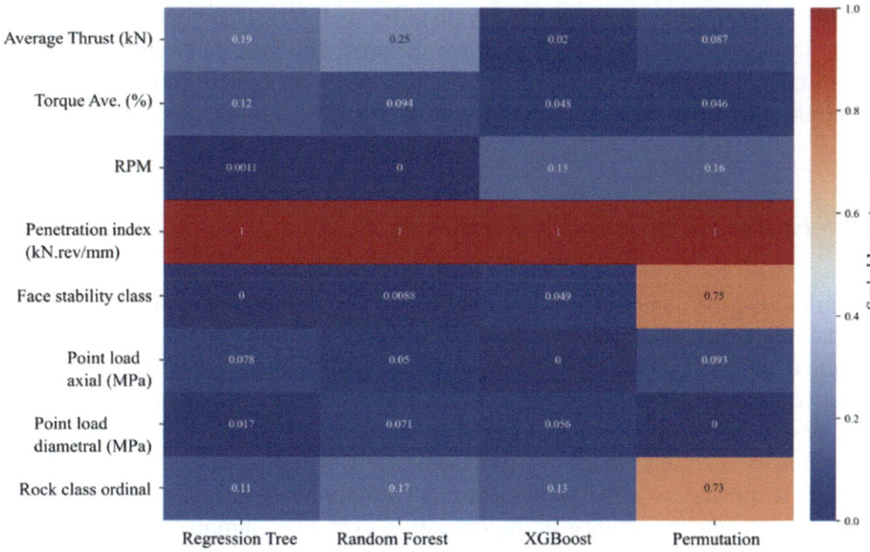

Fig. 7 Heatmap of each feature's contribution on the target variable net advance rate using four different regressor algorithms; note that each contribution type was scaled for comparison across all types and that the features were discretized using equal-width binning

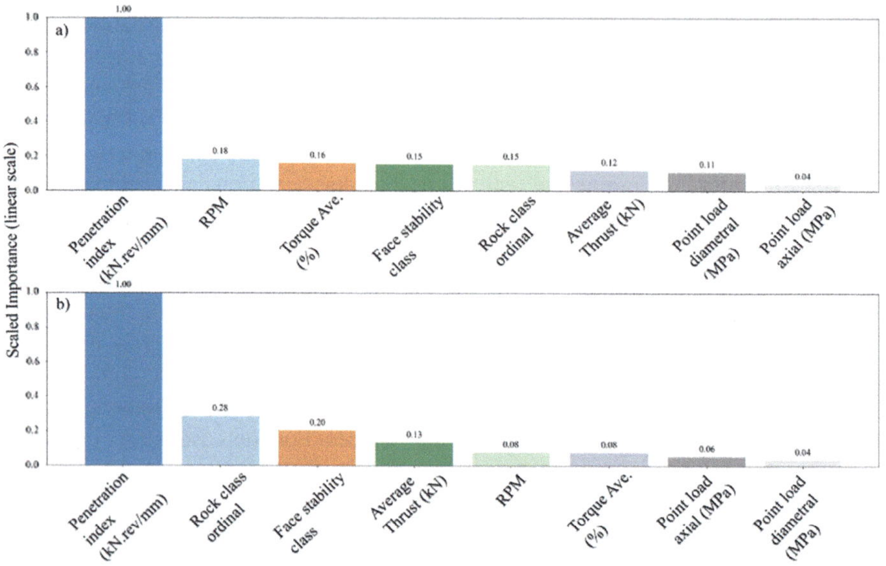

Fig. 8 Top ten features using regressors and permutation (average values); **a** raw values; **b** values discretized using equal-width binning

dataset discretized using equal-width binning, there is better differentiation between the features (Fig. 8b), and the top four features are consistent with the top four features identified by the wrapper and embedded methods. This highlights again that discretization can be a necessary step for features with high cardinality in order to obtain meaningful and consistent results. Based on the scaled importance metric, penetration index would be selected for modelling.

3.5 *Feature Extraction/reduction by PCA*

Feature reduction can also be accomplished while retaining many or all features in the dataset, especially where explicit reduction is not possible or straightforward. Principal component analysis (PCA) can be used for feature reduction in two ways: to derive new principal components (PCs), which can then be used for further analysis, or to seek highly correlated or dependent features to aid with explicit feature reduction. PCA results for the raw (not discretized) features of the case study dataset are first displayed as a PC versus cumulative explained variance plot (Fig. 9) to illustrate how much variance in the dataset is explained cumulatively as principal components (PCs) are added. For this dataset nearly 90% of the variance can be explained by 4 PC's, which could then be used for further analysis, for example for regression, while a higher number of PC's could cause problems with overfitting, especially for small datasets, such as this one.

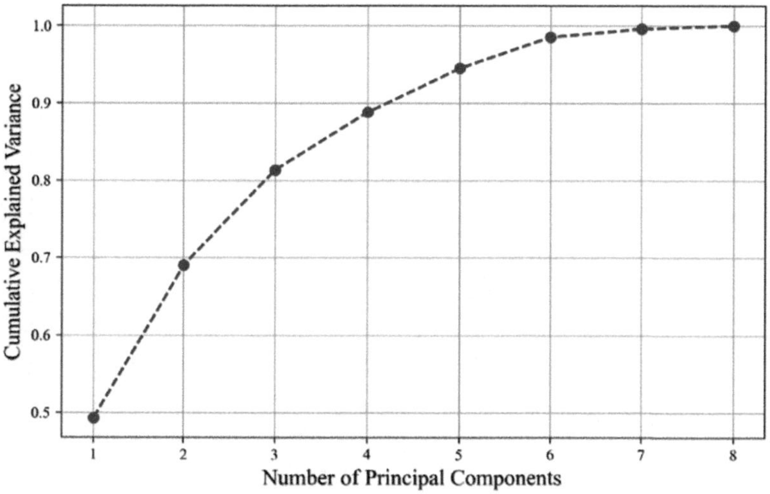

Fig. 9 Cumulative variance in the target variable explained by the principal components

The 3D scatter plot of the first three principal components from PCA (Fig. 10) provides a comprehensive visualization of the data structure in three dimensions. Each point is color coded according to the magnitude of the target variable. The principal components are directions of maximum variance; PC1 captures the largest variance, and PC2, orthogonal to PC1, captures the next largest, etc. Clusters of points suggest groupings within the data, with concentrations like those in the upper-left quadrant indicating similar characteristics among observations. When points of similar color cluster together, it might indicate that certain values of the dependent variable are associated with specific combinations of PC scores. This suggests inherent patterns or groupings within the original high-dimensional data which is captured by the PCA. There also appears to be a monotonic increase in the magnitude of the target variable with increasing PC magnitude, indicating that there are underlying structures or categories within the data that are preserved even after dimensionality reduction, potentially revealing latent factors or classes that could be crucial for further predictions.

Eigenvector plots allow visualization of the impact of each feature on the principal components, with the length and direction providing meaningful information (Fig. 11). In the PC1 versus PC2 plot, representing nearly 70% of the variance, the rock class and point load have the strongest, but opposing impacts, while the other features have weaker and orthogonal impacts. This exact pattern is not repeated for the other PC plots, with the eigenvector lengths and directions changing for each plot. Some patterns do emerge, however. Rock class is opposite to strength (as measured by point load) for the first three PCs, and both are typically orthogonal to torque. This suggests that strength is linked to rock characteristics, but both are not correlated to torque. The directions of the face stability class and penetration rate eigenvectors tend to coincide, suggesting that stable face is related to increased penetration index, as

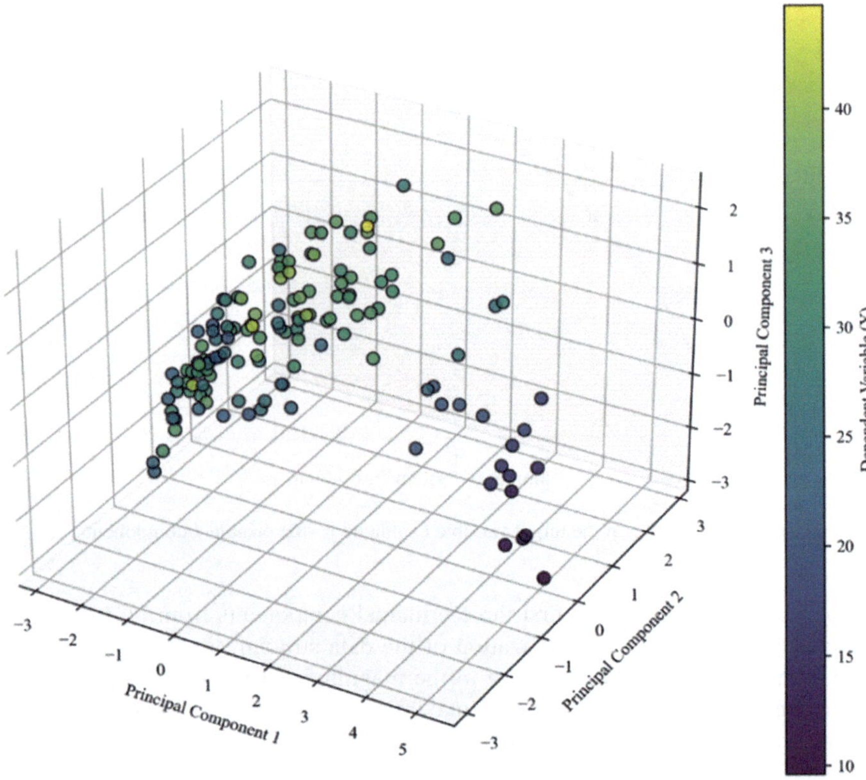

Fig. 10 3D scatter plot of the first three principal components color coded according to magnitude of the net advance rate

supported by Fig. 1. Thrust and RPM eigenvectors also tend to coincide, suggesting that the cutterhead is rotated faster when the rock is stronger, possibly to offset low penetration rates (as represented by the also often parallel penetration index eigenvector). Some relationships tend to hold over the first three PCs (accounting for over 80% of the variance), highlighting further potential for explicit feature reduction; for example, the point load (diametral) eigenvector has a consistently similar direction but smaller magnitude than the point load (axial) eigenvector, and could be excluded. Torque, which does not appear to have a high importance in the previous metrics, does appear to have a unique (often orthogonal) impact on the PCs, suggesting that it should not necessarily be excluded from a future model. However, torque is a feature that can be calculated from thrust and penetration rate, so it is not necessarily a fully independent parameter.

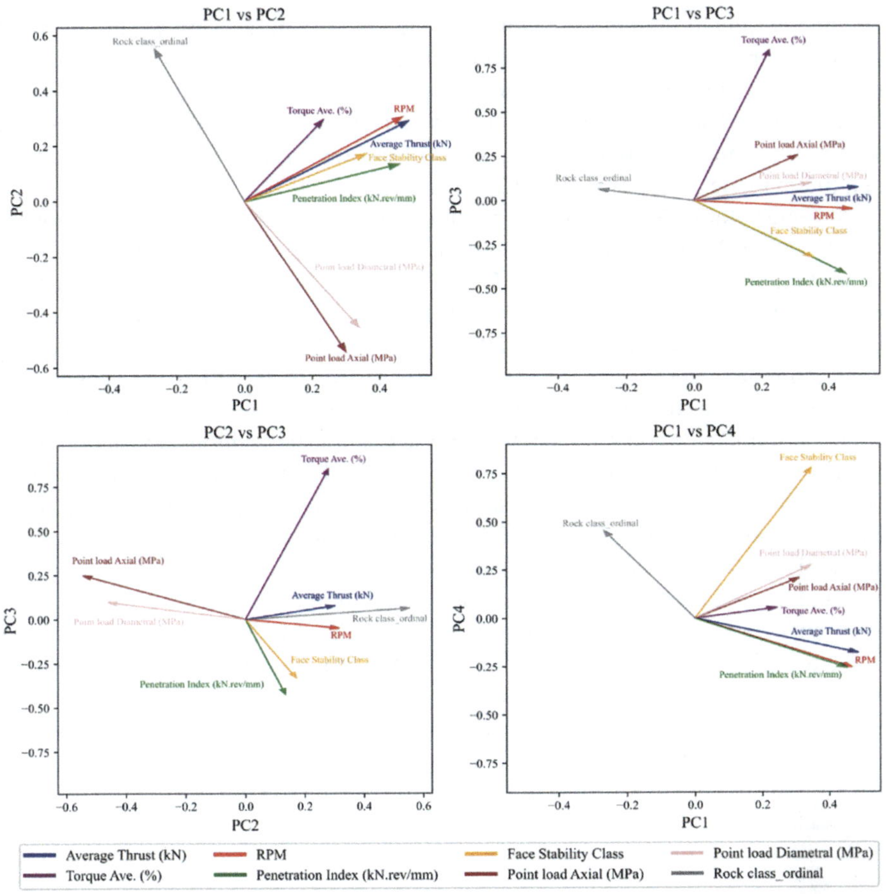

Fig. 11 Composite of selected biplots of feature eigenvectors of the first four PCs resulting from principal component analysis

3.6 Feature Selection Using Domain Expertise

In the tunneling industry, enriched domain expertise and expert judgment have provided valuable insights into the importance of various variables, leading to the development of engineered indexes and proposed metrics. In the presentation of feature optimization via this case study we have also highlighted aspects of feature engineering and explicit feature reduction, both of which require domain knowledge. The authors also strongly recommend leveraging domain knowledge to interpret the results of feature importance (FI) and feature reduction (as a part of feature extraction) analyses and to exclude unnecessary features within the framework of feature selection (FS).

The interpretation of these results, and the subsequent decision regarding which features to then use for model development rests on the domain expert. The feature importance, reduction and extraction provided a consistent top ranked feature, penetration index. The second and lower ranked features, however, are not consistent across all methods, and determining which of the lower ranked features are included in a future model requires expert judgement.

A scatter plot of NAR versus penetration index highlights the clear relationship visible at low NAR and high penetration index, dominated by stable face conditions (Fig. 12). Similarly to the NAR-speed relationship in Fig. 2, this relationship is related to ease of excavation, where higher strength leads to higher penetration index, as explored by Villeneuve (2017) (Fig. 13). Under these stable face conditions, a simple regression analysis should suffice to relate strength to both penetration index and NAR, as the relationship appears relatively consistent with low variability. Under unstable face conditions, however, the relationship exhibits a visually distinguishable non-linear inverse trend, with the range of NAR values decreasing as the penetration index increases. To accurately capture this complex relationship, more advanced models, such as polynomial regression or Random Forests, are necessary, as additional factors likely influence the trend.

The feature importance methods using correlation and regressors were not able to identify face stability as an important feature related to NAR, although permutation did select it as its second-most important feature. The feature selection methods ranked face instability in the middle, below penetration index, rock class and thrust. Strength, as measured by point load testing, consistently ranked low using all measures. While strength is a dominant feature for penetration index, it is not for NAR because of the NAR reduction under unstable face conditions.

Based on these metrics, a domain expert would suggest that a model should not be built with just the highest ranked feature, but the question remains: how many more

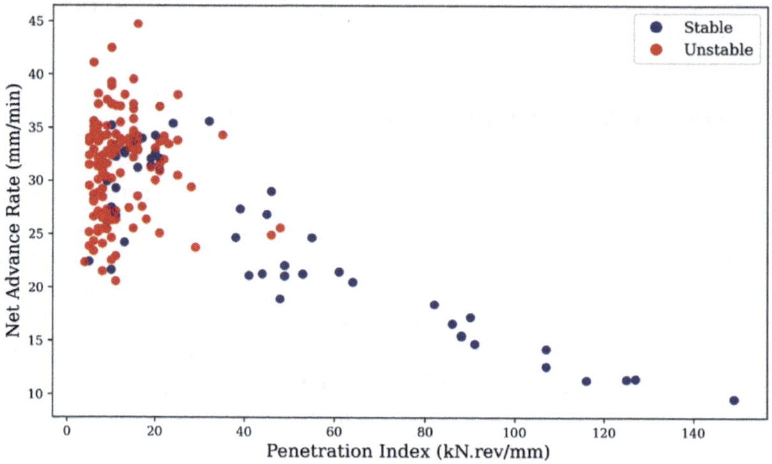

Fig. 12 Net advance rate versus penetration index, color-coded according to face stability

Fig. 13 Comparison of UCS versus penetration index from penetration tests in schist and granite (after Villeneuve 2017)

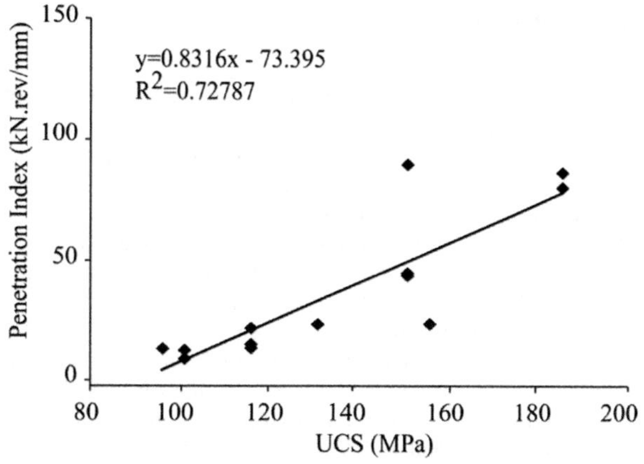

are valuable? The metrics are not fully in agreement, thus complicating the final feature selection. Here PCA can provide insight, by identifying correlated features, one of which could be excluded, for example point load (diametral). An argument could also be made for excluding one of thrust or RPM, as well as one of penetration index or face stability class. Here the domain expert must decide what question is to be answered by the model. Is the model meant to show that face instability causes decreased utilization, then the simplest model, which excludes features correlated to face stability (i.e. PI) could be useful. Another question could be whether decreased utilization can be predicted using only geological parameters, in which case a model excluding TBM performance parameters (thrust, torque, PI, RPM) would be best.

These results are based on the analysis of data for a single TBM and if the machine diameter and configuration were to change, the result of the analysis and exclusion of some features could be different. The results are not meant to be widely adopted, but rather to illustrate the feature optimization approach. Conducting feature optimization on a dataset including various machine types and sizes, as well as different ground conditions could reveal generalizable knowledge about which features are best for developing a predictive model applicable for a wide range of TBMs and ground conditions.

The case study demonstrates the iterative nature of feature optimization, and highlights the value gained, not only in reducing the number of features, but also of better understanding their relationships with each other. This was discussed within the context of previously published conventional analyses on the same dataset. The importance of domain expertise for the final feature selection should be stressed, especially in light of the potential conflicting results from the various optimization techniques, especially for limited datasets.

References

Abdi H, Williams LJ (2010) Principal component analysis. Wiley Interdiscip Rev: Comput Stat 2(4):433–459

Alishvandi A, Karimi J, Damari S, Moayedi Far A, Setodeh Pour M, Ahmadi M (2024) Estimating the compressive strength of plastic concrete samples using machine learning algorithms. Asian J Civ Eng 25(2):1503–1516

Alvo M, Cabilio P (1995) Rank correlation methods for missing data. Can J Stat 23(4):345–358

Bengio Y, Courville A, Vincent P (2013) Representation learning: a review and new perspectives. IEEE Trans Pattern Anal Mach Intell 35(8):1798–1828

Biesiada J, Duch W (2007) A Kolmogorov-Smirnov correlation-based filter for microarray data. In: International conference on neural information processing, pp 285–294

Biesiada J, Duch W (2008) Feature selection for high-dimensional data—a Pearson redundancy-based filter. In: Computer recognition systems, vol 2. Springer, pp 242–249

Breiman L (2001) Random forests. Mach Learn 45:5–32

Breiman L (2002) Manual on setting up, using, and understanding random forests v3. 1. Statistics Department University of California Berkeley, vol 1(58), CA, USA, pp 3–42

Brink H, Richards J, Fetherolf M (2016) Real-world machine learning. Simon and Schuster

Bruland A (2000) Hard rock tunnel boring. Fakultet for ingeniørvitenskap og teknologi.

Cawley G, Talbot N, Girolami M (2006) Sparse multinomial logistic regression via bayesian l1 regularisation. Adv Neural Inf Process Syst 19

Coates A, Ng A, Lee H (2011) An analysis of single-layer networks in unsupervised feature learning. In: Proceedings of the fourteenth international conference on artificial intelligence and statistics, pp 215–223

Cohen I, Huang Y, Chen J, Benesty J, Chen J, Huang Y, Cohen I (2009) Pearson correlation coefficient. Noise Reduct Speech Process 1–4

Cover TM (1999) Elements of information theory. John Wiley & Sons

Crysdian C (2022) The behaviour of rank correlation coefficients for incomplete data. Res Math 9(1):2107793

Deng J, Deng Y, Cheong KH (2021) Combining conflicting evidence based on Pearson correlation coefficient and weighted graph. Int J Intell Syst 36(12):7443–7460

Devi MS, Mathew RM, Suguna R (2019) Regressor fitting of feature importance for customer segment prediction with ensembling schemes using machine learning. Int J Eng Adv Technol 8(6):952–956

Duboue P (2020) The art of feature engineering: essentials for machine learning. Cambridge University Press

Farrokh E, Rostami J, Laughton C (2012) Study of various models for estimation of penetration rate of hard rock TBMs. Tunn Undergr Space Technol 30:110–123

Goh ATC, Zhang Y, Zhang R, Zhang W, Xiao Y (2017) Evaluating stability of underground entry-type excavations using multivariate adaptive regression splines and logistic regression. Tunn Undergr Space Technol 70:148–154

Guyon I, Weston J, Barnhill S, Vapnik V (2002) Gene selection for cancer classification using support vector machines. Mach Learn 46:389–422

Hamidi JK, Shahriar K, Rezai B, Rostami J (2010) Performance prediction of hard rock TBM using rock mass rating (RMR) system. Tunn Undergr Space Technol 25(4):333–345

Hansen TF, Erharter GH, Marcher T (2024) Towards reinforcement learning-driven TBM cutter changing policies. Autom Constr 165:105505

Hassanpour J, Rostami J, Zhao J (2011) A new hard rock TBM performance prediction model for project planning. Tunn Undergr Space Technol 26(5):595–603

Hassanpour J, Rostami J, Zhao J, Tarigh Azali S (2015) TBM performance and disc cutter wear prediction based on ten years experience of TBM tunnelling in Iran. Geomech Tunn 8(3):239–247

Heaton J (2016) An empirical analysis of feature engineering for predictive modeling. SoutheastCon 2016:1–6

Hosmer DW, Lemeshow S, Sturdivant RX (2013) Applied logistic regression. John Wiley & Sons

Hwang SW, Chung H, Lee T, Kim J, Kim Y, Kim JC, Kwak HW, Choi IG, Yeo H (2023) Feature importance measures from random forest regressor using near-infrared spectra for predicting carbonization characteristics of kraft lignin-derived hydrochar. J Wood Sci 69(1):1

Jović A, Brkić K, Bogunović N (2015) A review of feature selection methods with applications. In: 2015 38th International convention on information and communication technology, electronics and microelectronics (MIPRO), pp 1200–1205

Kendall MG (1938) A new measure of rank correlation. Biometrika 30(1–2):81–93

Kotsiantis S, Kanellopoulos D (2006) Discretization techniques: a recent survey. GESTS Int Trans Comput Sci Eng 32(1):47–58

Kraskov A, Stögbauer H, Grassberger P (2004) Estimating mutual information. Phys Rev E—Stat Nonlinear Soft Matter Phys 69(6):066138

Kuhn M, Johnson K (2019) Feature engineering and selection: a practical approach for predictive models. Chapman and Hall/CRC

Kurgan LA, Cios KJ (2004) CAIM discretization algorithm. IEEE Trans Knowl Data Eng 16(2):145–153

Kwak N, Choi CH (2002) Input feature selection by mutual information based on Parzen window. IEEE Trans Pattern Anal Mach Intell 24(12):1667–1671

Leek J (2015) The elements of data analytic style. Leanpub, Victoria, BC, Canada, p 93

Louppe G, Wehenkel L, Sutera A, Geurts P (2013) Understanding variable importances in forests of randomized trees. Adv Neural Inf Process Syst 26

Ma S, Huang J (2008) Penalized feature selection and classification in bioinformatics. Brief Bioinform 9(5):392–403

Macias FJ, Jakobsen PD, Seo Y Bruland A (2014) Influence of rock mass fracturing on the net penetration rates of hard rock TBMs. TunnLing Undergr Space Technol 44:108–120

Macias FJ, Dahl F, Bruland A (2016) New rock abrasivity test method for tool life assessments on hard rock tunnel boring: the rolling indentation abrasion test (RIAT). Rock Mech Rock Eng 49:1679–1693

May JO, Looney SW (2020) Sample size charts for Spearman and Kendall coefficients. J Biom Biostat 11(2):1–7

Moayedi Far A, Zare M (2025) Ensemble-based soil liquefaction assessment: leveraging CPT data for enhanced predictions. Civ Eng Des 7(1):23–35

Monedero I, Biscarri F, León C, Guerrero JI, Biscarri J, Millán R (2012) Detection of frauds and other non-technical losses in a power utility using Pearson coefficient, Bayesian networks and decision trees. Int J Electr Power Energy Syst 34(1):90–98

Morgenroth J, Unterlaß PJ, Sapronova A, Khan UT, Perras MA, Erharter GH, Marcher T (2022) Practical recommendations for machine learning in underground rock engineering–on algorithm development, data balancing, and input variable selection. Geomech Tunn 15(5):650–657

Muthukrishnan R, Rohini R (2016) LASSO: a feature selection technique in predictive modeling for machine learning. IEEE Int Conf Adv Comput Appl (ICACA) 2016:18–20

Nargesian F, Samulowitz H, Khurana U, Khalil EB, Turaga DS (2017) Learning feature engineering for classification. Ijcai 17:2529–2535

Osborne J (2010) Improving your data transformations: applying the box-cox transformation. Pract Assess Res Eval 15(1)

Pedregosa F, Varoquaux G, Gramfort A, Michel V, Thirion B, Grisel O, Blondel M, Prettenhofer P, Weiss R, Dubourg V (2011) Scikit-learn: machine learning in Python. J Mach Learn Res 12:2825–2830

Ross BC (2014) Mutual information between discrete and continuous data sets. PLoS ONE 9(2):e87357

Rostami J (2016) Performance prediction of hard rock tunnel boring machines (TBMs) in difficult ground. Tunn Undergr Space Technol 57:173–182

Saadati G, Javankhoshdel S, Mohebbi Najm Abad J, Mett M, Kontrus H, Schneider-Muntau B (2024) AI-powered geotechnics: enhancing rock mass classification for safer engineering practices. Rock Mech Rock Eng 1–31

Salimi A, Rostami J, Moormann C (2019) Application of rock mass classification systems for performance estimation of rock TBMs using regression tree and artificial intelligence algorithms. TunnLing Undergr Space Technol 92:103046

Salimi A, Rostami J, Moormann C, Hassanpour J (2022) Introducing tree-based-regression models for prediction of hard rock TBM performance with consideration of rock type. Rock Mech Rock Eng 55:4869–4891

Sandri M, Zuccolotto P (2006) Variable selection using random forests. In: Data analysis, classification and the forward search: proceedings of the meeting of the classification and data analysis group (CLADAG) of the Italian Statistical Society, University of Parma, June 6–8, 2005, pp 263–270

Sanio HP (1985) Prediction of the performance of disc cutters in anisotropic rock. Int J Rock Mech Min Sci Geomech Abstr 22(3):153–161

Spearman C (1904) The proof and measurement of association between two things

Stańczyk U, Jain LC (2014) Feature selection for data and pattern recognition: An introduction. In: Feature Selection for Data and Pattern Recognition. Springer, pp 1–7

van den Heuvel E, Zhan Z (2022) Myths about linear and monotonic associations: Pearson's r, Spearman's ρ, and Kendall's τ. Am Stat 76(1):44–52

Venkatesh B, Anuradha J (2019) A review of feature selection and its methods. Cybern Inf Technol 19(1):3–26

Verdonck T, Baesens B, Óskarsdóttir M, vanden Broucke S (2024) Special issue on feature engineering editorial. Mach Learn 113(7):3917–3928

Villeneuve M (2008) Examination of geological influence on machine excavation of highly stressed tunnels in massive hard rock

Villeneuve MC (2017) Hard rock tunnel boring machine penetration test as an indicator of chipping process efficiency. J Rock Mech Geotech Eng 9(4):611–622

Wang GJ, Xie C, Stanley HE (2018) Correlation structure and evolution of world stock markets: evidence from Pearson and partial correlation-based networks. Comput Econ 51:607–635

Wang Q, Geng P, Wang L, He D, Shen H (2024a) Machine learning-driven feature importance appraisal of seismic parameters on tunnel damage and seismic fragility prediction. Eng Appl Artif Intell 137:109101

Wang T, Fan H, Wang K, Wang L, Zhou G (2024b) Reliability analysis and risk assessment for settlement of cohesive soil layer induced by undercrossing tunnel excavation. Sustainability 16(6):2356

XGBoost: Scalable and Flexible Gradient Boosting (2024). https://xgboost.readthedocs.io/en/latest/index.html

Yagiz S (2008) Utilizing rock mass properties for predicting TBM performance in hard rock condition. Tunn Undergr Space Technol 23(3):326–339

Zhang L, Wang L (2023) Optimization of site investigation program for reliability assessment of undrained slope using Spearman rank correlation coefficient. Comput Geotech 155:105208

Zou H, Hastie T (2005) Regularization and variable selection via the elastic net. J R Stat Soc Ser B Stat Methodol 67(2):301–320

Other TBM Performance Prediction Methods

Jafar Hassanpour and Jamal Rostami

1 Introduction

To date, there have been substantial efforts to develop models for predicting the performance of Tunnel Boring Machines (TBM) and understanding the impact of various rock mass parameters on machine performance. These efforts have led to the creation of several machine performance prediction models, with some of the most prominent ones including (see previous chapters):

- The CSM model from the Colorado School of Mines (Rostami and Ozdemir 1993; Rostami 1997).
- The NTNU model from the Norwegian University of Science and Technology (Blindheim 1979; Bruland 1998; Macias 2016).
- The Q_{TBM} model (Barton 2000).
- UT model (Hassanpour et al. 2023a).

While these models provide valuable insights, they also have limitations that hinder their practical application. For example, the original CSM model has been developed based on the full-scale laboratory cutting tests using disc cutter and does not consider the impact of discontinuities on the machine penetration rate. Although some revised versions have attempted to address this issue by incorporating the effects of rock mass discontinuities (Cheema 1999; Yagiz 2002; Ramezanzadeh 2005), challenges remain. The NTNU model depends on input parameters that require specific laboratory equipment (Norwegian suite, DRI, AVS, CLI, …), and while these tests

J. Hassanpour (✉)
University of Tehran, Tehran, Iran
e-mail: hassanpour@ut.ac.ir

J. Rostami
Colorado School of Mines, Golden, CO, USA
e-mail: rostami@mines.edu

J. Hassanpour and J. Rostami (eds.), *Performance of Tunnel Boring Machines (TBM) in Rock*, Springer Series in Geomechanics and Geoengineering,
https://doi.org/10.1007/978-3-032-20424-0_8

are becoming more common, the pertinent information is not available in many of the projects. Furthermore, the Q_{TBM} model, which is based on the Q-system, involves numerous input parameters such as the Norwegian test suite to characterize the rock and joint properties to represent the rock mass, making its application complex. Meanwhile, the UT model, developed based on field data from various tunneling projects mainly in Iran, with limited range of geology and machine parameters, such as disc cutter diameter. In summary, while the models that are proposed for predicting machine performance offer useful insights, further refinement is necessary to overcome their limitations and improve their practical use in tunneling projects. Detailed descriptions of selected models were provided in previous chapters of this book.

In addition to existing models, researchers have been working on introducing other predictive models and improving existing formulas in last five decades. For example, Cassinelli et al. (1982) looked into how the RSR classification relates to actual machine performance to estimate the machine penetration rate. Sanio (1985) found a strong connection between rock strength and a specific energy parameter. This parameter measures the energy required to excavate a certain volume of rock in a mechanized tunneling project. Moreover, Sapigni et al. (2002) studied the relationship between RMR and penetration rate in three tunneling projects in Italy. Ribacchi and Fazio (2005) focused on the correlation between RMR and the performance of tunnel boring machines in a specific project in Italy. Gong and Zhao (2009) suggested an empirical relationship between certain rock mass properties and machine penetration rate, based on data from two excavated tunnels in granitic rock formations. Delisio (2014) in his doctoral dissertation at the École Polytechnique Fédérale de Lausanne (EPFL) developed a new model for predicting machine performance in blocky rocks, using data from the Lötschberg project in Switzerland. Macias (2016) worked on updating the NTNU model in his doctoral dissertation at the Norwegian University of Science and Technology. Additionally, Wilfing (2016) developed a new model called the Alpine model as part of an international project at the Technical University of Munich (TUM) by modifying the existing Gehring (1995) model.

More recently, Salimi (2021) developed and validated transparent, data-driven predictive models for estimating the penetration rate of Tunnel Boring Machines in hard rock. This was achieved by combining conventional regression approaches with modern machine learning techniques—with a particular focus on Classification and Regression Trees (CART). This marks the first application of such methods in this context, introducing a new era of transparent and interpretable TBM performance modeling. Salimi (2021) gave special attention to integrating rock mass classification systems, especially RMR, due to their simplicity, availability, and widespread use in mining and underground construction.

Dardashti (2024), in a section of her Ph.D. thesis, employed machine learning techniques to investigate the relationships between FPI, as the dependent variable of machine performance, and independent variables such as the four components of Basic RMR (UCS, RQD, joint spacing, Jc), as well as RTC (tock type), and developed an empirical equation for predicting FPI.

Some of the above-mentioned models are based on a single laboratory parameter related to the intact rock properties (Tarkoy 1975; Graham 1976; Farmer and Glossop

1980; Nelson et al. 1983, 1985), while others are more comprehensive and consider parameters such as rock mass characteristics and geological conditions in addition to the properties of the intact rock (e.g., NTNU, CSM, Q_{TBM} models). Table 1 provides a brief overview of these models.

As shown in Table 1, the most important rock mass parameters used in these models to predict penetration rate include UCS, TS, and the frequency and orientation of discontinuities. Additionally, key machine parameters crucial to these models include cutter spacing, cutter edge width, cutter diameter, thrust force, and cutterhead rotation speed. The main output parameter of all these models is the machine penetration rate during excavation cycle or stroke. Advanced models also include outputs such as advance rate or utilization factor, cutter life and replacement cost, tunnel completion time, and tunnel construction cost.

In developing predictive models, researchers aim to establish a connection between machine's performance and three sets of related parameters including geological parameters, operating parameters, and machine specifications.

Generally, data used to develop performance prediction models for machines are obtained from one or a combination of the following sources:

- Field observations and measurements during tunnel construction (field data);
- In situ tests like cutterhead penetration tests;
- Routine laboratory tests;
- Large-scale, special laboratory tests.

Predictive machine performance models can be broadly categorized into two types: theoretical and empirical models.

Theoretical models are developed based on the mechanisms of rock failure and equilibrium equations between cutting forces and machine operating parameters, incorporating findings from laboratory experiments. On the other hand, empirical models are derived from analyzing the relationship between geological conditions and actual machine performance parameters observed in various tunneling projects.

While laboratory tests on rocks are essential for predicting machine performance, they alone may not provide a comprehensive understanding of rock mass conditions and machine performance characteristics. In contrast, statistical and empirical approaches often do not prioritize the mechanism of rock mass failure. As a result, a realistic model for predicting TBM performance should combine data from laboratory tests with geological site characteristics and tunnel construction conditions.

In this chapter, aside from the commonly utilized and well-known models such as NTNU, CSM and Q_{TBM}, which were detailed in previous chapters, other theoretical and experimental models will be briefly introduced.

Table 1 Summary of existing models to predict TBM performance and their input(s) and output(s) parameters

No.	Model	Output parameter	Input parameters	
			Rock mass	Machine
1	Graham (1976)	P	UCS	F_n
2	Farmer and Glossop (1980)	P	TS	F_n
3	Nelson et al. (1983)	FPI, P	TH	F_n, RPM
4	Hughes (1986)	PR	UCS	F_n, RPM, N_{TBM}, D_{TBM}, CD
5	Gehring (1995)	P	UCS, JO, JS, σ	F_n, CD, N_{TBM}
6	Rostami (1997) (CSM)	P, PR, AR, CC	UCS, TS	Th, N_{TBM}, CD
7	Cheema (1999); Modified CSM model	PR	TS, E_i RMBI	Th, N_{TBM}, CD
8	Bruland (1998); NTNU model	P, PR, AR, CC	DRI, n, JS, JO, Jn	Th, N_{TBM}, CD, RPM
9	Barton (2000); Q_{TBM} model	PR, AR	Q, UCS, q, σ, CLI, n	F_n, D_{TBM}
10	Ramamurthy (2008)	PR	UCS, TS, DRI, E_i, E_{rm}, Jf, JC, JO, σ	Th, RPM, N_{TBM}
11	Gong and Zhao (2009)	P	UCS, Bi, Jv, JO	F_n, RPM
12	Hassanpour (2009, 2018); Hassanpour et al. (2009, 2010, 2011, 2014, 2015, 2016, 2021, 2022, 2023a,b); Samadi and Hassanpour (2025); UT model	FPI, P, PR, AR, CC	UCS, RQD, VHNR	F_n, RPM
13	Farrokh et al. (2012); Farrokh (2013)	P, PR	RT, UCS, RQD	D_{TBM}, F_n
14	Delisio (2014)	FPI_{blocky}, P	Jv, UCS	Th, RPM
15	Wilfing (2016); Alpine model	P	UCS, TS, JS, JO, LCPC, σ	F_n, CD, N_{TBM}

(continued)

Table 1 (continued)

No.	Model	Output parameter	Input parameters	
			Rock mass	Machine
16	Goodarzi et al. (2021)	P, PR	UCS	F_n, RPM
17	Pourhashemi et al. (2021a, b, c)	FPI, P, PR	UCS	F_n, RPM
18	Salimi (2021)	FPI, P, PR	UCS, RQD, RT	F_n, RPM
19	Dardashti (2024)	FPI, P, PR	UCS, RQD, JS, RT	F_n, RPM

RT: Rock type; UCS: Uniaxial compressive strength; TS: Tensile strength; DRI: Drilling rate index; CLI: Cutter life index; TH: Total hardness; q: Quartz content; B_i: Brittleness index; E_i: Intact rock elastic modulus; E_{rm}: Rock mass deformation modulus; n: Porosity; RQD: Rock quality designation; Q: Q system; RMBI: Rock Mass Boreability Index; JS: Spacing of discontinuities; JC: Joint conditions; JO: Joint orientation; Jv: Joint count number; Jf: Joint frequency; Jn: Number of joint sets; GW: Groundwater conditions; σ: Stress conditions; RPM: Cutterhead rotation speed; Po: Cutterhead power; Th: Thrust force; Tq: Cutterhead torque; F_n: Cutter load; D_{TBM}: TBM diameter; N_{TBM}: Number of disc cutters; CD: Disc cutter diameter; P: Cutterhead penetration; PR: Penetration rate; AR: Advance rate; CC: Cutter consumption

2 Theoretical Models

As previously mentioned, theoretical models that are developed based on equilibrium equations of force and energy during cutting process are crucial in understanding the interaction between rock and machine. However, one of the main drawbacks of these models is their failure to account for important factors that can influence the cutting process, such as geological conditions (presence of fractures and joints in the rock mass). The most significant of these models will be discussed below.

2.1 Roxborough and Phillips Model

Roxborough and Phillips (1975) employed theoretical principles to analyze the forces required to penetrate in rock for V-shape disc cutters. In this approach, the necessary force to penetrate the rock is calculated by considering the area of contact between the disc cutter and the rock surface, as well as the rock strength. Essentially, the normal force in this model depends on factors such as the depth of penetration, the rock strength, the cutter diameter, and the apex angle of the cutter ring.

2.2 CSM Model

One of the most recognized and comprehensive theoretical models is the CSM model, initially developed at the Colorado School of Mines (Ozdemir 1977) and later updated by Rostami in the 1990s (Rostami and Ozdemir 1993; Rostami 1997). Over the years

1999–2009, various corrections were suggested to enhance and refine this model (Cheema 1999; Yagiz 2002; Ramezanzadeh 2005; Hassanpour 2009). Due to its significance and widespread acceptance worldwide, a detailed explanation of the CSM model was provided in Chap. "Cutterhead Modeling of Hard Rock TBM".

2.3 Snowdon et al. Model

Snowdon et al. (1982), investigated the relationship between normal (F_n) and rolling (F_r) forces and cutterhead penetration (P) into the rock and proposed Eq. 1 to define this relationship:

$$\frac{F_n}{F_r} = 21.71\,P - 0.656 \tag{1}$$

They used a V-shape, small disc cutter in their studies and made an attempt to show that there is an optimum ratio of cutters spacing (S) to penetration depth (P) in which the minimum specific energy (SE) is required to break the rock and create chips.

This study also revealed that for each specific combination of cutter spacing and rock type, there is a critical penetration depth at which further decrease in specific energy doesn't occur. They also found that the required forces increase in a linear trend as the cutter spacing grows, reaching an optimum s/p ratio between 15 and 20.

Snowdon suggests that none of the strength parameters have a significantly greater influence on machine performance than others. This is because all mechanisms of compressive, shear, and tensile failure are involved in the excavating process. As a result, since UCS is typically the only strength parameter considered in projects, it is prioritized over other parameters.

2.4 Sanio Model

Sanio (1985) recommended the Eq. 2 for determining the force necessary for penetration into the rock and chip formation:

$$F_n = 2 \cdot h \cdot \mathrm{Tan}\left(\frac{\varepsilon}{2}\right) \cdot \sigma_0 \tag{2}$$

where ε is the tip angle of the disc cutter ring, h is penetration depth and σ_0 is the hydrostatic pressure in the crushed zone beneath the disc cutter.

2.5 *Sato et al. Model*

Sato et al. (1991) continued Sanio (1985) research by suggesting Eq. 3 for estimating the forces necessary for penetrating rock:

$$F = k \cdot ROP^a \cdot S^b \tag{3}$$

In the Eq. 3, F represents the force (either normal or rolling force), K is the cutting coefficient, ROP stands for the rate of penetration, S denotes the spacing between disc cutters, a is the penetration coefficient (0.5 for normal force and 1 for rolling force), and b is the spacing coefficient (0.43 for both normal and rolling forces).

2.6 *Boyd Model*

Boyd (1986) introduced a novel model that deviates significantly from existing frameworks. This model operates on the assumption that a certain quantity of energy (referred to as SE or specific energy) is required to break a rock mass. If the area of the tunnel face and the total power of the cutterhead are known, the penetration rate of the machine can be calculated using the Eq. 4:

$$PR = \frac{W_{CP}}{SE \cdot A} \tag{4}$$

where PR is the rate penetration rate (m/h), W_{CP} is the total power of the cutterhead (kW), SE is the specific energy (kWh/m^3), and A is the area of the tunnel face (m^2).

This methodology, known as the specific energy method, is applicable across various excavating systems, provided that the specific energy values for different rock types with distinct disc cutters can be computed. A more comprehensive formula for this approach has been put forth by Rostami et al. (1994). In the latest model, an additional parameter has been integrated into the formula—mechanical efficiency coefficient of the machine multiplied by the nominal power of the tunneling machine—to ascertain the effective power available for the cutterhead.

3 Empirical Models

Empirical models are developed from actual data of mechanized tunneling projects, identifying and proposing relationships between geological properties, machine specifications, operating parameters, and performance metrics like penetration rate. Their strength lies in naturally accounting for ground and machine factors, but they are limited by their dependence on specific machine technologies. As technology advances and new machines are introduced, these models' applicability diminishes and needs to be reviewed and corrected. Additionally, application of empirical models is generally limited to ground conditions similar to those in their underlying datasets, which can lead to inaccuracies when applied to projects with different or untested geological conditions. The next section provides a brief overview of some key empirical models.

3.1 Graham Model

In the Graham (1976) model, the cutterhead penetration in rock (P, in mm/rev) is dependent on the cutter load (F_L, in kN) as an operating parameter of machine and UCS (in kPa) as a rock mass parameter:

$$P = 3940\frac{F_L}{UCS} \tag{5}$$

Equation 5 has been proposed for rocks with UCS falling within the range of 140–200 MPa. It is evident that this model ignores the influence of joint characteristics within the rock mass, the type and spacing of disc cutters, as well as machine specifications. Consequently, it may not provide a reliable prediction of machine performance.

3.2 NTNU Model

Within empirical methods, the NTNU model is considered the most comprehensive and frequently employed model for assessing the performance of hard rock TBMs. Originating in 1979 at the Norwegian University of Science and Technology (NTNU), this model has undergone revisions and refinements by various researchers up to 2016, aligning with the completion of pertinent data sets. Similar to the CSM model, the NTNU model holds significant importance and more details about this model was provided in Chap. "The NTNU Prediction Model for Hard Rock Tunnel Boring: Performance Predictions and Cutter Life Assessments".

3.3 Farmer and Glossop Model

The Farmer and Glossop (1980) model shares many similarities with the Graham (1976) model, the primary difference being the use of the tensile strength of the rock (σ_t in kPa) instead of the compressive strength of the rock. Defined by Eq. 6, this relationship has been recommended for sedimentary rocks.

$$P = 624\frac{F_L}{\sigma_t} \tag{6}$$

3.4 *Cassinelli et al. Model*

Cassinelli et al. (1982) used the rock structure rating (RSR) system, initially designed for tunnel support assessment, to develop an empirical equation to predict machine penetration rate (PR, m/h):

$$PR = -0.0059\,RSR + 1.59 \tag{7}$$

Their study revealed that as the RSR increases—which signifies improved rock quality and fewer discontinuities—the machine's penetration rate decreases (see Fig. 1). The proposed model doesn't account for key machine factors such as diameter, cutter type and specifications, spacing, thrust force, which undermines its reliability in accurately predicting machine performance. The model at best can be considered a site and machine specific model.

This relationship was developed based on data from a tunnel in the Western Alps (Italy), which was excavated in rocks with a uniaxial compressive strength ranging from 50 to 150 MPa.

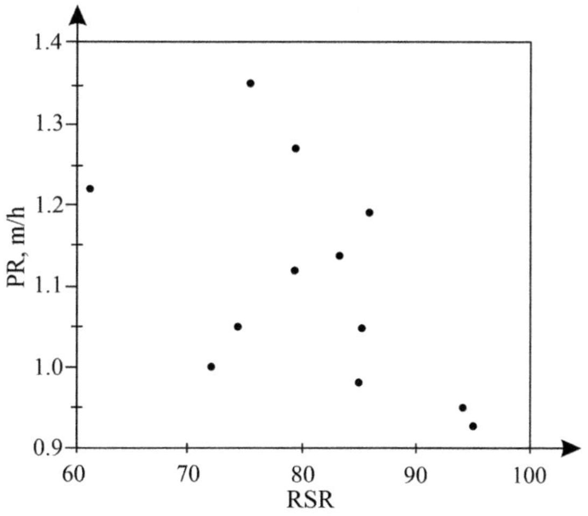

Fig. 1 The relationship between RSR and PR (Cassinelli et al. 1982)

3.5 Nelson et al. Or FPI Model

Nelson et al. (1983) proposed Eq. 8 as a method for predicting the cutterhead penetration (P, in mm/rev), by analyzing data gathered from four excavated tunnels in sedimentary rocks:

$$P = 10.45 - 1.19\,H_A \tag{8}$$

where, H_A is the Taber abrasion hardness as introduced by Tarkoy (1975). Moreover, they suggested Eq. 9 in their research to anticipate machine performance while considering the impact of thrust force.

$$R_f = 5.95 + 0.18H_T \tag{9}$$

In this equation, R_f stands for the field penetration index (FPI, in kN/cutter/mm/rev), or the ratio of the cutter load to the cutterhead penetration, whereas H_T denotes the total hardness calculated with Eq. 10:

$$H_T = H_R\sqrt{H_A} \tag{10}$$

H_R signifies the Schmidt hammer hardness.

3.6 Bamford Model

Bamford (1984) proposed Eq. 11 to calculate the penetration rate (PR, in m/h) by correlating machine performance data from two tunnels with diverse rock properties:

$$PR = 0.5325\,S - 8049 - 0.00344\,T - 0.00823\,N + 0.00137\,\varphi \tag{11}$$

In this equation S is the Schmidt hammer hardness, T (in tons) is the machine thrust force, N (or NCB in N/mm) is the cone indenter hardness (NCB 1977), and φ is the internal friction angle.

3.7 Hughes Model

Hughes (1986) introduced Eqs. 12 and 13 to estimate the penetration rate (V, in m/h) and the power requirement (PW, in kW) for machines working in coal-bearing layers. Here, P is the cutter load (kN), N is the cutterhead rotational speed (rev/s), h is the number of disc cutters, Fc is the rock's uniaxial compressive strength (MPa), R is the cutter ring radius (m), and D is the TBM diameter (m).

$$V = \frac{6P^{1.2}N * h}{Fc^{1.2}R^{0.6}} \tag{12}$$

$$PW = 28.45\,D + 9.07\,D^2 \tag{13}$$

3.8 *Innaurato et al. Model*

Innaurato et al. (1991) introduced a model similar to Cassinelli et al. (1982), based on data from 112 tunnel sections. It is based on the RSR system and includes a key modification by incorporating the rock strength (UCS), as shown in Eq. 14:

$$P = 40.41\,UCS^{-0.44} + 0.047\,RSR + 3.15 \tag{14}$$

In this Equation, P denotes the cutterhead penetration (in mm/rev).

3.9 *Gehring Model*

Gehring (1995) model was initially developed based on data gathered from four tunneling projects in South Africa and South Korea. Since then, it has undergone several updates to refine its accuracy. While the fundamental principles of the model have remained consistent, there have been new recommendations for adjustment factors introduced. In this model, the penetration in rock (P in mm/rev), is influenced by two primary factors: the cutter load (F_n) and the rock strength as shown in Eq. 15:

$$P = \frac{F_n}{UCS} \cdot K_0 \cdot K_1 \cdot K_2 \cdot K_3 \cdot K_4 \cdot K_5 \tag{15}$$

The model incorporates various correction factors, denoted as k_0 to k_5, each serving a specific purpose such as correcting for basic penetration, specific failure energy, rock mass fabric, state of stress in rock mass, and cutters diameter and spacing. Gehring (1995) assigned a value of 4 to k_0 and suggested the use of Eq. 16 to determine k_1, which is an empirical equation obtained from data collected during tunnel excavation by tunnel boring machines (Fig. 2).

$$K_1 = 0.475\,wf^{-0.56} \tag{16}$$

$$wf = \frac{WF}{UCS} \tag{17}$$

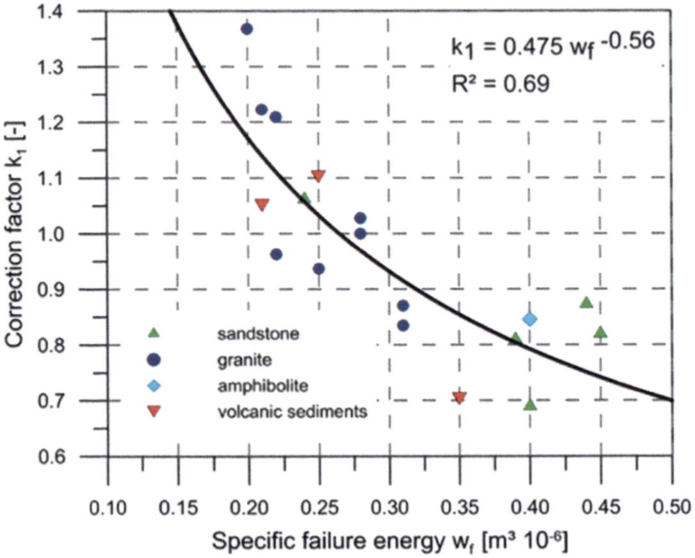

Fig. 2 The proposed plot for determining the correction factor of specific failure energy, (K_1) (Gehring 1995)

In these equations, WF and wf denote failure energy and specific failure energy, respectively, obtained from laboratory uniaxial compressive strength tests. Failure energy indicates the energy needed to cause rock failure in uniaxial test.

Gehring (1995) recommends using Table 2 to adjust the rock mass fabric, which provides guidance on selecting a correction factor based on joint set spacing and its average orientation relative to the tunnel axis (α angle). The α angle is determined using Eq. 18.

$$\alpha = \sin^{-1}\left[\sin\alpha_f \cdot \sin(\alpha_t - \alpha_s)\right] \tag{18}$$

where, α_s and α_f are strike and dip angle of joint set and α_t is the azimuth of tunnel axis.

Table 2 K_2 correction factor for different ranges of α angle and spacing

Spacing of discontinuities (cm)	α angle			
	0	30	60	90
> 50	1.0	1.0	1.0	1.0
10–50	1.2	1.3	1.6	1.3
5–10	1.4	1.8	2.3	1.6
< 5	1.7	2.3	3.0	2.0

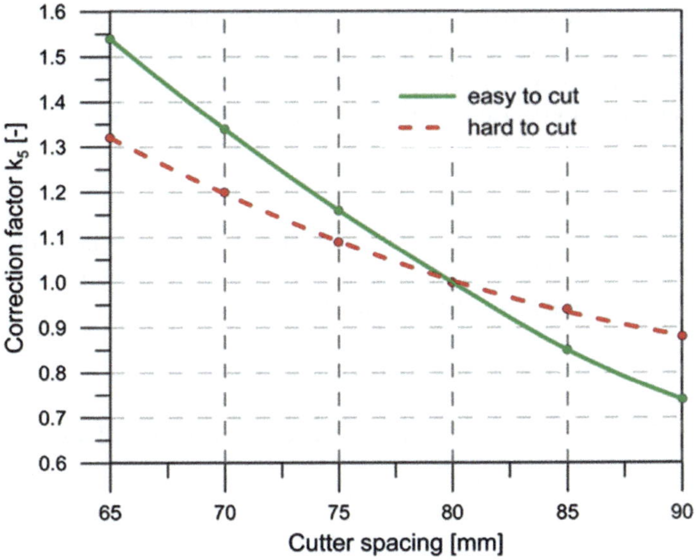

Fig. 3 The proposed plot for determining the correction factor of disc cutters spacing (K_5) (Gehring 1995)

Gehring's model is based on 17-inch disc cutters with a cutter tip width of 15.875 mm and a spacing of 80 mm. To account for varying cutter sizes and spacings, correction factors k_4 and k_5 are necessary.

Research by the Norwegian University of Science and Technology (NTH 1983) indicates that the cutterhead diameter correction factor (k_4) decreases linearly with increasing disc cutter diameter (dc), falling below one for diameters over 17 inches as shown in Eq. 19:

$$K_4 = \frac{430}{dc} \tag{19}$$

The correction factor for disc cutter spacing (k_5), shown in Fig. 3, exceeds one for spacings under 80 mm due to increased cutter penetration. This graph, based on research by the Norwegian University of Science and Technology (NTH 1983), was utilized by Gehring in his model.

3.10 Palmström Model

Presented by Palmström (1995), this method uses RMi classification parameters to estimate TBM performance. It is based on the NTNU prediction model but replaces laboratory tests with alternative relationships, such as substituting DRI with UCS.

Figure 4 shows the workflow for predicting TBM performance with the Palmström model. Palmström (1995) argues that the NTNU model does not fully account for the joint spatial distribution, represented by K_S, and recommends using block volume (V_b) to determine a more accurate K_S. He proposed Eq. 20 to relate V_b to K_S:

$$K_s = 1.6c_o \cdot V_b^{-0.33} \quad (20)$$

In Eq. 20, c_o is a coefficient related to the orientation of the main discontinuity set, estimated from Table 3. This equation can also be expressed using the jointing parameter (JP) introduced by Palmström (1995), as Eq. 21:

$$K_s = 0.43c_o \cdot JP^{-1} \quad (21)$$

In the NTNU model, it is necessary to convert the parameter K_S to an equivalent fracturing factor or k_{ekv} (Eq. 22):

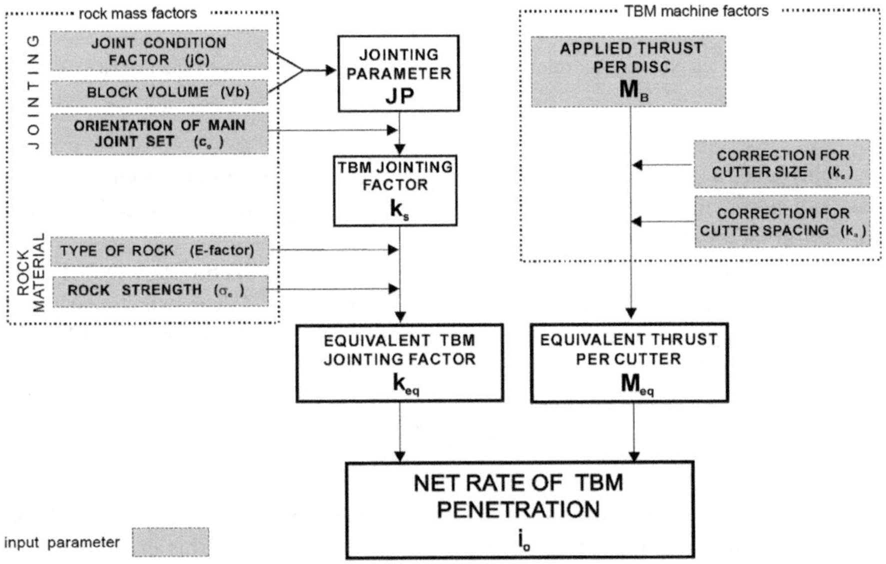

Fig. 4 Flowchart to predict TBM performance using RMi model (Palmström 1995)

Table 3 Determining the C_0 value, a coefficient related to the orientation of the discontinuity planes (Palmström 1995)

The angle between tunnel axis and the main joint set or α angle (°)	0–15	15–30	30–45	45–75	75–90
Average value of C_0	1	1.25	1.5	1.75	1.5 ($C_0 = 1.75$ for $V_b < 0.1\ m^3$)

$$K_{ekv} = K_s \cdot K_{DRI} \tag{22}$$

According to Palmström (1995):

$$K_{DRI} = 0.14\sqrt{DRI} \tag{23}$$

$$DRI = E \cdot UCS^{-0.6} \tag{24}$$

Therefore:

$$K_{ekv} = \frac{0.06c_o\sqrt{E}}{JP \cdot UCS^{0.3}} \tag{25}$$

The parameter E indicates the rock group and has the following values:

- For most non-schistose, hard rocks (UCS > 40): E = 1000.
- For metamorphic schists (30 < UCS < 150): E = 750.
- For argillaceous rocks (10 < UCS < 100), E = 500.

To predict TBM performance using this model, the remaining stages are exactly the same as the NTNU model, (See NTNU model, Chap. "The NTNU Prediction Model for Hard Rock Tunnel Boring: Performance Predictions and Cutter Life Assessments").

3.11 Barton Model

In the late 1990s, Barton expanded the Q system to introduce the Q_{TBM} model for estimating tunnel boring machine performance (Barton 1999, 2000). Recent publications highlight its increasing use in projects. In Chap. "AR, Time, Tunnel Length and Geology Using Q_{TBM}", Professor Barton, the model's developer, details its key characteristics.

3.12 Alber Model

The model proposed by Alber (2000) offers an empirical approach for predicting TBM performance using statistical analyses. According to Alber's model, the advance rate of the machine is influenced by two key factors: the boreability and the stability of the rock mass. The boreability is directly linked to the penetration rate, while the stability of the rock mass (which dictates the type and amount of support system) contributes to a reduction in the utilization factor.

This model defines the penetration rate based on the rock mass uniaxial strength (σ_{cm}) according to the Hoek and Brown (1980) failure criterion. In this model, to compare different machines with varying rotational speeds, cutter numbers, and thrust forces, the specific penetration metric, described by Eq. 26, is used:

$$SP = \frac{P}{RPM * F_n} \tag{26}$$

In this equation, P is cutterhead penetration (cm), RPM is the cutterhead rotation speed (rev/min), and F_n is the cutter load (kN). The specific penetration (SP) is inversely related to the field penetration index (FPI).

The relationship between the specific penetration and the uniaxial compressive strength of the rock mass is illustrated in Fig. 5a. As expected, the specific penetration of a boring machine increases as the uniaxial compressive strength of the rock mass decreases. This increase continues until a certain point (σ_{cm} = 15 MPa). In such scenarios, where the rock mass is highly fragmented with minimal discontinuities spacing, the presence of rock chips between disc cutters significantly impedes the penetration rate in these weaker rock formations.

The utilization factor of Tunnel Boring Machines (TBMs) depends on several variables, but in this specific model, it is primarily linked to the installation of support systems within the tunnel. The type of support system required is determined by the behavior of the rock mass during excavation, which is influenced by its resistance to shear stresses on the tunnel wall, and can be expressed by a metric known as the factor of safety (FS). Alber (2000) established a relationship between the factor of safety at the roof centerline and the TBM utilization factor, based on actual data from over 100 km of tunnel excavated by mechanized methods. This relationship is illustrated in Fig. 5b.

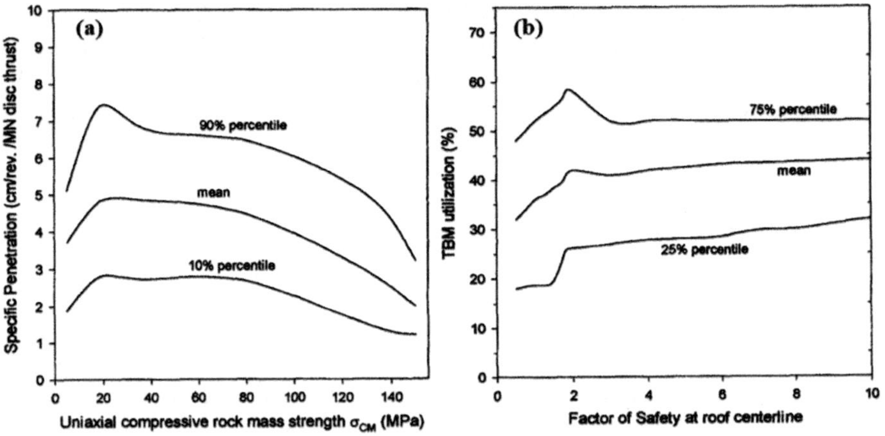

Fig. 5 **a** Relationship between specific penetration and rock mass compressive strength, **b** relationship between utilization factor and confidence coefficient in the tunnel roof (Alber 2000)

In the graph, it is evident that in stable rock mass conditions (FS > 2), the utilization factor of the TBM varies between 25 and 50% (with an average of 40%). For situations where 1.25 < Fs < 2, the average utilization factor decreases to 35%. When FS < 1.25, the utilization factor of the TBM falls below 35% and may even reach zero in tunnel sections where FS is extremely low.

By understanding the relationship between rock mass strength and penetration rate, as well as the correlation between the factor of safety and utilization factor, it is possible to predict the advance rate. Alber proposes classifying rock masses into three broad classes based on strength: High, medium, and low rock mass strength. Additionally, he suggests categorizing rock mass behavior in terms of stability into three classes: Stable, friable, and squeezing rock masses. Hence, the rock masses excavated within the tunnel using TBM can be described in terms of three strength classes and three behavioral classes (refer to Table 4).

By utilizing these two parameters, a total of 9 distinct categories of advance rate can be identified. As illustrated in Fig. 6, Alber (2000) has established a probability distribution function (PDF) for the advance rates within each of these nine categories. He notes that categorizing rock masses into these nine groups is generally sufficient for predicting the performance of Tunnel Boring Machines.

It is important to note that in stable rock masses, there is no failure on the tunnel walls, and the rock behaves elastically without the formation of any plastic zones. In such scenarios, only a minimal support system is required. Conversely, in friable rock masses, some failure may occur on the tunnel wall. In such cases, the radius of the plastic zone is generally less than 1.5 times the radius of the tunnel.

In rock masses with the potential for squeezing, deformations occur at depth and a substantial amount of rock is influenced. Here, the radius of the plastic zone often exceeds 1.5 times the tunnel radius, leading to significant convergence in the tunnel. Therefore, a heavy and flexible support system is necessary for such rock masses. The discussions by Alber shows some general trends in TBM performance, but it does not take into account many of the machine and operational parameters such as machine diameter, power, type of cutters, etc.

Table 4 Classification of rock masses into different classes based on strength and behavior (Alber 2000)

Rock mass strength		Rock mass behavior	
Description	σ_{cm}	Stability class	FS
I, Low rock mass strength	< 20 MPa	A, Rock mass with squeezing potential	FS < 1.25
II, Medium strong rock mass strength	20–80 MPa	B, Friable and unstable rock mass	1.25 < FS < 2
III, High rock mass strength	80–140 MPa	C, Stable rock mass	FS > 2

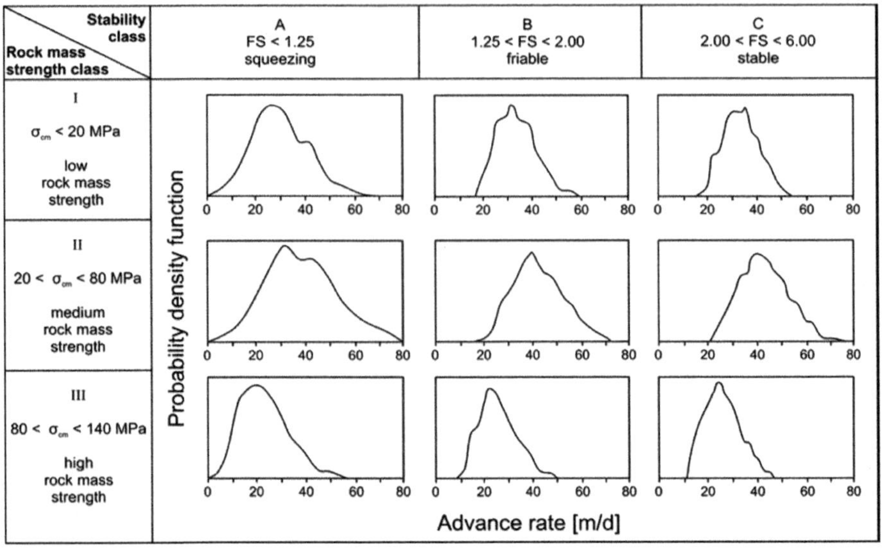

Fig. 6 The probability density function of the advance rate in 9 separated categories (Alber 2000)

3.13 RME Model

Bieniawski et al. (2008) delved into data from various tunnels in Spain, Germany, Switzerland and Ethiopia to propose Eqs. 27–32 for predicting the average advance rate of various types of TBMs:

Open TBM: Based on data from 49 tunnel sections with total length of 1724 m:

$$ARA_T = 0.839RME_{07} - 40.8 \quad \text{UCS} > 45\text{MPa} \quad \text{R} = 0.763 \tag{27}$$

$$ARA_T = 0.324RME_{07} - 6.8 \quad \text{UCS} < 45\text{MPa} \quad \text{R} = 0.729 \tag{28}$$

Single Shield TBMs: Based on data from 62 tunnels sections studied at the Tunnels of Guadarrama in Spain and at the Katzenberg Tunnel in Germany, all totaling 3620 m in length:

$$ARA_T = 23\left(1 - 242^{((45-RME)/17)}\right) \quad \text{UCS} > 45 \text{ MPa} \tag{29}$$

$$ARA_T = 10\ln(RME) - 13 \quad \text{UCS} < 45 \text{ MPa} \quad \text{R} = 0.784 \tag{30}$$

Double Shield TBMs: Based on data from 225 sections from the tunnels of Guadarrama and Abdalajís West in Spain and Gibe II Inlet and Gibe II Outlet in Ethiopia, all totaling 20.7 km in length:

$$ARA_T = 0.422RME_{07} - 11.6 \quad \text{UCS} > 45 \text{ MPa} \quad \text{R} = 0.658 \tag{31}$$

$$ARA_T = 0.661RME_{07} - 20.4 \quad \text{UCS} < 45\text{MPa} \quad \text{R} = 0.867 \tag{32}$$

In the above equations, ARA_T represents the machine's theoretical advance rate (m/day), while RME_{07} indicates the Rock Mass Excavability based on its latest version (Bieniawski et al. 2008). The RME_{07} index is calculated by summing the ratings of five parameters: UCS and DRI (Drilling Rate Index) of intact rock, discontinuity conditions, tunnel stand-up time, and groundwater inflow, with detailed ratings for each parameter provided in Table 5. The sum of the ratings of the above parameters varies between 0 and 100 and it is expected that the higher the RME value, the easier and more productive the excavation of the tunnel.

Bieniawski et al. (2008) proposed Eq. 33 with three adjustment factors (F_E, F_A and F_D) to convert ARA_T to the real advance rate (ARA_R) of the machine in a given tunnel section:

$$ARA_R = ARA_R \cdot F_E \cdot F_A \cdot F_D \tag{33}$$

where F_E, F_A and F_D are adjustment factors for crew efficiency, team adaptation to the terrain, and tunnel diameter, respectively.

The TBM crew who handles the tunneling machine every day has an important influence on the performance achieved. Grandori (2007) proposed the Eq. 34 and Table 6 to calculate adjustment factor for crew efficiency:

$$F_E = 0.7 + F_{E1} + F_{E2} + F_{E3} \tag{34}$$

Experiences indicate that after excavating a number of kilometers of tunnel, even with a similar ground condition, the rate of machine advancement often exceeds what one would anticipate from a typical "learning curve" effect. This increase is clearly linked to the length of tunnel already completed, suggesting that the crew is adapting more effectively to the project over time. The quantitative effect of this adaptation factor (F_A) is given in Table 7.

Equations 27–32 were derived for tunnels with diameters close to 10 m. To account for the effect of tunnel diameter on the TBM advance rate, the F_D adjustment factor has been introduced as shown in Eq. 35:

$$F_D = \frac{10}{D} \tag{35}$$

where D is the tunnel diameter (in m). $F_D = 1.0$ for $D = 10$ m

It is important to note that a characteristic section utilized for obtaining the ARA must meet certain conditions (Bieniawski et al. 2008):

Table 5 Ranking score for calculation of RME_{07} (Bieniawski et al. 2008)

Uniaxial compressive strength of intact rock [0–25 points]

UCS (MPa)	< 5	5–30	30–90	90–180	> 180
Rating	4	14	25	14	0

Drillability [0–15 points]

DRI	> 80	65–80	50–65	40–50	< 40
Rating	15	10	7	3	0

Discontinuities in front of the tunnel face [0–30 points]

Homogeneity		Number of joints per meter					Orientation with respect to tunnel axis		
Homogeneous	Mixed	0–4	4–8	8–15	15–30	> 30	Perpend	Oblique	Parallel
Rating 10	0	2	7	15	10	0	5	3	0

Stand up time [0–25 points]

Stand up time (h)	< 5	5–24	24–96	96–192	> 192
Rating	0	2	10	15	25

Groundwater inflow [0–5points]

Water inflow (lit/s)	> 100	70–100	30–70	10–30	< 10
Rating	0	1	2	4	5

Table 6 Criteria for evaluation of coefficients F_{E1}, F_{E2} and F_{E3} (Grandori 2007)

Contractor's TBM experience	No experience	1–5 tunnels built	6–10 TBM tunnels built	11–20 TBM tunnels built	> 21 TBM tunnels built
Value of F_{E1}	0	0.05	0.1	0.15	0.2
Qualifications of the tunnelling crew	Little trained and none with TBMs	Trained but none with TBMs	Trained overall and with TBMs		
Value of F_{E2}	0	0.1	0.15		
Resolutions of disputes	TBM manufacturer rep on site	No TBM manufacturer rep on site	Time to resolve problems: < 1 month	Time to resolve problems: > 1 month	
Value of F_{E3}	0.075	0	0.075	0	

Table 7 Adjustment Factor for team adaptation to the terrain (F_A) based on excavated length of tunnel

Tunnel length excavated (km)	Adaptation factor (F_A)
0.5	0.68
1.0	0.80
2.0	0.90
4.0	1.00
6.0	1.08
8.0	1.12
10.0	1.16
12.0	1.20

- The section length should be more than 30 m;
- The section should not have significant variations in the value of RME and have a representative rock mass quality RMR, that is, not varying by more than 10 points;
- The section should not feature extraordinary repairs to the TBM;
- During the period of excavating the tunnel section the TBM utilization should be within 30% to 60% of the cycle.

As one can observe, the model is heavily impacted by the ground conditions and there are many of the machine and backup settings that are not incorporated in the calculations. In other words, the model seems to be an elevated site-specific model with limited influence of machine specifications. Also, when a project is laid out, the utilization rate estimation should include probable repair time and disruptions to offer a realistic picture of the potential downtimes and related lower utilization rates.

Table 8 The values of n based on orientations of joints, β (Ramamurthy 2008)

Orient. of joint, β°		0	10	20	30	40	50	60	70	80	90
n	U-shaped	0.82	0.46	0.11	0.05	0.09	0.30	0.46	0.64	0.82	0.95
	Shoulder-shaped	0.85	0.60	0.20	0.06	0.12	0.45	0.80	0.90	0.95	0.98

3.14 Ramamurthy Model

Ramamurthy (2008) proposed Eq. 36 to calculate the penetration rate of the machine, accounting for various rock mass properties, including deformation modulus:

$$PR = \frac{\left(\frac{T}{A}\right)\left(\frac{\sigma_{ci}}{\sigma_t}\right) \cdot R \cdot N \cdot \left(\frac{DRI}{100}\right) \cdot s}{P_o \cdot M_{rj}} \tag{36}$$

where, T is net thrust, A is the total area of the cutterhead face (m^2), σ_{ci} is the intact rock compressive strength (MPa), σ_t is the intact rock tensile strength (MPa), R is the number of rotations of the cutterhead per hour (rpm), N is the number of disc cutters per cutterhead face area (#/m^2), DRI is the drillability rate index (determined based on the rock strength), s is the length of the tunnel section (m), P_o is the mean biaxial stress at the tunnel face (t/m^2), and M_{rj} ($= E_{tj}/\sigma_{ci}$) is the modulus ratio of the rock mass derived from Eq. 37:

$$\frac{M_{rj}}{M_{ri}} = \exp(-0.0035 J_f) \tag{37}$$

$$J_f = \frac{J_n}{n \cdot r} \tag{38}$$

In Eqs. 37 and 38, M_{ri} (E_{ti}/UCS) is the modulus ratio of the intact rock, J_n is joint frequency (joints/m), n is a coefficient representing the critical joint inclination (β) with the direction of drilling or loading giving lowest strength of rock mass (Table 8), and r represents the joint surface strength parameter (Tables 9 and 10).

3.15 Gong and Zhao Model

Gong and Zhao (2009) developed a model to estimate penetration rate of TBM based on data collected from two sewer tunnels, excavated in granite rocks with varying levels of weathering, in Singapore—T05 (D = 3.6 m and L = 12.6 km) and T06 (D = 3.3 m and L = 9.6 km). By measuring key parameters of the rock mass and TBMs, and applying statistical analyses, they proposed Eq. 39 for estimating the specific rock mass boreability index (SRMBI):

Table 9 The proposed values for r based on intact rock compressive strength (Ramamurthy 2008)

Uniaxial compressive strength, σ_{ci} (MPa)	joint strength parameter, r	Remarks
2.5	0.30	Fine-grained to coarse-grained
5	0.45	
15	0.60	
25	0.70	
45	0.80	
65	0.90	
100	1.00	

Table 10 The proposed values for "r" based on the filling materials of joints (Ramamurthy 2008)

Gouge material	Friction angle, σ_j°	Joint strength param, r = tan ϕ_j
Gravelly sand	45	1.00
Coarse sand	40	0.84
Fine sand	35	0.70
Silty sand	32	0.62
Clayey sand	30	0.58
Clayey silt: 25% clay	25	0.47
50% clay	15	0.27
75% clay	10	0.18

$$BI_{(1)} = 37.06UCS^{0.26}Bi^{-0.1}\left(0.84e^{-0.05J_V} + e^{0.09\sin(\alpha+30)}\right) \tag{39}$$

In this equation, $BI_{(1)}$ is the specific rock mass boreability index (kN/cutter/mm/rev), UCS is the rock uniaxial compressive strength (MPa), Bi is the rock brittleness index ($=\sigma_{ci}/\sigma_t$), J_V is the volumetric joint count and angle α is the orientation of the joint planes with respect to the tunnel axis.

The specific rock mass boreability index, $BI_{(1)}$ defined as the boreability index at the penetration equal to 1mm per revolution and its relationship with boreability index (BI) and cutterhead penetration (p) is defined by Eq. 40.

$$BI \approx BI_{(1)}P^{-0.75} \tag{40}$$

The model utilizes the boreability index (BI) as an equivalent of the field penetration index (FPI) to account for machine specifications and operational parameters. Gong and Zhao (2009) used the developed model to explore how various rock mass parameters affect the cutterhead penetration. They found that UCS and J_v have a greater influence than brittleness index and discontinuity orientation (See Fig. 7a-d).

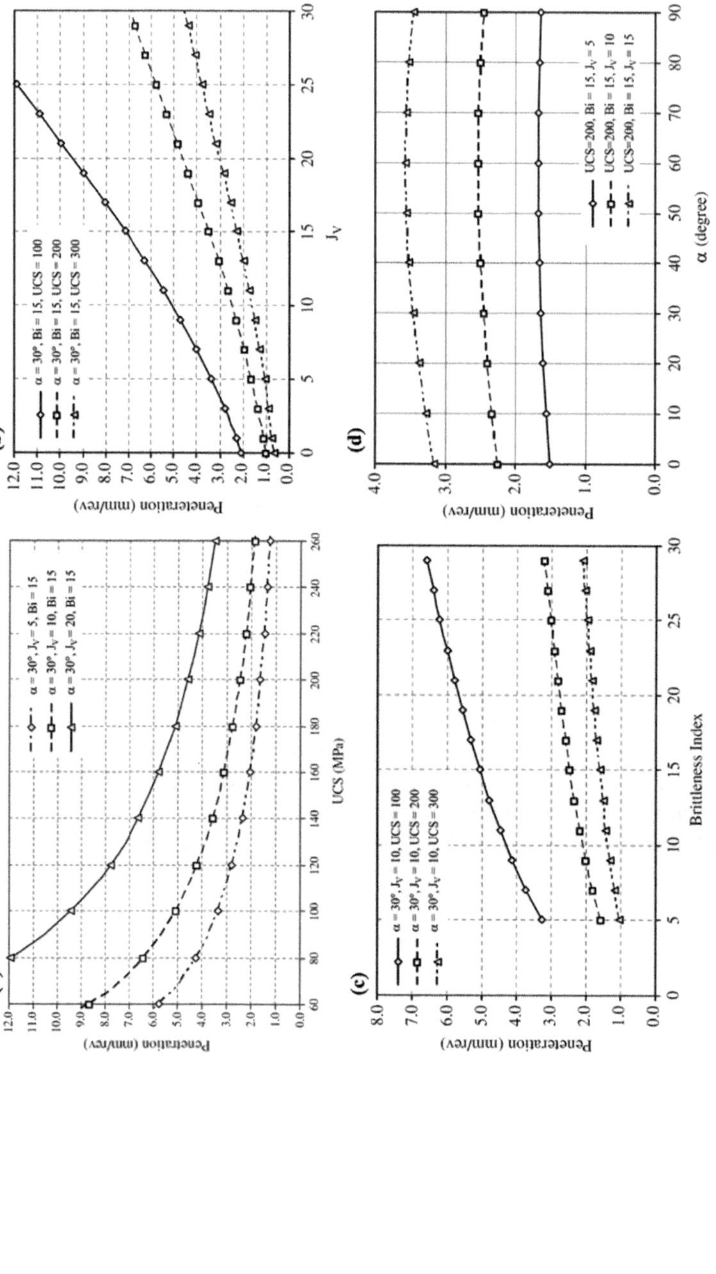

Fig. 7 The effect of **a** UCS, **b** Jv, **c** Bi and **d** α angle on cutterhead penetration at a constant cutter load of 200 kN (Gong and Zhao 2009)

Table 11 Coding of RQD values (RQD_C) in Farrokh et al. (2012) database

Code	RQD	Number of discontinues in m (CFF)	Description
S or 1	100–90	< 8	Low frequency
M or 2	90–60	12–8	Medium frequency
H or 3	< 60	16–12	High frequency

Table 12 Coding of rock types (RT_C) in Farrokh et al. (2012) database

Code	Numerical code (RTc)	Rock type
C	5	Mudstone, claystone, marl, slate, phyllite, argillite
S	3	Sandstone, conglomerate, quartzite, limestone
L	3	Limestone, chalk, dolomite, marble
K	3	Karstic limestone
M	2	Metamorphic rocks such as gneiss and schist
G	1	Intrusive rocks such as granite and diorite
V	2	Volcanic rocks such as basalt, andesite, tuff

3.16 *Farrokh et al. Model*

Farrokh et al. (2012) and Farrokh (2013) conducted a statistical analysis of data from various tunnels worldwide, organized in a database. They proposed Eqs. 41 and 42 for estimating cutterhead penetration and machine penetration rate.

$$P = e^{0.41+(0.404D)-\left(0.027D^2\right)+(0.0691RTc)-(0.00431UCS)+(0.0902RQDc)+(0.000893F_n)} \tag{41}$$

$$PR = \frac{F_n^{0.186} * RQDc^{0.133} * RTc^{0.183} * RPM^{0.363} * D^{5.47} * e^{\left(0.046D^2\right)}}{5.64 * UCS^{0.248} * e^{(1.58D)}} \tag{42}$$

In these equations, D is the tunnel diameter (millimeters in Eq. 41 and meters in Eq. 42), while RQD_C and RT_C are codes indicating RQD and rock type from Tables 11 and 12, respectively. The other parameters have been previously introduced.

3.17 *Delisio Model*

Delisio (2014), using data from the Lötschberg railway tunnel in Switzerland, developed a model in his PhD thesis at EPFL (École polytechnique fédérale de Lausanne) to predict TBM performance in blocky rocks. He focused on tunnel sections with minor face instabilities that reflect the complexities of unstable blocky grounds.

His studies showed that there is a significant variability in machine performance in blocky grounds, caused by irregular thrust force distribution on disc cutters due to incomplete contact with the rock face. To address this, Delisio introduced the blocky FPI, defined by Eq. 43, in his model (Delisio et al. 2013).

$$FPI_{blocky} = \frac{Th}{P} \tag{43}$$

where Th is the total machine thrust force (kN) and P is the penetration (mm/rev). In the updated FPI formula, cutter load is replaced by the total cutterhead thrust force (Th), preserving the influence of thrust force as a key factor regardless of its distribution across the disc cutters.

Through extensive statistical analyses and exploration of the relationships between the new FPI and various geological parameters (Figs. 8 and 9), Delisio (2014) proposed Eq. 44 as a predictive model for machine performance in such rock conditions. With a regression coefficient of 0.78, this model effectively captures the complex interactions between thrust force, cutter load, and geological factors, providing valuable insights for improving machine efficiency in such demanding conditions. The use of total thrust and not accounting for the diameter of the TBM and the number of cutters could make this formula limited to the range of diameters in the database and not scalable.

$$FPI_{blocky} = 5952 - 1794Ln(J_V) + UCS \tag{44}$$

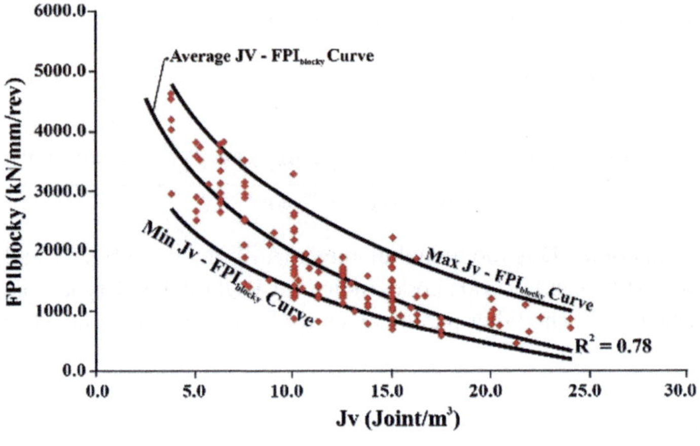

Fig. 8 Relationship between FPI_{blocky} and Jv (Delisio 2014; Delisio et al. 2013)

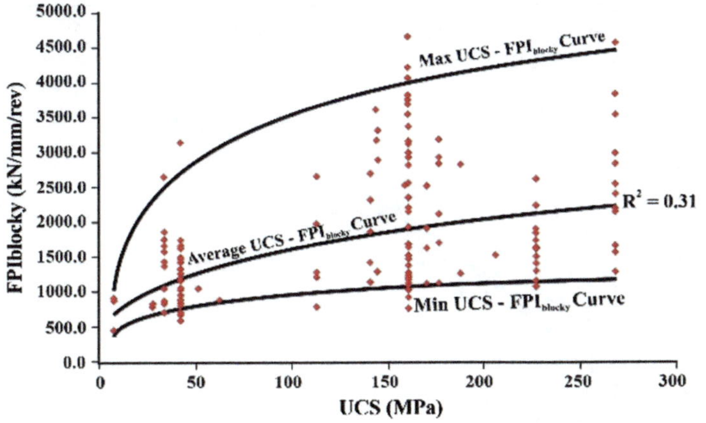

Fig. 9 Relationship between FPI_{blocky} and UCS (Delisio 2014; Delisio et al. 2013)

3.18 Wilfing Model

Wilfing (2016), in her PhD thesis at the Technical University of Munich (TUM) as part of the "Abrock" project, evaluated Gehring's (1995) model by comparing its results with data from cutterhead penetration tests in Koralm tunnel (Austria) and Røssåga tunnel (Norway). Her findings indicated the need for an update. Figure 10 displays an example of the cutterhead penetration test results used in this study, comparing them with the Gehring and CSM models.

The result of Wilfing's modifications and corrections on the Gehring's model, which has been named the Alpine model, was introduced as Eq. 45, which is still very similar to Gehring's original model (See Eq. 15):

$$P = \frac{F_n - b_{BTS3mm}}{UCS} * K_0 * K_2 * K_i + 3 \tag{45}$$

where, P is the cutterhead penetration (in mm/rev), F_n is cutter load (in kN), k_0 is basic penetration factor (= 4.0), k_2 is the correction factor for discontinuity pattern and k_i represents the further correction factors for geotechnical/machine parameters.

A key aspect of Wilfing's work was her focus on the load-penetration (P-F_n) curve of the disc cutters. She proposed that a penetration of 3 mm marks the threshold at which chips effectively form beneath the disc cutters, identifying the load corresponding to this threshold as a critical load. By selecting a subset of cutterhead penetration tests conducted in massive intact rock, Wilfing correlated the critical load values from these tests to laboratory rock properties such as LCPC or LBC and Brazilian tensile strength (σ_t) using Eqs. 46 and 47 (see Fig. 11). Wilfing modified the Gehring model by eliminating the portion of the curve below 3 mm, utilizing parameters b_{BTS3mm} (or b_{LBC3mm}) derived from the established relationships. These

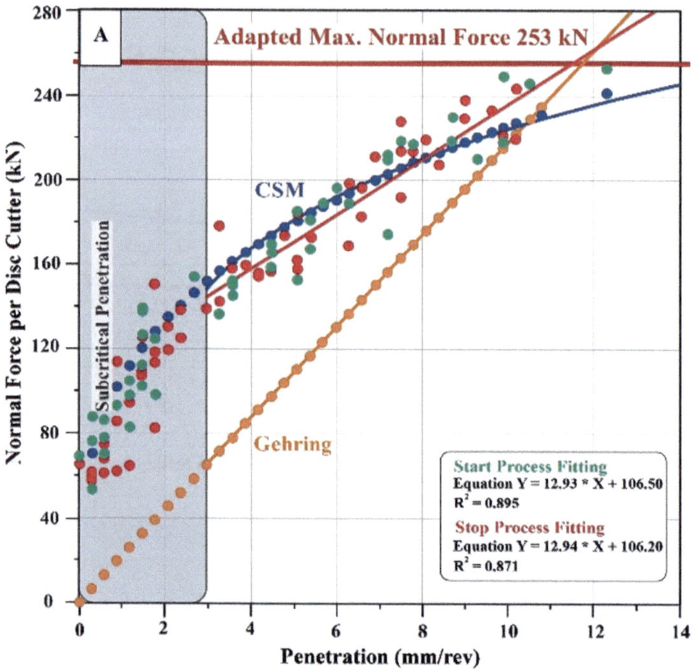

Fig. 10 An example of the load-penetration curve in the cutterhead penetration test along with a comparison of the results obtained from the Gehring and CSM models (Wilfing 2016)

parameters represent the cutter load at 3 mm of penetration. She also added the numeral 3 into the equation.

Wilfing made an attempt to explore adjustments related to rock toughness and discontinuity conditions, but she faced challenges due to limited data.

$$b_{BTS3mm} = e^{0.08\sigma_t + 4.1} \tag{46}$$

$$b_{LBC3mm} = -1.3LBC + 194.7 \tag{47}$$

3.19 Salimi Model

The primary goal of the study done by Salimi (2021) was to develop and validate transparent, data-driven predictive models for estimating the penetration rate of TBMs in hard rock. This was achieved by combining conventional regression approaches with modern machine learning techniques—with a particular focus on Classification and Regression Trees (CART). This marks the first application of

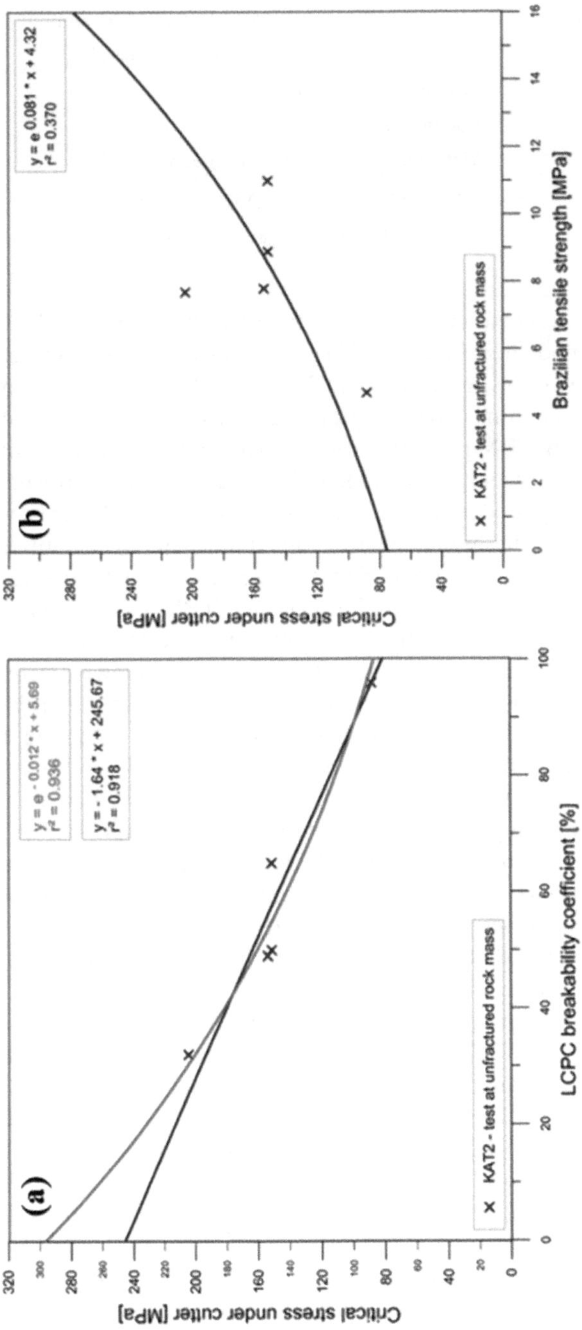

Fig. 11 The relationship between the critical stress under cutter and laboratory parameters in penetration tests: **a** LCPC, **b** Brazilian tensile strength (Wilfing 2016)

Table 13 Rock type categorization (RTC) in database (Salimi 2021)

Rock type	Code
Claystone, mudstone, marl, slate, phyllite, argillite	C
Sandstone, siltstone, conglomerate, quartzite	S
Limestone, chalk, dolomite, marble	L
Karstic Limestone	K
Metamorphic rocks such as gneiss and schist	M
Coarse igneous such as granite and diorite	G
Fine volcanic such as basalt, tuff, and andesite	V

such methods in this context, introducing a new era of transparent and interpretable TBM performance modeling. Salimi (2021) gave special attention to integrating rock mass classification systems, especially RMR, due to their simplicity, availability, and widespread use in mining and underground construction. He also concluded that most inputs influencing TBM performance—such as UCS, Brazilian tensile strength (BTS/σ_t), joint frequency and orientation, and rock abrasivity—correspond closely with parameters already used in standard classification systems. Notably, while abrasivity primarily affects cutter wear and utilization, the remaining features are directly relevant to penetration rate prediction. Salimi compiled a comprehensive and structured database using data from seven international TBM projects, encompassing 580 tunnel sections over 61 km. These sections were selected based on reliability and data verifiability. To ensure robustness, he conducted a stepwise statistical analysis, starting with Principal Component Analysis (PCA) to detect multicollinearity among variables, followed by tolerance diagnostics (VIF). Considering the distinct textural and mechanical behavior of various rock types under cutting forces, the study incorporated Rock Type Classification (RTC) to improve boreability estimation (Table 13). Additionally, groundwater conditions were found—through bivariate analysis and PCA—to primarily influence the advance rate via machine utilization, rather than the penetration rate itself. Consequently, he excluded groundwater from the PR prediction model, and included RTC instead.

Salimi (2021) conducted both deterministic and probabilistic sensitivity analyses to identify the most influential input parameters. As illustrated in the tornado graph (Fig. 12), the results of the probabilistic sensitivity analysis confirmed that parameters such as UCS, RQD, and RTC exert a strong influence on the FPI—highlighting its role as a robust and transparent performance indicator in hard rock TBM tunneling.

Later, Salimi et al. (2022) expanded the database to include 666 datasets covering a total tunnel length of 93 km, with TBM diameters ranging from 3.6 to 10.5 m. Using regression tree algorithms, they developed multiple regression tree models (also referred to as tree graphs), each calibrated for different rock type classifications (Fig. 13).

Salimi et al. (2022) discuss how the regression tree (RT) models developed in their study provide a more accurate and transparent alternative to traditional multiple regression methods for predicting tunnel boring machine (TBM) performance. These

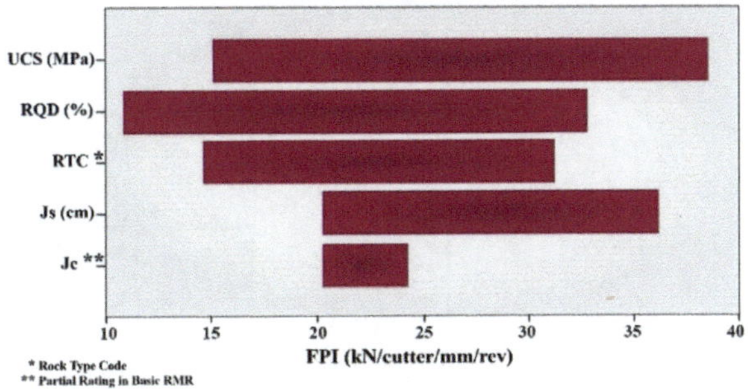

Fig. 12 Tornado graph resulted of sensitivity analysis of CART model (Salimi 2021)

models were designed as decision graphs, categorizing rock types based on similarities in rock texture, including factors such as cementation, grain size, and grain shape. The findings indicate that incorporating rock type as a categorical input significantly enhances model performance, particularly when combined with other important parameters like unconfined compressive strength (UCS) and joint frequency (Jv or RQD). Tree-based models also offer the advantage of visualizing decision paths, which adds valuable interpretability for practical engineering applications. The research proposes a hybrid modeling framework that combines the transparency of empirical models with the flexibility of machine learning. By integrating probabilistic sensitivity analysis with tree-based methods, the study ensures both predictive robustness and model interpretability.

His study highlighted the effectiveness of the FPI as a project-independent and scalable indicator for evaluating TBM performance under various geological and operational conditions. The FPI's simplicity, ease of measurement, and strong correlation with cutting parameters make it a reliable tool for both pre-construction planning and on-site monitoring.

Additionally, Salimi (2021) concluded that rock mass classification systems are essential tools in tunneling and underground construction. Incorporating these systems into performance prediction models not only utilizes their established value for assessing ground behavior and designing support but also enhances model adaptability across various geotechnical contexts.

3.20 *Dardashti Models*

In her PhD thesis, Dardashti (2024) aimed to develop empirical performance prediction models for Tunnel Boring Machines (TBMs) in hard rock conditions. She utilized modified parameters from the Rock Mass Rating (RMR) system specifically designed for mechanized tunneling. By employing machine learning techniques such as Regression Tree (RT) and Random Forest (RF) algorithms, the model adjusts the RMR input parameters and the rock type to provide a more accurate estimation of the FPI, which is a critical measure of machine performance.

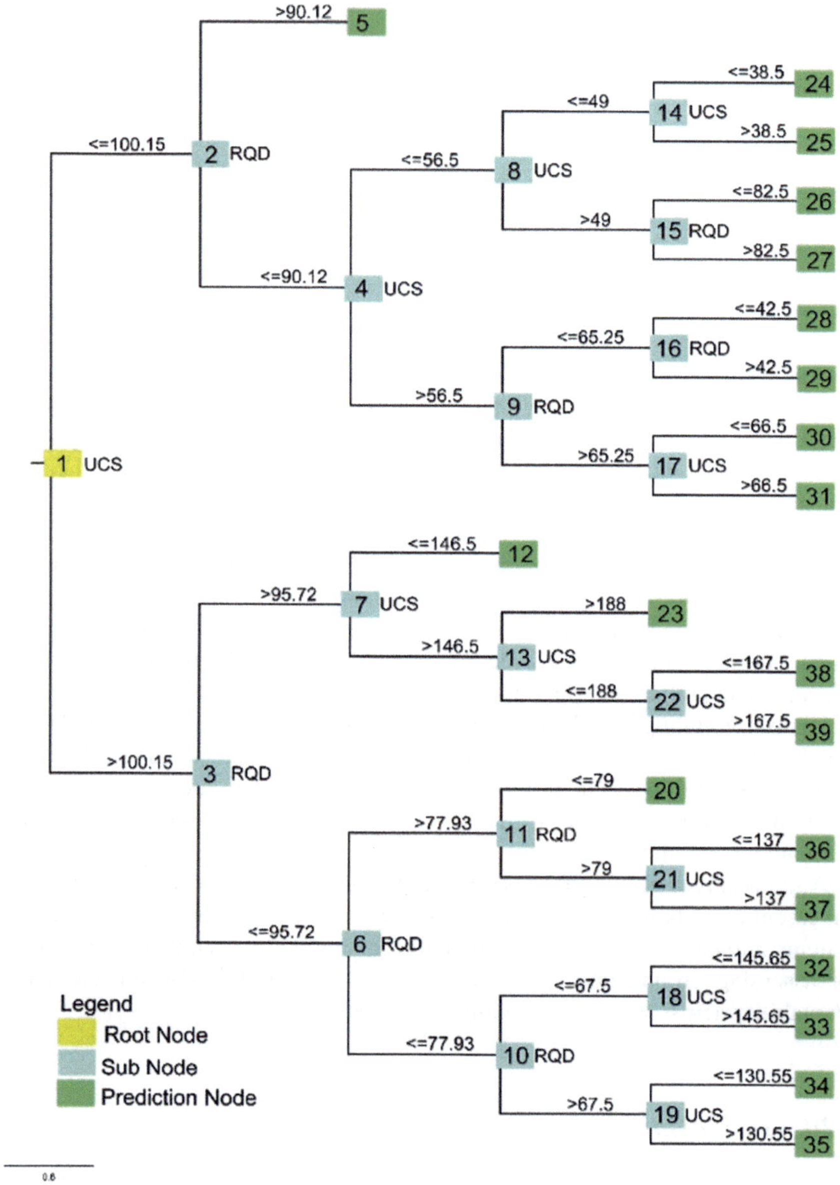

Fig. 13 Regression tree developed for estimation of FPI in rock type classes of M and V in Table 13 (Salimi et al. 2022)

This study shows that the RMR system exhibits a strong correlation with TBM performance parameters compared to other rock classification systems. However, modifications to the RMR parameters and their weighting were necessary to improve the prediction of TBM performance, leading to the development of the RMR based prediction models.

Dardashti (2024) collected data from 10 tunneling projects, totaling approximately 82 km with various geological conditions, and compiled them into a database. This database includes 523 tunnel sections of limited length, corresponding to excavation shifts involving multiple strokes. The database includes operational data such as total thrust, RPM, cutterhead torque and power, section length, and excavation time. From these parameters, additional calculations were performed to determine values such as PR, AR, U, FPI, and SE for each section. The database also contained geological and geotechnical information regarding the intact rock and rock mass properties encountered along the tunnel.

The tunneling projects included the 2nd lot of Zagros water conveyance tunnel (ZWCT-2), the 1st lot of Karaj Water Conveyance Tunnel (KWCT-1), the 3rd and 4th lot of Ghomrood Water Conveyance Tunnel (GWCT-3,4), southern lot of Kerman Water Conveyance Tunnel (KrWCT-S), and Golab Water Conveyance Tunnel in Iran, Yinsong water supply tunnel (lot 4), and Yinhan Water Conveyance Tunnel in China, MINAS Water Conveyance Tunnel in Ecuador, Manapouri second tailrace tunnel in New Zealand, and Maroshi–Ruparel water supply tunnel in Mumbai, India.

In her study, the key input parameters for predicting TBM performance included 4 out of 6 RMR input parameters including UCS, RQD, spacing (JS) and conditions (JC) of critical joint set, as well as lithology or rock type.

Dardashti (2024), based on her own analyses and previous researches, showed that Basic RMR exhibits a stronger correlation with TBM performance parameters compared to Adjusted RMR, indicating that the joint orientation does not significantly improve the correlation. Although joint orientation, play an important role in the propagation of cracks during rock cutting process, establishing a clear relationship between joint orientation and machine performance is a challenging task (Gong et al. 2005; Hassanpour et al. 2011, 2022; Sapigni et al. 2002; Salimi 2021). Therefore, she excluded adjustment rating of joint orientation from her analyses.

She also removed the rating of groundwater condition, as it is not a directly influential parameter on the machine penetration rate. According to Hassanpour (2009), the mere presence of water, regardless of its quantity, can weaken the intact rock and rock mass due to alteration and swelling phenomena, and plays a positive role in increasing the machine penetration rate. On the other hand, according to experiences, the quantity of groundwater inflow into the tunnel affects the advance rate of the machine rather than its penetration rate. So, it seems that just considering the effect of groundwater presence on the strength of intact rock and rock mass is enough for estimating the penetration rate of the machine.

The texture of the rock, including factors such as cement type, degree of cementation, and grain size and shape, significantly affects the machine penetration rate. Sensitivity analysis and parametric studies conducted by Fatemi et al. (2016) also confirmed the significant impact of lithology on rock mass boreability. Therefore,

Dardashti (2024), to account for the impact of lithology on the machine performance, used Rock Type Code (RTC), introduced by Farrokh et al. (2012) and Salimi et al. (2019), as an input parameter in her models. The seven Rock Type Code categories, were used by Dardashti (2024) in her analyses, has been listed in Table 14.

Among TBM performance parameters, Dardashti (2024) selected the FPI, to analyze interaction between rock mass and machine. Using this parameter, the effects of penetration rate, cutter load, and cutterhead rotation speed are combined into a single factor.

Dardashti (2024), in a part of her PhD thesis, examined the relationships between FPI as the dependent variable of machine performance and independent variables, including four components of Basic RMR, namely UCS, RQD, joint spacing (JS) and partial rating of joint conditions in RMR system (JC), as well as RTC by machine learning (ML) techniques. Notably, ML-based regression analysis enables the modeling of categorical parameters in regression algorithms. Thus, Dardashti (2024) used the rock type category (RTC), a qualitative and categorical parameter, in the modeling. The result of the analyses was an empirical equation with $R^2 = 0.81$ (Eq. 48):

$$FPI = 5.173(UCS)^{0.22} \cdot (RQD)^{0.19} \cdot (JS)^{0.41} \cdot (RTC) \tag{48}$$

where, FPI is in kN/cutter/mm/rev, UCS is in MPa, RQD is in percent, JS is in m and RTC is defined as codes presented in Table 14.

Based on the T-test, Jc has been determined no statistically significant, and was removed from the model. In Eq. 48, FPI is estimated accurately by utilizing the three input parameters of Basic RMR and the RTC coefficient relevant to the lithology, as outlined in Table 14. According to Dardashti (2024), in the G and V categories of RTC, which encompass a wide range of rock strengths, the tunnel sections were analyzed separately based on strength (less than 150 MPa and more than 150 MPa)

Table 14 Classification of rock types (RTC) in the database used for developing of the models by Dardashti (2024)

Rock type		Code	RTC
Claystone, mudstone, marl, slate, phyllite, argillite		C	1
Sandstone, siltstone, conglomerate, quartzite		S	1.598
Limestone, chalk, dolomite, marble		L	1.325
Karstic Limestone		K	-
Metamorphic rocks such as schist		M	1.202
Coarse igneous such as granite, diorite, and gneiss	G	G_1 (UCS < 150 MPa)	1.177
		G_2 (UCS ≥ 150 MPa)	1.848
Fine volcanic such as basalt, tuff, andesite, and pyroclastic	V	V_1 (UCS < 150 MPa)	0.8
		V_2 (UCS ≥ 150 MPa)	1.876

and the corresponding RTC coefficients were obtained for each strength range, separately. Therefore, when using Eq. 48 for the G and V categories, attention should be given to the UCS value, and the appropriate RTC coefficient should be used to predict FPI. Figure 14 illustrates the measured FPI versus the predicted FPI using Eq. 48 for both the Train and Test datasets, providing a visual representation of the model's performance.

Dardashti (2024) in another part of her PhD thesis, by modifying the ratings of each parameter in the Basic RMR system and adjusting the internal weighting of each parameter, developed an optimized RMR, named RMR_{TBM}, specifically tailored for predicting TBM performance in hard rock. The approach adopted for developing this optimized RMR system involves nonlinear regression algorithms within the machine learning domain, such as Regression Tree and Random Forest algorithms.

The Random Forest algorithm introduces an index that reveals the relative importance of variables in the model's predictions. Known as Variable Importance, this index can be calculated as a percentage. In Dardashti's study, this index was employed to determine optimal RMR input parameter ratings for estimating TBM performance. The modified RMR system is termed RMR_{TBM}. Ultimately, based on the results of the machine learning algorithms, the optimized ratings for the input parameters were proposed as follows: UCS = 40, RQD = 30, Js = 25, and Jc = 5. Following the assignment of ratings to each input parameter of the RMR_{TBM}, considering the range of variations of each parameter and the predicted FPI corresponding to that range in the developed regression trees, a trial-and-error approach was employed by Dardashti (2024) to determine the optimal internal weighting of each input parameter. Ultimately, she obtained the diagrams in Fig. 15 for internal weighting of the RMR_{TBM} input parameters.

Consequently, the RMR_{TBM} system is proposed with the objective of predicting TBM performance in hard rock and estimating the FPI according to Table 15. In this system, each of the UCS, RQD, JS, and JC parameters is assigned a score based on Table 15 and the diagrams in Fig. 15. The sum of these parameter ratings determines the RMR_{TBM} value on a scale of 0–100. Given this value, the RMR_{TBM} class is determined, and based on Table 16, the FPI range, boreability class, excavability description, and tunnel stability conditions are predicted.

Given the variety of approaches that have been discussed in this chapter, the last two models seem to offer a reasonable way to estimate FPI from rock mass parameters through the use of frequently available rock and rock mass properties in many tunneling site investigations. This refers to UCS, RQD, JS, JC, and Rock Type that are widely used in geotechnical site investigations for tunneling projects as part of the rock mass classification effort. Using the proposed transparent tables by Salimi (2021) or the modified RMR_{TBM} by Dardashti (2024), one can estimate FPI and by using the anticipated cutter load and estimation of the cutterhead speed, based on machine diameter, offer a reasonable estimate of penetration rate of machine (PR). While these models lack input on joint orientation, they still offer a reasonable estimate of PR and in cases where the joint spacing is very low or in schistose rock, proper adjustment can be made by using NTNU models or by using expert judgement to account for joint orientation.

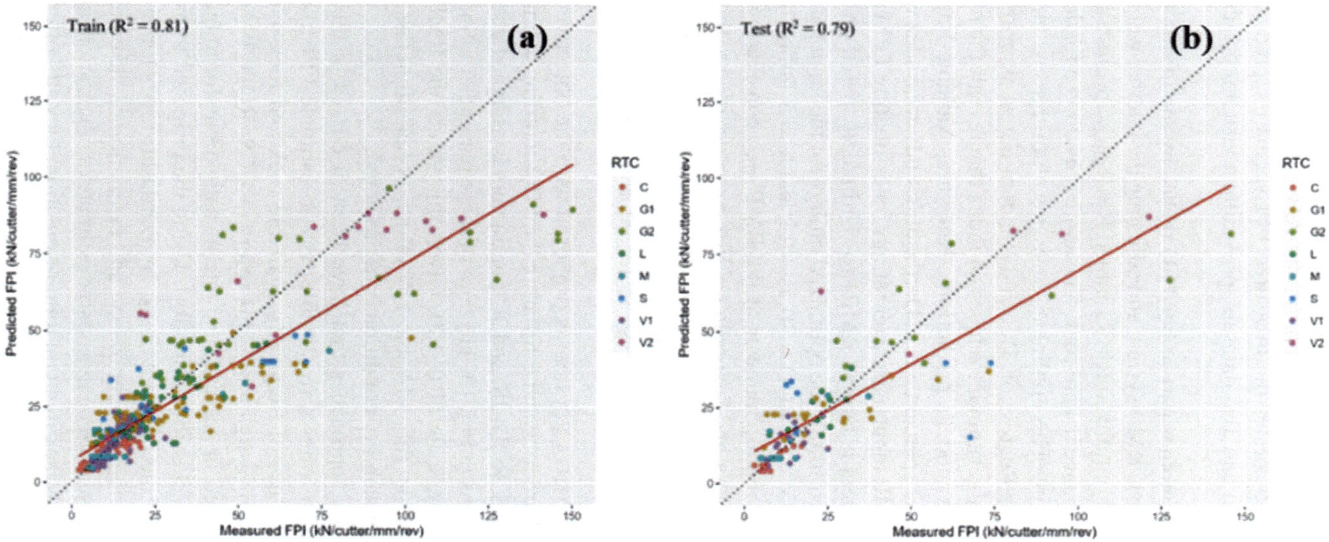

Fig. 14 The measured FPI versus the predicted FPI via Eq. 48 for: **a** Train and **b** test datasets (Dardashti 2024)

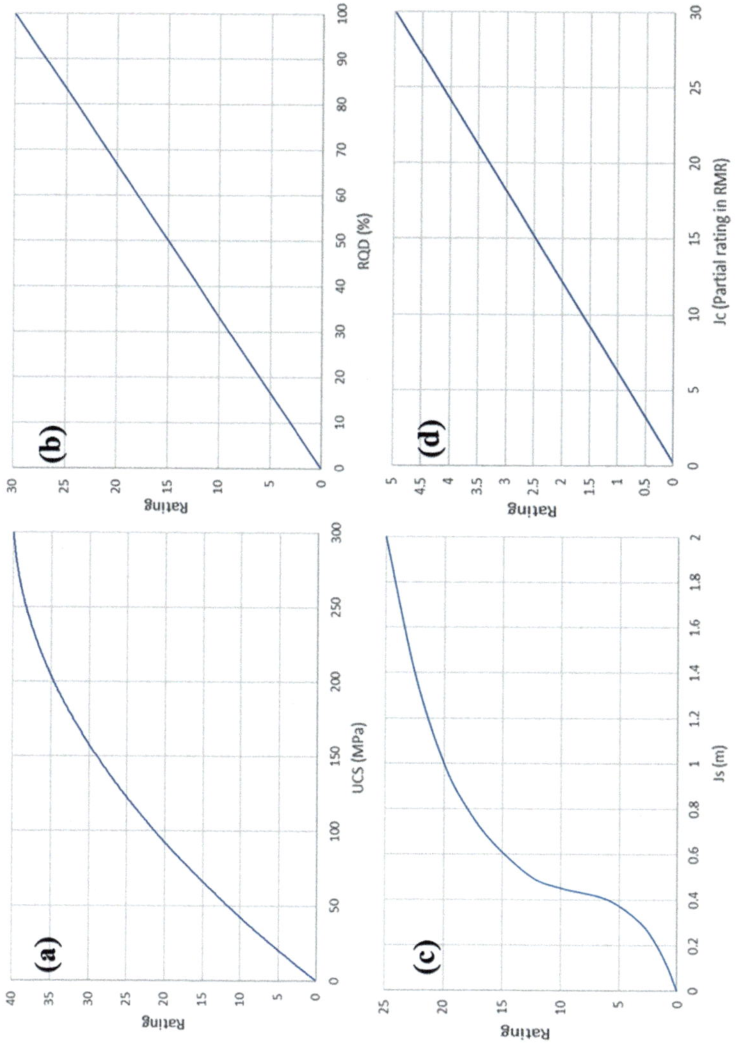

Fig. 15 Diagrams used for internal weighting of four RMR_{TBM} input parameters (Dardashti 2024)

Table 15 The developed RMR_{TBM} system for TBM performance prediction (Dardashti 2024)

UCS (MPa)	Range	> 250	200–250	150–200	100–150	50–100	25–50	< 25
	Rating	40	37	32	25	17	9	3
RQD (%)	Range	> 99	90–99	70–90	50–70	30–50	10–30	< 10
	Rating	30	28	24	18	12	6	1.5
Spacing (m)	Range	> 2	0.6–2	0.2–0.6		0.06–0.2	< 0.06	
	Rating	25	22	6		1.2	0	
Joint conditions	Description	Very rough, non-continuous, tight, un-weathered	Slightly rough, aperture < 1 mm, slightly weathered	Slightly rough, aperture < 1 mm, highly weathered		Slickensided, gouge < 5 mm, aperture = 1–5 mm, continuous	Soft gouge > 5 mm, apperture > 5 mm, continuous	
	Rating	5	3.3	2.5		1.6	0	

Table 16 RMR_{TBM} system and the range of FPI values and Boreability conditions (Dardashti 2024)

Class	I	II	III	IV	V
Total rating of RMR_{TBM}	81–100	61–80	41–60	21–40	0–20
FPI (kN/cutter/mm/rev)	≥ 70	40–70	20–40	10–20	< 10
Boreability class	B-0	B-I	B-II	B-III	B-IV & B-V
Rock mass boreability	Tough	Fair-Tough	Good-Fair	Very Good-Good	Extremely Good
TBM excavatability	Difficult boring	Fair boring	Good boring	Very good and easy boring, potential support problems	Shielded tunneling, potential support problems, potential soft rock problems
Stability condition	Completely stable	Stable	Minor instabilities	Local or major instabilities	Collapse, squeeze, gripper problems
Rock types	Very strong and massive intrusive and metamorphic rocks	Strong igneous and metamorphic rocks	Blocky and jointed tuffs, limestones	Alternations of sandstones, limestones and shales, Flysch type sediments, most low strength rocks and jointed rock masses	Highly foliated and schistose rocks, shale, marlstone, thick fault zones

References

Alber M (2000) Advance rates of hard rock TBMs and their effects on projects economics. Tunn Undergr Space Technol 15(1):55–64

Bamford WF (1984) Rock test indices are being successfully correlated with tunnel boring machine performance. In: Proceedings of 5th Australian tunneling conference, vol 2. Melbourne, Australia, pp 19–22

Barton N (2000) TBM tunneling in jointed and faulted rock. A. A. Balkema, Brookfield

Barton N (1999) TBM performance estimation in rock using QTBM. Tunn Tunn Int 31(9)

Bieniawski ZT, Celada B, Galera JM, Tardáguila I (2008) New applications of the excavability index (RME) for selection of TBM types and predicting their performance. In: Proceedings of world tunnel congress. Delhi, India

Blindheim OT (1979) Boreability predictions for tunneling. Ph.D. Thesis, Department of Geological Engineering, The Norwegian Institute of Technology

Boyd RJ (1986) Hard rock continuous mining machine: mobile miner MM-120. In: Rock excavation engineering seminar, edited by Howarth et al., Department of Mining and Metallurgical Engineering, University of Queensland
Bruland A (1998) Hard rock tunnel boring, vol 1–10. Ph.D. Thesis, Norwegian University of Science and Technology (NTNU), Trondheim (Norway)
Cassinelli F, Cina S, Innaurato N, Mancini R, Sampaolo A (1982) Power consumption and metal wear in tunnel boring machines: analysis of tunnel-boring operation in hard rock. In: Proceedings of tunneling '82. Institution of Mining and Metallurgy, London, pp 73–81
Cheema S (1999) Development of a rock mass boreability index for the performance of tunnel boring machines. Ph.D. Thesis, Colorado School of Mines (CSM), Golden (CO)
Dardashti A (2024) Application of rock mass classification systems for improved performance estimation of rock TBMs. Ph.D. Thesis, University of Isfahan
Delisio A, Zhao J, Einstein HH (2013) Analysis and prediction of TBM performance in blocky rock conditions at the Lötschberg Base Tunnel. Tunn Undergr Space Technol 33:131–142
Delisio A (2014) Field and numerical studies on the causes and effects of blocky rock conditions on TBM tunnelling. Ph.D. Thesis, EPFL, Lausanne (Switzerland)
Farmer IW, Glossop NH (1980) Mechanics of disc cutter penetration. Tunnels Tunneling 12(6):22–25
Farrokh E, Rostami J, Laughton C (2012) Study of various models for estimation of penetration rate of hard rock TBMs. Tunn Undergr Space Technol 30:110–123
Farrokh E (2013) Study of utilization factor and advance rate of hard rock TBMs. Ph.D. Thesis, The Pennsylvania State University, University Park (USA)
Fatemi SA, Ahmadi M, Rostami J (2016) Evaluation of TBM performance prediction models and sensitivity analysis of input parameters. Int J Bull Eng Geol Environ. https://doi.org/10.1007/s10064-016-0967-2
Gehring K (1995) Leistungs-und Verschleissprognosen in maschinellen Tunnelbau. Felsbau 13(6)
Gong QM, Zhao J, Jiao YY (2005) Numerical modeling of the effects of joint orientation on rock fragmentation by TBM cutters. J Tunnell Undergr Space Technol 20(2):183–191. https://doi.org/10.1016/j.tust.2004.08.006
Gong QM, Zhao J (2009) Development of a rock mass characteristics model for TBM penetration rate prediction. Int J Rock Mech Min Sci 46(1):8–18
Goodarzi S, Hassanpour J, Yagiz S, Rostami J (2021) Predicting TBM performance in soft sedimentary rocks, case study of Zagros mountains water tunnel projects. Tunn Undergr Space Technology 109:103705
Graham PC (1976) Rock exploration for machine manufacturers. In: Proceedings of the symposium on exploration for rock engineering, vol 1. Johannesburg (South Africa), pp 173–180
Grandori R (2007) TBM performances and RME classification system. Jornada sobre 'Experiencias recientes con tuneladoras'. CEDEX, Madrid
Hassanpour J (2009) Investigating the impact of engineering geological parameters on TBM performance and modifying existing TBM performance prediction models. Ph.D. Thesis, Tarbiat Modares University, Tehran, Iran
Hassanpour J, Rostami J, Khamehchiyan M, Bruland A (2009) Development new equations for performance prediction. Geomech Geoeng Int J 4(4):287–297
Hassanpour J, Rostami J, Khamehchiyan M, Bruland A, Tavakoli HR (2010) TBM performance analysis in pyroclastic rocks, a case history of Karaj Water Conveyance Tunnel (KWCT). J Rock Mech Rock Eng 4:427–445
Hassanpour J, Rostami J, Zhao J (2011) A new hard rock TBM performance prediction model for project planning. Tunn Undergr Space Technol 26:595–603
Hassanpour J, Rostami J, Zhao J, Tarigh Azali S (2014) Introduction of an empirical TBM cutter wear prediction model for pyroclastic and mafic igneous rocks; a case history of Karaj water conveyance tunnel, Iran. Tunn Undergr Space Technol 43:222–231

Hassanpour J, Rostami J, Zhao J, Tarigh Azali S (2015) TBM performance and disc cutter wear prediction based on ten years experience of TBM tunnelling in Iran. Geomech Tunn 8(3):239–247

Hassanpour J, Ghaedi Vanani AA, Rostami J, Cheshomi A (2016) Evaluation of common TBM performance prediction models based on data from the second lot of Zagros water conveyance tunnel (ZWCT2). Tunn Undergr Space Technol 52:147–156

Hassanpour J (2018) Development of an empirical model to estimate disc cutter wear for sedimentary and low to medium grade metamorphic rocks. Tunn Undergr Space Technol 75:90–99

Hassanpour J, Firouzei Y, Hajipour G (2021) Actual performance analysis of a double shield TBM through sedimentary and low to medium grade metamorphic rocks of Ghomrood water conveyance tunnel project (lots 3 and 4). Bull Eng Geol Env 80:1419–1432

Hassanpour J, Saljooghi Khoshkar A, Ghasemi Farasani M, Hashemnejad A (2022) Investigating the relationships between rock mass classification systems based on data from mechanized tunneling projects in Iran. Bull Eng Geol Environ 81:147. https://doi.org/10.1007/s10064-022-02641-y

Hassanpour J, Rostami J, Khamehchiyan M (2023a) Engineering geology and tunneling: evaluation of TBM performance in rock tunnels. University of Tehran Press (in Persian)

Hassanpour J, Goodarzi S, Yagiz S, Rostami J, Davari Sarem M (2023b) Soft sedimentary rocks and their boreability characteristics in mechanised tunnelling projects. Geomechanik und Tunnelbau 16(1):28–37

Hoek E, Brown ET (1980) Underground excavations in rock. The Institute of Mining and Metallurgy, London

Hughes HM (1986) The relative cuttability of coal-measures stone. J Min Sci Technol 3(2):98–109

Innaurato N, Mancini R, Rodena E, Zaninetti A (1991) Forecasting and effective TBM performances in a rapid excavation of a tunnel in Italy. In: Proceedings of 7th International congress on rock mechanics. Aachen, Germany, pp 1009–1014

Macias FJ (2016) Hard rock tunnel boring: performance predictions and cutter life assessments. Ph.D. Thesis, Norwegian Science and Technology University (NTNU), Trondheim (Norway)

NCB (1977) NCB cone indenter. MRDE handbook No. 5. Staffordshire: Mining Research and Development Establishment. National Coal Board, Bretby, UK

Nelson P, O'Rouke TD, Kulhawy FH (1983) Factors affecting TBM penetration rates in sedimentary rocks. In: Proceedings of 24th U.S. symposium on rock mechanics. USA

Nelson P, Ingraffea AR, O'Rouke TD (1985) TBM performance prediction using rock fracture parameters. Int J Rock Mech Min Sci Geomech Abstr 22(3):189–192

Norwegian Institute of Technology (NTH) (1983) Hard rock tunnel boring, project report 1–83. Norwegian Institute Technology, Division of Construction Engineering, Trondheim, Norway

Ozdemir L (1977) Development of theoretical equations for predicting tunnel boreability. Ph.D. Thesis, Colorado School of Mines, Golden, CO (USA)

Palmström A (1995) RMi—a rock mass characterization system for rock engineering purposes. Ph.D. Thesis, Oslo University, Oslo (Norway)

Pourhashemi SM, Ahangari K, Hassanpour J, Eftrekhari SM (2021a) TBM performance analysis in very strong and massive rocks; case study: Kerman water conveyance tunnel project, Iran. Geomech Geoeng 17(4):1110–1122. https://doi.org/10.1080/17486025.2021.1912410

Pourhashemi SM, Ahangari K, Hassanpour J, Eftrekhari SM (2021b) Evaluating the influence of engineering geological parameters on TBM performance during grinding process in Limestone strata. Bull Eng Geol Env 80:3023–3040

Pourhashemi SM, Ahangari K, Hassanpour J, Eftrekhari SM (2021c) Analysis of grinding and chipping processes beneath disc cutters of hard rock TBMs (Case study: the Uma-Oya water conveyance tunnel, Sri-Lanka). J Min Environ (JME) 12(1):281–297

Ramamurthy T (2008) Penetration rate of TBMs. In: Proceedings of world tunnel congress. India

Ramezanzadeh A (2005) Performance analysis and development of new models for performance prediction of hard rock TBMs in rock mass. Ph.D. Thesis, INSA, Lyon (France)

Ribacchi R, Lembo Fazio A (2005) Influence of rock mass parameters on the performance of a TBM in a Gneissic formation (Varzo tunnel). Rock Mech Rock Eng 38(2):105–127

Rostami J (1997) Development of a force estimation model for rock fragmentation with disc cutters through theoretical modeling and physical measurement of crushed zone pressure. Ph.D. Thesis, Colorado School of Mines, Golden, CO (USA)

Rostami J, Ozdemir L (1993) A new model for performance prediction of hard rock TBM. In: Bownerman and Monsees (ed) Proceedings of rapid excavation and tunneling conference (RETC). Boston, MA, Chapter 50, pp 793–809. Society for Mining, Metallurgy, and Exploration Inc., Littleton, CO

Rostami J, Ozdemir L, Neil DM (1994) Performance prediction: a key issue in mechanical hard rock mining. Min Eng 46(11):1263–1267

Roxborough F, Phillips HR (1975) Rock excavation by disc cutter. Int J Rock Mech Min Sci Geomech Abstr 12:361–366

Salimi A, Rostami J, Moormann C (2019) Application of rock mass classification systems for performance estimation of rock TBMs using regression tree and artificial intelligence algorithms. Tunn Undergr Space Technol 92:103046. https://doi.org/10.1016/j.tust.2019.103046

Salimi A (2021) Investigation and evaluation of rock mass characteristics for development of new TBM performance prediction model. Ph.D. Thesis, Institut für Geotechnik der Universität Stuttgart

Salimi A, Rostami J, Moormann C, Hassanpour J (2022) Introducing tree-based-regression models for prediction of hard rock TBM performance with consideration of rock type. Rock Mech Rock Eng 55:4869–4891

Samadi H, Hassanpour J (2025) Application of the stacking-ensemble machine learning algorithm to investigate the rock mass cuttability index in different lithological units using tunnel boring machine performance data. Rock Mech Rock Eng 58(7):8445–8475

Sato K, Gong F, Itakura K (1991) Prediction of disc cutter performance using a circular rock cutting ring. In Proceedings of the International symposium on mine mechanization and automation. Colorado School of Mines, Golden (Colorado)

Sanio HP (1985) Prediction of the performance of disc cutters in anisotropic rocks. Int J Rock Mech Min Sci 22(3):153–161

Sapigni M, Berti M, Bethaz E, Busillo A, Cardone G (2002) TBM performance estimation using rock mass classification. Int J Rock Mech Min Sci 39:771–788

Snowdon AR, Ryley DM, Temporal J (1982) Study of disc cutting in selected British rocks. Int J Rock Mech Min Sci Geomech Abstr 19:107–121

Tarkoy PJ (1975) Rock hardness index properties and geotechnical parameters for predicting tunnel boring machine performance. Ph.D. Thesis, University of Illinois

Wilfing LSF (2016) The influence of geotechnical parameters on penetration prediction in TBM tunneling in hard rock. Ph.D. Thesis, Technical University of Munich (TUM), Munich (Germany)

Yagiz S (2002) Development of rock fracture and brittleness indices to quantify the effects of rock mass features and toughness in the CSM model basic penetration for hard rock tunneling machines. Ph.D. Thesis, Department of Mining and Earth Systems Engineering, Colorado School of Mines, Golden, CO (USA)

Application of Machine Learning (ML)-Based Algorithms in Hard Rock TBM Performance Analysis Using Explainable Artificial Intelligence (XAI)

Hanan Samadi, Jafar Hassanpour, and Jamal Rostami

1 Introduction

Successful planning of tunneling activities involving tunnel boring machines (TBMs) depends on making well-informed assumptions and judgments of TBM performance, as this assessment influences critical aspects such as project planning (Hassanpour et al. 2011), cost management (Einstein 2020), and the utilization factor (Rostami et al. 2013) and feasibility of tunneling using a certain boring machine type. Gross errors or substantial miscalculations in TBM performance can lead to unexpected expenses, project delays, and even the viability of the whole project cost and schedule (Bieniawski et al. 2006; Hassanpour et al. 2015, 2021; Samadi et al. 2024a). Numerous researchers have highlighted the significance of intact rock properties, rock mass parameters, operational parameters of boring machines, and machine specifications for accurate estimation of TBM performance in hard rock conditions. To achieve this objective, studies have focused on two techniques: theoretical and empirical (Rostami and Ozdemir 1993; Bruland 1998; Barton 2000; Bieniawski et al. 2006; Hoek et al. 2008; Hassanpour et al. 2009, 2019; Samadi and Hassanpour 2025). Theoretical models employ cutting forces and laboratory cutting tests to establish formulas based on force equilibrium. While these models provide a fundamental

H. Samadi · J. Hassanpour (✉)
University of Tehran, Tehran, Iran
e-mail: hassanpour@ut.ac.ir

H. Samadi
e-mail: samadi.hanan1996@ut.ac.ir; hsamadi@systra.com

H. Samadi
R&D Department, SYSTRA Co., Trento, Italy

J. Rostami
Colorado School of Mines, Golden, CO, USA
e-mail: rostami@mines.edu

J. Hassanpour and J. Rostami (eds.), *Performance of Tunnel Boring Machines (TBM) in Rock*, Springer Series in Geomechanics and Geoengineering,
https://doi.org/10.1007/978-3-032-20424-0_9

understanding of rock fragmentation, they do not entirely capture the complexities of real-case rock conditions (Rostami and Ozdemir 1993; Bruland 1998). The second approach, based on field performance observations, considers the site's geomechanical conditions. Researchers have proposed various statistical methods, based on regression analysis, for predicting TBM performance (Hassanpour et al. 2009, 2010, 2016; Farrokh et al. 2012; Pourhashemi et al. 2022; Dardashti et al. 2024a, Samadi and Hassanpour 2024).

There are also some empirical models that are based on soft computing techniques and artificial intelligence (AI) principles. These models type have gained significant popularity in recent years to address the complexity of predicting TBM performance. The identification of factors influencing TBM performance and its parameters is a common challenge that can be effectively addressed using smart learning methods. These techniques have become extensively employed in tunnelling projects to predict TBM performance and have shown promising and reliable outcomes. This chapter takes a closer look at machine learning (ML)-based algorithms (one of the techniques for prediction based on AI principles) for TBM performance prediction.

2 Introduction to Learning Algorithms

In regression analysis, there are generally two methodologies commonly used. The first method involves simple regression analysis, and the second method involves more complex mathematical functions. The first method is used for linear or nonlinear regression analysis, generally without specific steps for normalizing and dividing the database into training, testing, and validation subsets. In comparison to traditional regression analysis, newer algorithms are based on advanced mathematical and computational techniques, known as smart soft computing techniques that have been under development in recent years (Japkowicz and Shah 2011). These models are classified in Fig. 1. More detailed explanations of these methodologies can be found in the literature (Japkowicz and Shah 2011; Sarker 2021) and will not be repeated here.

Machine learning algorithms, as one of the widely used learning techniques in TBM performance prediction, can be classified into supervised, unsupervised, reinforcement algorithms (Fig. 2), and semi-supervised (Fig. 3). The description of these learning algorithms was presented in the literature (Bonaccorso 2018; Sarker 2021). There are further categories within each group of algorithms (except supervised and some unsupervised learning techniques) that are not commonly utilized in tunnelling and TBM performance analysis, so a detailed description is beyond the scope of this chapter. Supervised learning is a machine learning algorithm that utilizes labeled datasets to train models or algorithms. The objective is to establish a mapping from input data to output labels, enabling predictions or classifications on new, unseen data. This learning method can be categorized into nine main divisions: linear model, k-nearest neighbor (KNN), support vector machine (SVM), stochastic gradient descent (SGD), decision tree (DT), ensemble learning, generative model,

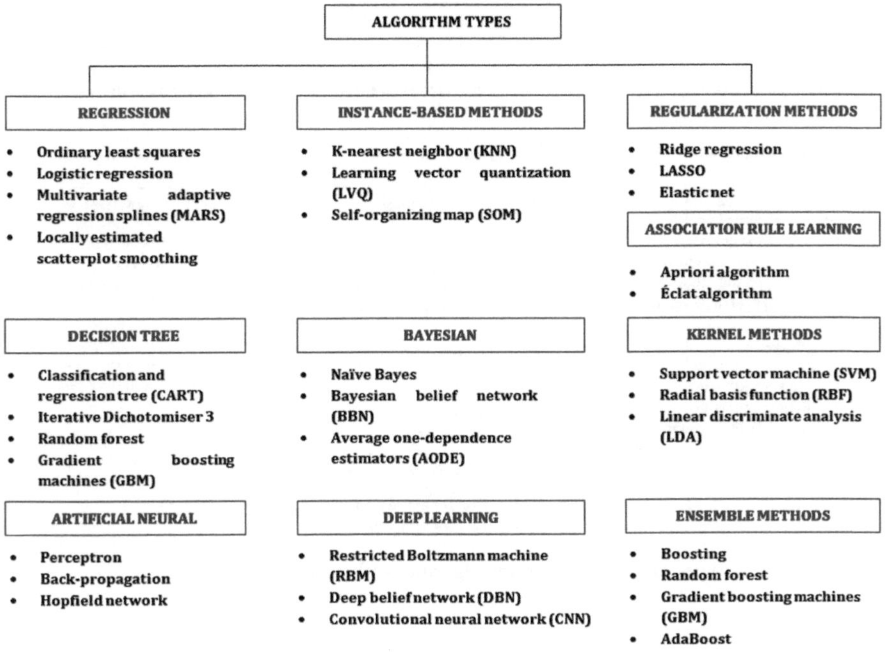

Fig. 1 The systematic classification of learning algorithms

time series forecasting, and supervised dimensionality reduction (SDR) techniques (Fig. 4).

3 Application of ML-Based Models for Predictive Purposes in Hard Rock Tunnelling

In the initial phase of the study, it is crucial to collect a database consisting of information gathered from tunnelling projects. This dataset should include the geomechanical properties of intact rock, such as the uniaxial compressive strength (UCS) (Bieniawski et al. 2006; Rostami 2008), as well as the geomechanical properties of the rock mass, such as rock quality designation (RQD) (Hoek et al. 2008; Hassanpour et al. 2009, 2010, 2024; Barton 2015; Rostami 2016; Samadi and Hassanpour 2025). Additionally, it should include TBM operational factors, including cutterhead thrust force, rotation speed, torque, and cutterhead power, as well as the specification of the TBM, such as the number and diameter of disc cutters and the cutterhead diameter. Moreover, TBM performance indicators like penetration rate and advance rate should be included (Rostami 2008; Barton 2009; Hassanpour et al. 2021; Samadi and Hassanpour 2022; Samadi et al. 2024b).

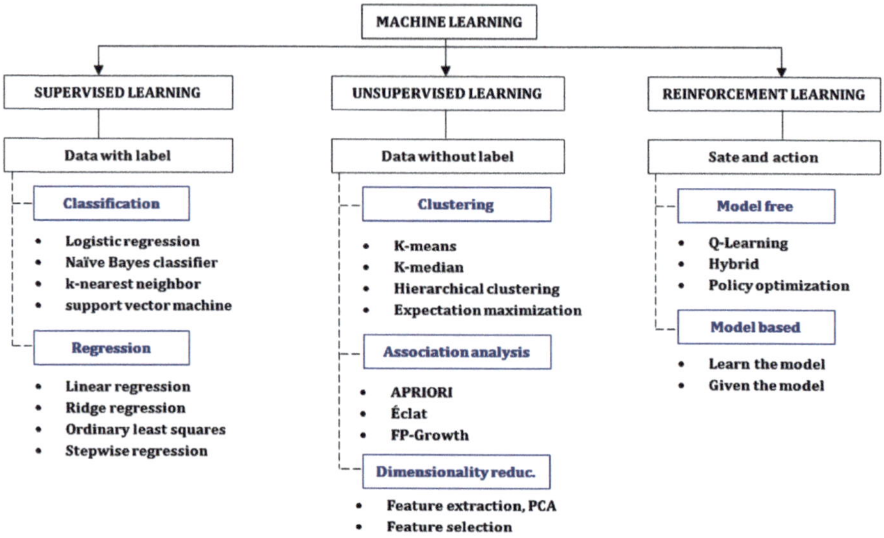

Fig. 2 Different types of machine learning algorithms (three major classes)

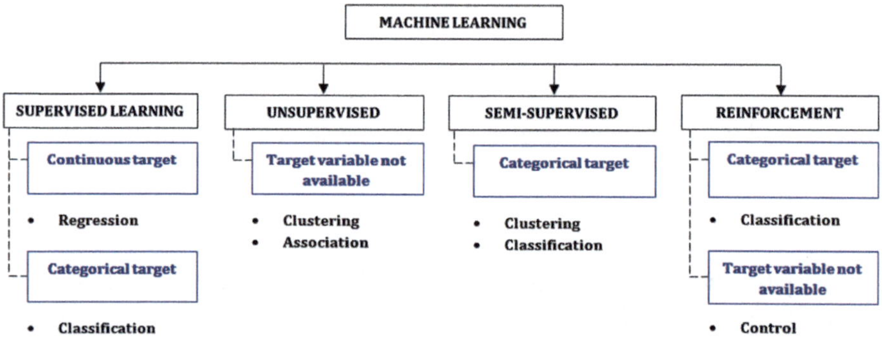

Fig. 3 Different types of machine learning algorithms (four major classes)

Subsequently, the input and output parameters within the database need to be identified, following which the dataset should be divided into two subsets—a training set and a testing set. The consideration of 75, 80, 85, or 90% of the total database for the training set, and the remaining percentage for the testing set, is a commonly employed approach. Deviations from these percentages may occur due to errors encountered during model training, leading to lower accuracy. Normalization of the dataset using a standardized formula is essential to scale all variables to a common range of 0–1. This normalization process becomes crucial when the parameters exhibit variations in units and ranges. Ensuring uniform fluctuation of data points across both the training and testing subsets is crucial. The database should capture the full distribution of data

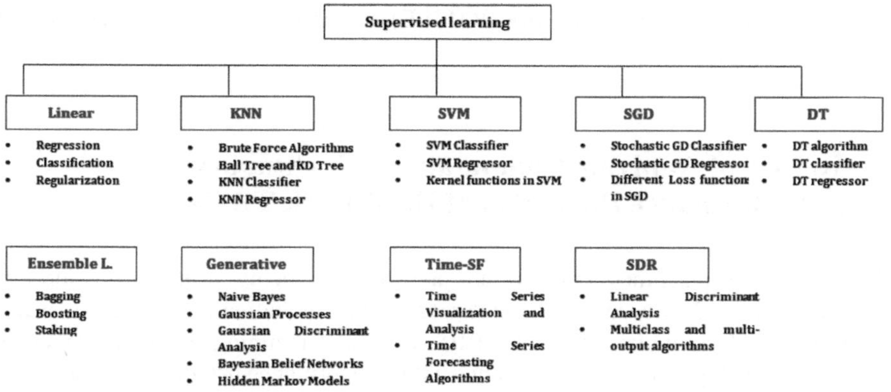

Fig. 4 Different types of supervised machine learning algorithms

points within all ranges. Subsequently, suitable ML-based models must be chosen for predictive analyses.

When selecting the most suitable model for a particular dataset, several factors must be taken into consideration. Firstly, the size of the dataset plays a crucial role in determining which model will yield the best results. SVM networks and kernel function-based models perform well with small datasets, whereas RNNs are better suited for larger datasets. Furthermore, the type of data being used also influences the choice of model. Numerical data can be effectively handled by most AI-based models for regression tasks, while non-numerical data may benefit from classification-based models such as decision trees, k-nearest neighbors (KNN), and Naïve Bayes. Another important consideration is the number of features present in the dataset (Bonaccorso 2018; Sarker 2021).

Additionally, the nature of the data points themselves is an important factor to consider. For instance, gated recurrent unit (GRU) networks may be more suitable for time series prediction tasks where the sequential nature of the data is significant. It is worth noting that while simple regression models can achieve high accuracy levels with minimal error, more advanced algorithms generally yield better results. However, the specific characteristics of the dataset and the chosen neural network architecture should always be taken into account when selecting the most appropriate model (Sarker 2021).

Following the training and testing of ML algorithms to obtain predictive results, the accuracy of these outcomes is typically assessed using various metrics and statistical measurements. One commonly employed approach involves the utilization of loss functions and evaluation metrics to analyze the accuracy of predictions. Notable loss functions for this purpose include but are not limited to, root mean square error (RMSE), mean square error (MSE), normalized root mean square error (NRMSE), root relative squared error (RRSE), relative standard error (RSE), mean absolute deviation (MAD), mean absolute percentage error (MAPE), relative root mean square error (RRMSE), mean bias error (MBE), relative absolute error (RAE), root mean squared logarithmic error (RMSLE), log cash loss (LCL), and coefficient of determination (R^2).

4 Review of Applying Computational Techniques in Performance Prediction of Hard Rock TBMs

Tunnel boring machines (TBMs) are increasingly being used worldwide for various projects (Jakobsen and Babendererde 2017), including transportation, water conveyance (Hassanpour et al. 2009, 2010, 2011, 2015), mining, etc. Various models have been developed to assess and predict TBM performance in hard rock conditions, based on theoretical, empirical regression, and empirical ML-based techniques. The proposed models based on theoretical and empirical-regression-based techniques were discussed in detail in previous chapters and will not be repeated here. This chapter focuses on research using empirical ML-based models to predict TBM performance. Table 1 presents a summary of recent works that employed ML algorithms for TBM performance prediction from the year 2000 to the present.

Figure 5 presents the visualization of scientific landscapes in the field of TBM performance prediction using ML-based algorithms, utilizing data sourced from the Web of Science from 2020 to the present. From 2020 to 2023, the focus of studies lay in the utilization of various traditional and simple ML algorithms for TBM performance prediction. However, a new scientific trend in scholarly papers has emerged after 2023, focusing on the correlation between TBM performance and rock geomechanical parameters using new hybrid or stacking algorithms. The current necessity lies in developing and utilizing hybrid and stacking smart models, formulating new equations and conducting analyses to uncover the relationship between rock mass geomechanical parameters and TBM performance, drawing insights from observed and generated databases of mechanized tunneling projects.

The number of papers released from 2000 to the present using data collected from Web of Science databases about TBM performance in hard rock environments, using various machine learning, neural networks, decision trees, fuzzy systems, stacked models, genetic approaches, and other AI-based algorithms, is illustrated in Fig. 6. The continuous rise in publications focusing on the application of AI-based techniques in TBM performance analysis reflects the growing interest in utilizing advanced AI algorithms to enhance the efficiency and reliability of AI-based, ML or deep learning (DL)-based algorithms in TBM performance prediction.

Figure 7 shows the quantity of papers from 2000 to the present in TBM performance prediction with smart ML algorithms per year. The data shows a rising trend in the publication rate within this field. In recent years, researchers have increasingly recognized the potential of smart ML algorithms in improving the accuracy and efficiency of performance prediction in TBM projects. This growing body of literature shows a shift towards more advanced data-driven approaches. Figure 8 depicts

Table 1 Summary of recent works in the field of TBM performance prediction using ML-based algorithms from 2000 to the present

References	Inputs/outputs	Method	Train-data	Test-data
Grima et al. (2000)	UCS, RPM, TF, D_c	ANN	512	128
	UCS, RPM, TF, D_c	ANFIS	512	128
Benardos and Kaliampakos (2004)	RQD, RMR, k, N, WTS, WZ	ANN	–	–
Simoes and Kim (2006)	D, RQD, RMR, WIR	FL	Queens, Milyang, Manapouri projects	
Benardos (2008)	RQD, RMW, SF, RMR, UCS, H, WT, K	ANN	642 data points	
Mikaeil et al. (2009)	UCS, BTS, PSI, DPW, α	FS	151 data points	
Gholamnejad and Tayarani (2010)	UCS, RQD, DPW	ANN	185 data points	
Oraee et al. (2012)	RQD, UCS, DPW	ANFIS	177 data points	
Ge et al. (2013)	UCS, BTS, PSI, DPW, α	LS-SVM	151 data points	
Ling et al. (2013)	UCS, BTS, PSI, DPW, α	PLS-FNN	151 data points	
Martins and Miranda (2013)	UCS, PSI, DPW, α	ANN	151 data points	
	UCS, PSI, DPW, α	SVR	151 data points	
Salimi and Esmaeili (2013)	UCS, BTS, PSI, DPW, β	ANN	46 data points	
Shao et al. (2013)	DPW, α, UCS, PSI	ELM	133	20
	DPW, α, UCS, PSI	LS-SVM	133	20
	DPW, α, UCS, PSI	PLS	133	20
	DPW, α, UCS, PSI	GP	133	20
Salimi et al. (2015)	UCS, BTS, Js, RQD, α, RMR, Q, GSI	ANN	75 data points	
	UCS, BTS, Js, RQD, α, RMR, Q, GSI	ANFIS	75 data points	

(continued)

Table 1 (continued)

References	Inputs/outputs	Method	Train-data	Test-data
	UCS, BTS, Js, RQD, α, RMR, Q, GSI	SVR	75 data points	
Tao et al. (2015)	UCS, BTS, BI, DPW, α	RF	151 data points	
Fattahi (2016)	UCS, BI, DPW, α	ANFIS	151 data points	
Salimi et al. (2016)	UCS, BTS, Js, RQD, α, RMR, Q, GSI	ANFIS	75 data points	
	UCS, BTS, Js, RQD, α, RMR, Q, GSI	SVR	75 data points	
Minh et al. (2017)	UCS, BTS, BI, DPW, α	FL	151 data points	
Liu et al. (2019)	UCS, BI, DPW, α	SVR	150	30
Salimi et al. (2019)	UCS, RQD, DPW, JC	CART	Various tunnels worldwide	
	UCS, RQD, DPW, JC	GP	Various tunnels worldwide	
Feng et al. (2021)	RPM, CT, TF, P	DBN	1173 + 3367 + 3575	
Xu et al. (2021)	AS, RPM, CT, TF	BR	86,400, 20 GB	
	AS, RPM, CT, TF	KNN	86,400, 20 GB	
	AS, RPM, CT, TF	RF	86,400, 20 GB	
	AS, RPM, CT, TF	GTB	86,400, 20 GB	
	AS, RPM, CT, TF	SVM	86,400, 20 GB	
	AS, RPM, CT, TF	CNN	86,400, 20 GB	
	AS, RPM, CT, TF	LSTM	86,400, 20 GB	
Ma et al. (2022)	TF, RPM, UCS, VJC	MR	–	–
	TF, RPM, UCS, VJC	BPNN	–	–

(continued)

Table 1 (continued)

References	Inputs/outputs	Method	Train-data	Test-data
	TF, RPM, UCS, VJC	SVR	–	–
Mahmoodzadeh et al. (2022)	UCS, BTS, PSI, DPW, α, RFC	LSTM	900	225
	UCS, BTS, PSI, DPW, α, RFC	GPR	900	225
	UCS, BTS, PSI, DPW, α, RFC	SVR	900	225
	UCS, BTS, PSI, DPW, α, RFC	KNN	900	225
	UCS, BTS, PSI, DPW, α, RFC	DT	900	225
Yang et al. (2022)	CP, TF, CT, RPM, C, φ, UCS, RQD	GWO-FW-MKL-SVR	503 data for training and testing	
Huang et al. (2023)	190 features	KNN	80% of data	20% of data
	190 features	SVR	80% of data	20% of data
	190 features	ANN	80% of data	20% of data
	190 features	RF	80% of data	20% of data
	190 features	DT	80% of data	20% of data
Latif et al. (2023)	RPM, CT, TF, AS, GP, Q, RPM	SVR	1285 data for training and testing	
	RPM, CT, TF, AS, GP, Q, RPM	ANN	1285 data for training and testing	
Wang et al. (2023)	141 digital and 232 analogue features	PSO	Nov 2020–Mar 2021 data collection	
	141 digital and 232 analogue features	BPSO	Nov 2020–Mar 2021 data collection	
	141 digital and 232 analogue features	BPSO-XGBoost	Nov 2020–Mar 2021 data collection	
	141 digital and 232 analogue features	PSO-XGBoost	Nov 2020–Mar 2021 data collection	

(continued)

Table 1 (continued)

References	Inputs/outputs	Method	Train-data	Test-data
Xu et al. (2023)	RPM, TF, TD, GL, ES	LSTM	115,397	44,875
Dardashti et al. (2024b)	Ten tunnel projects, input parameters of RMR	SR	523 tunnel sections	
Monthanopparat and Tanchaisawat (2024)	PR, UCS, RMT, RPM, CT, TF	DFF	Mae Tang–Mae Ngad Project	
	PR, UCS, RMT, RPM, CT, TF	LSTM	Mae Tang–Mae Ngad Project	
Yu et al. (2024)	RPM, TF, CT, ES, BC	ResNet-LSTM	8:1:1	8:1:1
Zhang et al. (2024)	IoT application	GCN	Singapore's Circle Line 6 C885 project	
	IoT application	LSTM	Singapore's Circle Line 6 C885 project	
	IoT application	MOO	Singapore's Circle Line 6 C885 project	
Kullarkar and Thote (2025)	UCS, I_{50}, RQD, RMR, n, BTS	RF	235 data points	
	UCS, I_{50}, RQD, RMR, n, BTS	KNN	235 data points	
	UCS, I_{50}, RQD, RMR, n, BTS	XGBoost	235 data points	
	UCS, I_{50}, RQD, RMR, n, BTS	SVR	235 data points	
	UCS, I_{50}, RQD, RMR, n, BTS	ANN	235 data points	
Samadi and Hassanpour (2025)	UCS and RQD for FPI	FSR-OPCB	80% of data	20% of data
	UCS and RQD for FPI	SCG-BPNN	80% of data	20% of data
	UCS and RQD for FPI	AO-RNN	80% of data	20% of data
	UCS and RQD for FPI	CD-KNN	80% of data	20% of data

(continued)

Table 1 (continued)

References	Inputs/outputs	Method	Train-data	Test-data
	UCS and RQD for FPI	HBA-KNN	80% of data	20% of data
	UCS and RQD for FPI	AdaD-LSTM	80% of data	20% of data
	UCS and RQD for FPI	AdaG-GRU	80% of data	20% of data
	UCS and RQD for FPI	Stacking	80% of data	20% of data

Note AdaD: AdaDelta, AdaG: AdaGrad, ANFIS: adaptive neuro-fuzzy inference system, ANN: artificial neural network, AO: Adam optimization, BPNN: back propagation neural network, BPSO: beetle antennae search particle swarm optimization, BR: Bayesian ridge regression, CART: classification and regression tree, CD: city block, CNN: convolution neural network, DBN: deep belief network, DT: decision tree, ELM: extreme learning machine, FL: fuzzy logic, FNN: feed-forward neural network, FS: fuzzy system, FSR: feed-forward stepwise regression, GBoost: gradient boosting, GCN: graph convolutional network, GP: genetic programming, GRU: gated recurrent units, GTB: gradient tree boosting, GWO-FW-MK-SVR: grey wolf optimizer-feature weighted-multiple kernel-support vector regression, HBA: honey badger algorithm, KNN: k-nearest neighbor, LR: linear regression, LSSVM: least-square support vector machine, LSTM: long short term memory, MOO: online multi-objective optimization, MR: multiple regression model, OPCB: overfit prevention criterion boosting technique, PSO: particle swarm optimization, RF: random forest, RNN: recurrent neural network, RT: regression tree, SR: stepwise regression, SCG: scaled conjunction gradient, SVR: support vector regression, XGBoost: extreme gradient boosting

the number of papers in TBM performance utilizing ML algorithms across various research categories. The majority of articles fall into research and review papers.

5 Application of ML Techniques in Hard Rock TBM Tunnelling (Case Study)

Numerous researchers have highlighted the significance of predicting metrics such as the rate of penetration (PR, m/h), penetration per revolution (P, mm/rev), field penetration index (FPI = Fn/P, kN/mm/rev), and cutter load (Fn, kN) for accurate estimation of TBM performance in hard rock conditions. To achieve this objective, studies have focused on three prediction techniques: theoretical, empirical, and soft computing approaches (as presented in Table 2). The advantages and disadvantages of these models, along with solutions to address the disadvantages, are presented in Table 3.

Precise evaluation of the performance of TBM is essential to achieve the suitable advance and utilization of the machine, and also controls the challenges associated with the considerable costs and complex planning involved in tunnel construction. To address this issue, a database containing mainly the Iranian mechanized tunnelling projects, complemented by other projects from around the world (Fig. 9) has been

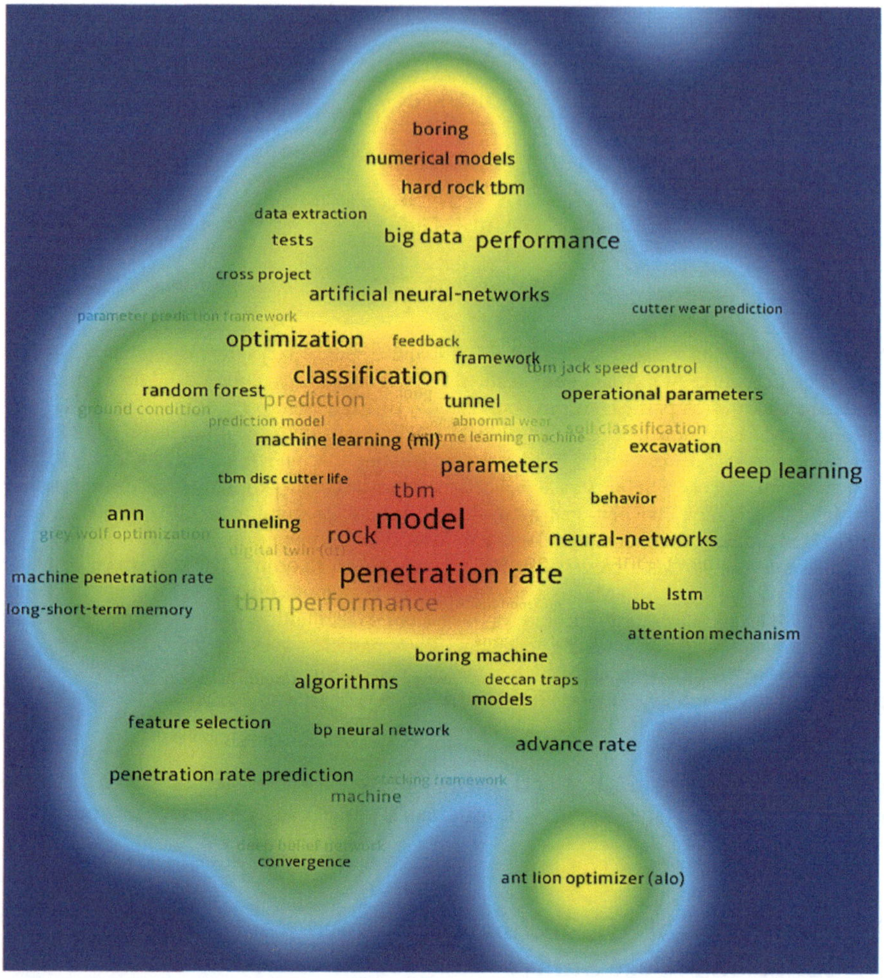

Fig. 5 Visualization of scientific landscapes in the field of TBM performance using AI-based algorithms from 2020 to the present

compiled, which contains the geotechnical and engineering geological information, along with the machine's operational parameters and recorded performance.

The first step, which is an important step for the accuracy of the analysis is to understand the data set. This can be done by a preliminary qualitative and descriptive statistical analysis of the data to see the ranges of data. In this stage, the objective function or parameter should be clear and relevant, while other input parameters should be identified. A common mistake in TBM performance analysis is to look at TBM torque, whereas torque is not an independent parameter, rather a function of the thrust and the rock type, where increased thrust will lead to increased penetration, hence leading to higher rolling forces and torque. So the strong correlation between

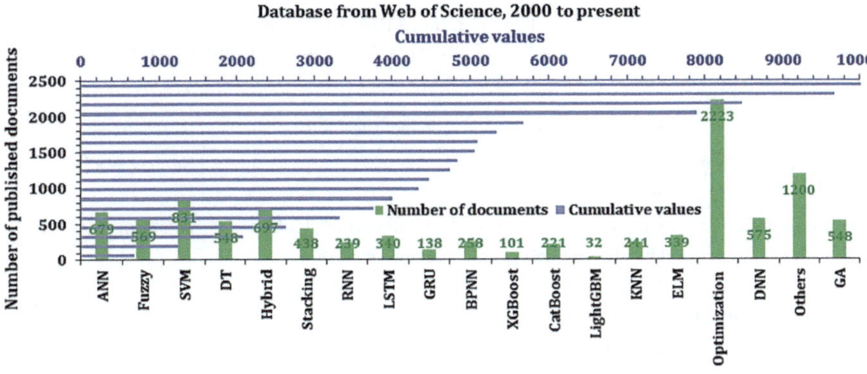

Fig. 6 Number of published documents in the field of TBM performance using AI-based algorithms, considering different models from 2000 to the present

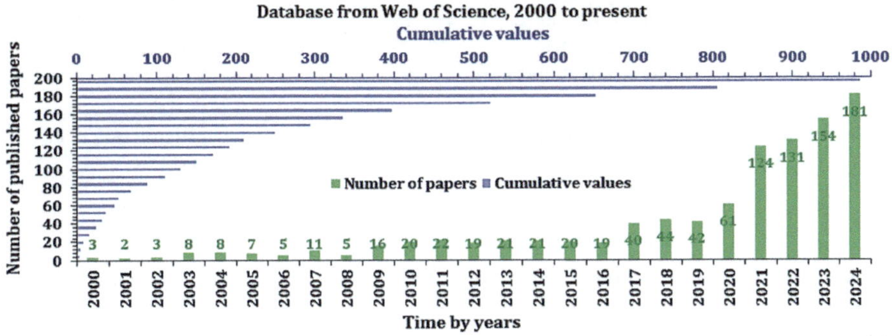

Fig. 7 Number of published papers in the field of TBM performance prediction using ML-based algorithms by year from 2000 to the present

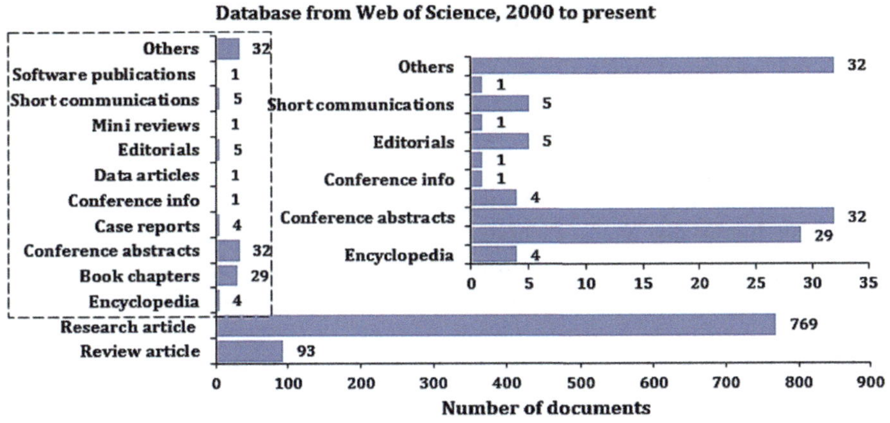

Fig. 8 Number of papers in the field of TBM performance prediction using ML-based algorithms, considering different research types from 2000 to the present

Table 2 Summary of widely used models to predict TBM performance based on three techniques: theoretical, empirical (regression-based), and empirical (computing machinery approaches)

Target parameter	Rock mass factors	Machine factors	References
(A) Theoretical model			
Cutter load, Rolling cutter force	Uniaxial compressive strength	–	Sanio (1985)
Penetration rate	Uniaxial compressive strength, Tensile strength	Cutter force, Cutter rotation speed	Rostami (1997)
(B) Empirical model based on simple regression			
Penetration rate, Advance rate	Drilling rate index, Angle between tunnel axis and joint, Joint spacing	–	Bruland (1998)
Penetration rate, Advance rate	Q system parameters, Cutter life index, Quartz content, Uniaxial compressive strength	Cutter force	Barton (2000)
Advance rate	Drilling rate index, Uniaxial compressive strength, Stand-up time, Joint condition, Groundwater	TBM diameter	Bieniawski et al. (2006)
Field penetration index	Uniaxial compressive strength, Rock quality designation	Cutter force	Hassanpour et al. (2009, 2011)
Penetration rate	Uniaxial compressive strength, Rock type	TBM diameter, Cutter rotation speed	Farrokh et al. (2012)
Field penetration index	Uniaxial compressive strength, Rock quality designation	FPI = Fn/P	Samadi and Hassanpour (2025)
(C) Empirical model based on soft computing techniques			
Presented in Table 1			

torque and penetration rate should not mask the fact that torque is a dependent variable. One should also check co-linearity or regression between input parameters should be examined to avoid overfitting by the models that is related to the interrelationship between the input parameters that are supposed to be independent of each other.

Bi-lineal statistical analysis between the parameters shows the general trends between the input parameters and the objective function/parameter. The individual set of parameters and their relationship may not be clear, but at least general trends and the sensitivity of the output to individual parameters can be seen in this step.

Table 3 The advantages and disadvantages of the described models in Table 2

	Advantages	Disadvantages	Solutions
Theoretical model	(a) Based on extensive research and analysis from small-scale cutting performance laboratory studies (b) Customizable to meet specific project needs and geological conditions, especially when there is no prior experience with real cases	(a) Laboratory cutting force analysis may not reliably predict TBM performance due to uncertainties in geological conditions, machine behavior, and other challenging geological factors (b) The predictions may not apply to all tunneling projects, as they are often based on specific assumptions and constraints that may not be acceptable in all geological and lithological conditions	(a) Collecting data from actual tunneling projects would be advantageous for developing a model based on empirical analysis and real scenarios (b) It can be beneficial to combine theoretical analysis from small-scale laboratory studies with empirical analysis from real tunneling project data
Empirical (regression-based)	(a) Clearly develops the relationship between TBM performance, geomechanical data, and machine specifications (b) Easily applies with minimal computational resources and time, unlike AI-based predictive modelling techniques (c) Effectively identifies trends and patterns in data (d) Offers statistical measures to indicate the significance and strength of variable relationships	(a) A linear relationship between independent and dependent variables may not reflect the true nature of the data (b) The model may lack the complexity needed to accurately predict outcomes influenced by various geomechanical and machine factors on TBM performance (c) It may be prone to overfitting by capturing noise or underfitting by oversimplifying relationships (d) The approach considers only the main effects of individual variables, ignoring potential interactions or complex relationships in the data	(a) This problem can be addressed using advanced nonlinear functions in machine learning analysis (b) This problem can be solved using AI-based smart computing techniques (c) This problem can be addressed using advanced machine learning techniques to prevent overfitting (d) This problem can be addressed with robust functions and optimization techniques

(continued)

Table 3 (continued)

	Advantages	Disadvantages	Solutions
Empirical (computing machinery approaches)	(a) Efficiently analyze extensive geomechanical and machine data to extract complex patterns between rock mechanics and TBM specifications, surpassing traditional statistical methods (b) Automatically identify and extract relevant rock geomechanical and machine operating features from raw data collected from field studies, TBM PLCs, laboratory investigations, and work shifts, minimizing manual feature engineering and associated errors (c) Scale effectively to handle large, diverse datasets with varying units, making them ideal for applications that demand predictions on a large, standardized, and normalized scale (d) Utilize knowledge from one domain to enhance predictions in another through transfer learning, particularly in stacking and bagging learning when labelled data is limited	(a) Frequently considered as "black box" models, making it difficult to detect their prediction processes (b) Sensitive to the quality of training data (c) Demand considerable computational resources, including high-performance GPUs and ample memory (d) Prone to adversarial attacks, where malicious actors alter input data to deceive the model into making errors	(a) This study addresses gaps in previous research by examining network structures, hyper-parameter ranges, SHAPE analysis, and classifying results into sub-datasets based on lithology (b) An extensive, high-quality dataset from various global projects across different lithological units was generated for this study (c) Additionally, a specific system with 32 GB of RAM and a 2.40 GHz Intel Core 17-131700H CPU was employed to solve the issue (d) The paper presents a comprehensive approach that includes input preprocessing, a detection mechanism, and an appropriate model architecture to resolve the problem

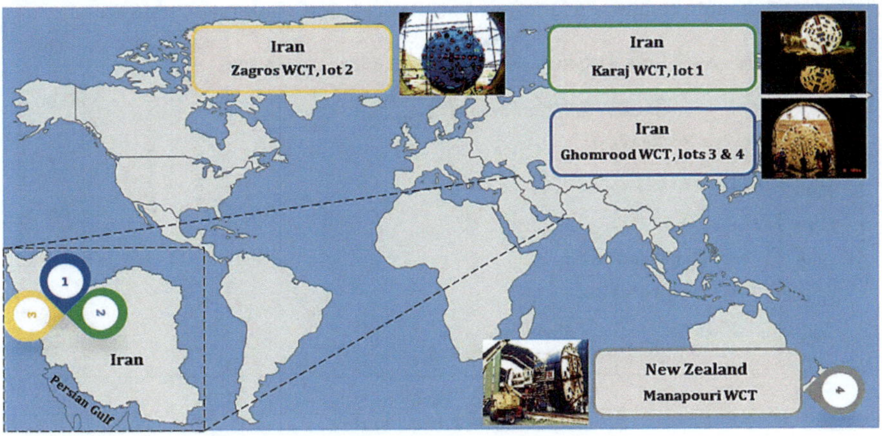

Fig. 9 Geographical distribution of selected TBM tunnelling projects in this work

This should be followed by an analysis of the impact of input variables on the output production.

One of the most common approaches is to perform a principal component analysis (PCA-unsupervised learning method), where the software can examine interdependence between the output parameter and different subsets of input parameters, leading the users to see what is the best single parameter for predicting the outcome, followed by the best set of two, three, ... parameters to be used as independent input parameters (Fig. 10). The Kaiser–Meyer–Olkin (KMO) measure of sampling adequacy was 0.798, indicating that the dataset is adequately suitable for PCA. Rock engineering parameters load strongly on Component 1 (PC-1), especially UCS with 0.984. This analysis identifies the optimal regression that can be achieved with a subset of input parameters, thereby highlighting the importance of each parameter and the sequence of selecting critical input parameters for modelling. PCA or similar analysis also quantitatively show that by adding the next parameter, what is the gain in improving the correlations/regressions. In some cases, adding additional parameters may have little to no impact on the accuracy of predictions. In such cases, the user should decide how many input parameters to use and which ones are more important.

The analysis of input parameters should be followed by simple linear and nonlinear regressions to establish a benchmark for the prediction models (Fig. 11). This allows the end users to see formulas for the prediction of the output parameter based on selected input parameters (as presented in Hassanpour et al. 2011). Such formulas can easily be programmed by other users to see the output of the model and to see how they measure up against the more complex, black box models offered by the ML algorithms. The formulas and models used in this way are typically most useful for practical applications by practitioners and operators/engineers who do not have access to source codes for the ML programs, hence often cited and used in tunneling applications. Once these steps are done, the data can be used in follow-up steps in the ML modelling process.

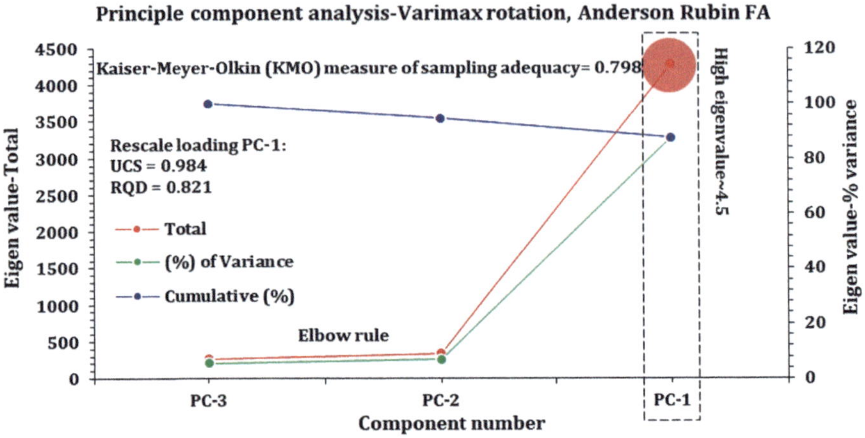

Fig. 10 The PCA analysis for the assessment of FPI in this study

Several ML-based algorithms have been developed, including hybrid-optimized, and stacking-ensemble supervised learning models for TBM performance assessment in different lithological types (ET-I to ET-IV) (Hassanpour et al. 2023). The flowchart of this study is presented in Fig. 12. The detailed description of geological formations, engineering geological and geotechnical properties, and machine specifications were presented in literature and published documents by authors of this chapter (Hassanpour et al. 2009, 2010, 2011, 2015, 2021, 2023; Samadi et al. 2024a, b; Samadi and Hassanpour 2025) and summarized in Tables 4 and 5 and Fig. 13. Laboratory tests, daily operating records, and TBM data logger were recorded for each stroke and engineering geological unit, resulting in a dataset of 156 data points (see Chap. "University of Tehran Model, A Geological-based Model for Rock Mass Boreability Classification and TBM Performance Prediction" for more details) that included seven parameters (Fig. 14). The dataset was divided into 85% (132 data points) for training and 15% (24 data points) (Fig. 15) for testing the hybrid-optimized supervised learning models to predict FPI based on two rock mechanical parameters, including UCS and RQD (the input rock mechanical parameters for introducing the University of Tehran or UT model (Hassanpour et al. 2023). The coding, training, and evaluation of these learning algorithms were carried out using the Python programming language and MATLAB software.

The results are presented in Table 6, showing the performance of models including tree seed algorithm-extreme gradient boosting (TSA-XGBoost), salp swarm algorithm-extreme gradient boosting (SSA-XGBoost), honey badger algorithm-k-nearest neural network (HBA-KNN), equilibrium slime mould algorithm-k-nearest neural network (ESMA-KNN), Gaussian-fuzzy inference system (G-FIS), least square-support vector machine (LSSVM), radial basis function (RBF), scaled conjugate gradient-deep neural network (SCG-DNN), scaled conjugate gradient-back propagation neural network (SCG-BPNN), scaled conjugate gradient-feed forward

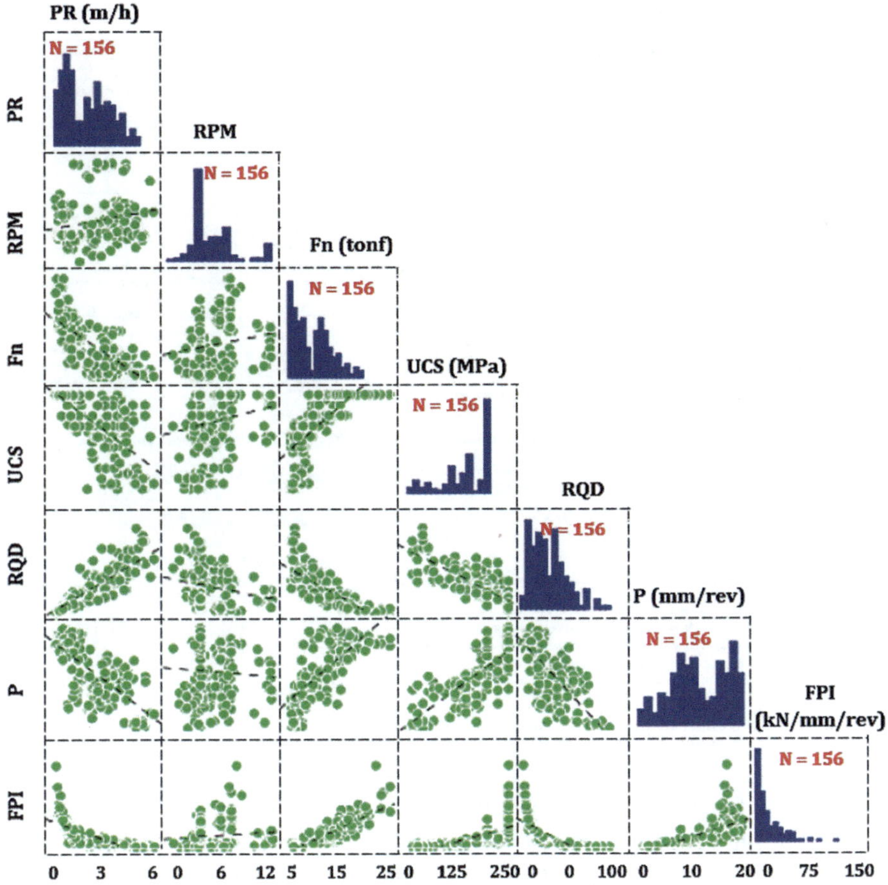

Fig. 11 The scatterplot matrix-regression for FPI assessment

neural network (SCG-FFNN), stochastic gradient descent-support vector machine (SGD-SVM), and stochastic gradient descent-support vector regression (SGD-SVR).

Among these, the SCG-DNN model demonstrated exceptional accuracy exceeding 97% in both the training ($R^2 = 0.9752$) and testing ($R^2 = 0.9713$) phases. Notably, the SCG-DNN model outperformed other optimized neural network models with minimal prediction error in training and testing phases (Fig. 16a). The comparison of SCG-FFNN and SCG-BPNN (Fig. 16b, c) and G-FIS (Fig. 16d) models showed the superiority of G-FIS in both training and testing performance. Furthermore, comparing the performance, the SCG-DNN (Fig. 16a) model surpassed the fuzzy model (Fig. 16d), and SCG-BPNN (Fig. 16c) has better performance than SCG-BPNN (Fig. 16b) in prediction.

Additionally, the comparison between SGD-SVR (Fig. 17a), SGD-SVM (Fig. 17b), LSSVM (Fig. 17c), and RBF (Fig. 17d) models indicated that SGD-SVR provided higher accuracy in predicting TBM performance compared to other

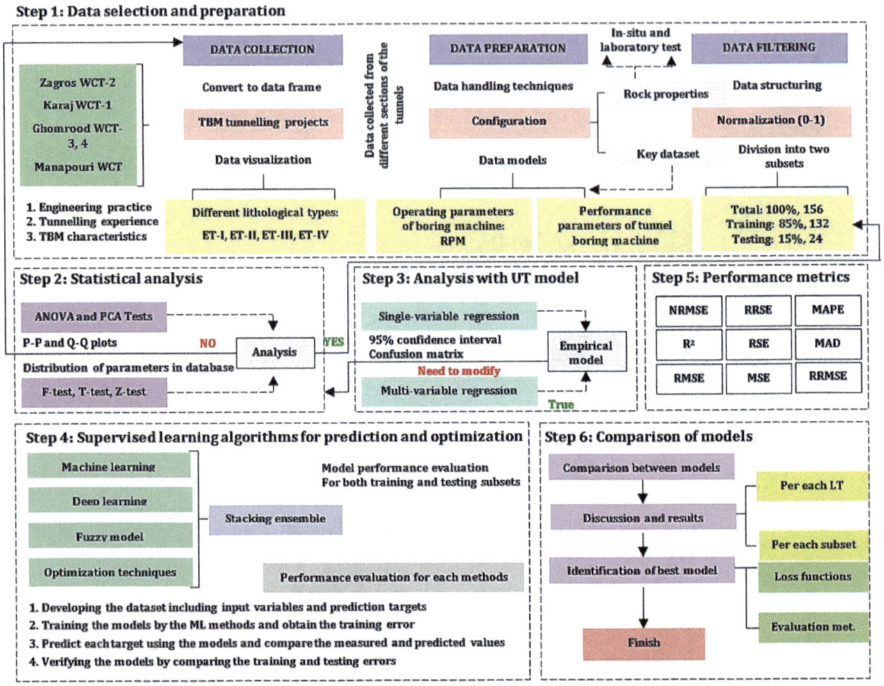

Fig. 12 The flowchart of the current study to predict FPI

kernel-based functions. In summary, deep learning models, particularly SCG-DNN (Fig. 17a), outperformed other machine learning models due to their ability to handle complex problems requiring numerous neurons and a one-directional computational process.

Also, this research employs other hybrid-optimized learning models by utilizing the k-nearest neighbors (KNN) and extreme gradient boosting (XGBoost) approaches. To enhance the model performance, advanced optimization techniques such as tree seed algorithm (TSA), salp swarm algorithm (SSA), honey badger algorithm (HBA), and equilibrium slime mould algorithm (ESMA) were used to optimize and develop TSA-XGBoost (Fig. 18a), SSA-XGBoost (Fig. 18b), HBA-KNN (Fig. 18c), and ESMA-KNN (Fig. 18d) models. The results showed that the TSA-XGBoost (Fig. 18a) model has achieved higher performance in both phases compared to SSA-XGBoost. In comparing the performance of KNN-optimized HBA and ESMA techniques, the ESMA-KNN model outperforms another KNN-optimized model.

Performance results have been analyzed by conducting various tests, including score analysis, importance ranking, intuitive display, and Taylor chart evaluation. The analysis of scores involves a comparison of the efficacy of the top architectural models using statistical methods. Each model is assigned a score of n (in this study, n = 108 because of the 12 by 9 matrix) to determine the optimal value for every performance

Table 4 Geological characteristics of tunnelling projects

No.	Projects	Geol. zone	Eng. Geol. unit	Formation	Lithology
1	ZWCT-2	Zagros simply folded zone	S1, S2, S3, S4, S5, LS1, LS2, LS3, LS4, SL1	Carbonate-argillaceous rocks of Pabdeh, Gurpi and Ilam Formations	Limestone, shale and limy shales
2	GWCT-3,4	Sanandaj-Sirjan metamorphic belt	–	Jurassic metamorphic rocks (low to medium grade) and Cretaceous limestone	Limestone, shale and sandstone, slate, phyllite, schist with quartzitic veins
3	KWCT-1	Central Alborz	Gta1, Sts1, Gta2, Tsh-1, Gta3,	Pyroclastic rocks of Karaj formation	Tuffs, shaly and sandy tuffs, agglomerate
4	MSTT (MWCT)	Fiordland complex	–	Paleozoic metamorphic and igneous rocks	Gneiss, calc-silicate and quartzite and the intrusive rocks

Table 5 Mechanical and operating properties of TBMs in the database

Parameters	ZWCT-2	GWCT-3,4	KWCT-1	MSTT
Manufactured by	Herrenknecht	Wirth	Herrenknecht	Robbins
TBM type	Double shield TBM			Open TBM
Weight (ton)	573	–	170	–
Diameter of machine (m)	6.73	4.525	4.665	10.05
Diameter of disc cutters	432 mm (17 in)			
Ave. cutter spacing (mm)	90	90	75	90
Nu. of cutters	42	35	31	68
Cutter load (kN)	250			280
Cutter width	18–19 (mm)			
Max. thrust force (kN)	28,134	18,000	16,913	–
Power (kW)	2100	1120	1250	3463
RPM	0–11	0–12	0–11	5.0
Cutterhead torque (kN.m)	4450	802	1029	9859

Fig. 13 Collecting required data during the construction phase, **a**, **b** tunnel face observation, **c** evaluating muck materials during construction; Examples of outcrops and drill cores representing the various rock mass categories observed in the projects under investigation, **d**, **e** thick-bedded of Karaj formation, **f** foliated rocks in GWCT project

No.	Parameters	FPI	P	PR	RPM	Fn	UCS	RQD
	Name of projects	(kN/mm/rev)	(mm/rev)	(m/h)	-	(tonf)	(MPa)	-
1	Karaj WCT-lot 1	17.48	7.47	3.68	8.20	13.06	60.00	40.00
.		.	.	.	.	.	.	.
.		.	.	.	.	.	.	.
40		11.63	10.83	3.83	5.90	12.59	40.00	25.00
41		15.45	8.96	3.28	6.10	13.84	50.00	30.00
1	Zagros WCT-lot 2	20.29	5.82	1.33	5.20	11.81	100.00	70.00
.		.	.	.	.	.	.	.
.		.	.	.	.	.	.	.
34		21.05	4.56	1.04	3.80	9.60	60.00	72.00
35		21.78	7.20	2.24	5.20	15.62	60.00	72.00
1	Ghomrood WCT-lots 3 and 4	9.28	9.27	5.68	10.22	8.60	40.00	60.00
.		.	.	.	.	.	.	.
.		.	.	.	.	.	.	.
30		10.89	6.22	4.18	11.21	6.77	50.00	70.00
31		20.89	4.85	2.18	7.50	10.14	60.00	75.00
1	Manapouri project	48.40	4.61	1.20	4.30	22.30	128.00	80.00
.		.	.	.	.	.	.	.
.		.	.	.	.	.	.	.
48		41.60	5.22	1.43	4.60	21.70	55.00	70.00
49		67.80	3.53	1.08	5.10	23.93	102.00	100.00

Fig. 14 The developed database from laboratory tests, daily operating records, TBM data logger, and statistical analysis

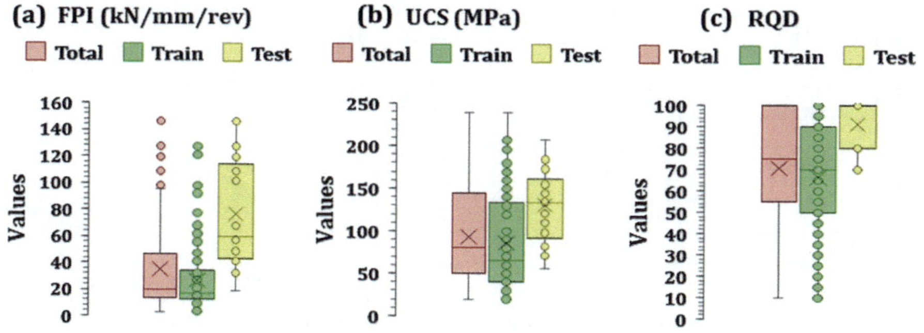

Fig. 15 The distribution of data points for each parameter in three categories: train, test, and total

Table 6 The results of the loss functions of ML-based models in the prediction of FPI

Models	RMSE	MSE	MAPE	MAD	RRMSE	NRMSE	RSE	RRSE
SCG-DNN	0.034	0.000	0.000	0.158	0.000	0.000	0.000	0.000
SCG-BPNN	0.203	0.023	0.001	0.185	0.022	0.001	0.000	0.021
SCG-FFNN	0.058	0.003	0.003	0.218	0.057	0.000	0.001	0.025
SGD-SVM	0.062	0.041	0.001	0.182	0.025	0.000	0.001	0.026
G-FIS	0.008	0.000	0.001	0.182	0.059	0.000	0.003	0.055
HBA-KNN	0.059	0.005	0.006	0.184	0.128	0.000	0.147	0.384
SGD-SVR	0.178	0.000	0.002	0.176	0.025	0.001	0.000	0.021
ESMA-KNN	0.023	0.041	0.000	0.165	0.031	0.000	0.001	0.025
RBF	0.437	0.020	0.001	0.212	0.111	0.003	0.005	0.074
LSSVM	0.032	0.002	0.001	0.183	0.049	0.000	0.002	0.046
TSA-XGBoost	0.005	0.007	0.001	0.180	0.035	0.000	0.001	0.037
SSA-XGBoost	0.123	0.001	0.001	0.190	0.033	0.001	0.001	0.037

indicator, with a focus on the best architectural models for soft computing considered in the analysis.

The score analysis reveals the relative performance of models, with higher and lower values of performance indicators indicating superior and inferior training and testing examples, respectively. Subsequently, the scores from each phase of training and testing, considering nine evaluation metrics (as presented in Table 6) are used to calculate the final model score, which includes the validation score as well. The findings from the score analysis are presented in Fig. 19, which highlights that hybrid models achieved scores exceeding 90 in both the training and testing phases. In contrast, RBF obtained the lowest score of 26. Notably, Fig. 19 underscores the superior predictive capabilities of hybrid models, particularly SCG-DNN, with a high score of 99 in predicting TBM performance (FPI).

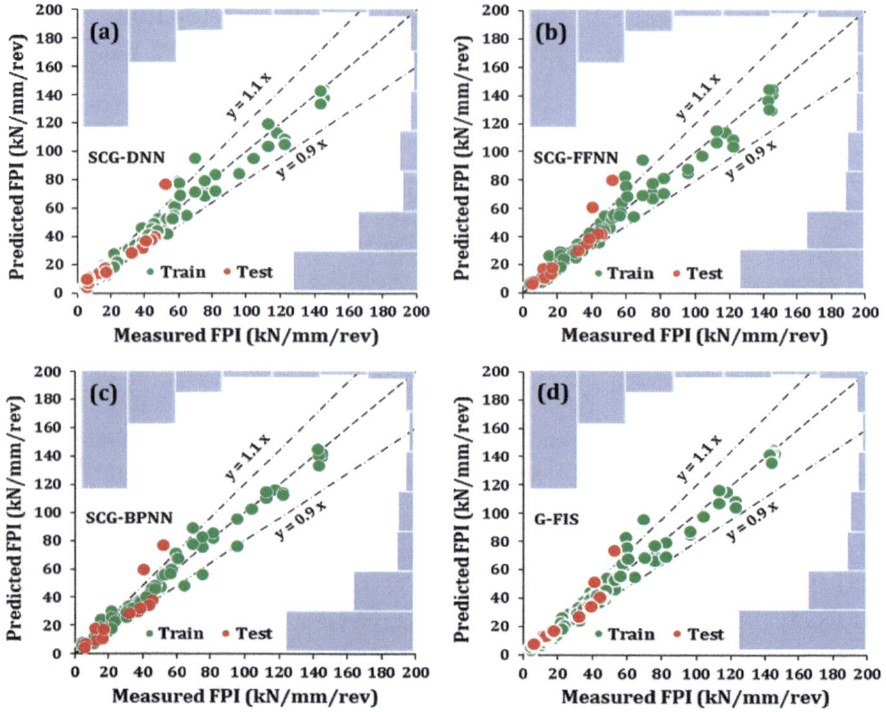

Fig. 16 The correlation between the measured and predicted results of FPI based on optimized neural networks and fuzzy approaches

Further comprehensive evaluations of the prediction results of multiple soft computing techniques for FPI were conducted using normalized error analysis, variance, and normalized standard deviation, as depicted in Fig. 20. This Taylor diagram assesses the performance of the applied algorithms based on the method's prediction results, normalized standard deviation, normalized root mean square error, and variance. A comparative analysis using calculated relative error rates of models in prediction was done. The boxplot graph in Fig. 21a presents a comparative analysis of the relative error rates of models applied to the training set for predictive modelling. Figure 21b shows the calculated error rate for each model in the testing phase. It can be concluded that all models show an acceptable accuracy rate in prediction in both training and testing sets.

Algorithm evaluation involves assessing metrics like the area under the receiver operating characteristic (ROC) curve (AUC) and accuracy when comparing different algorithms to identify the most optimal model in prediction. In this study, the assessment of various algorithms for TBM performance prediction indicated that the SCG-DNN and ESMA-KNN algorithms displayed superior performance, achieving an accuracy rate of 90% and an AUC-ROC value of 98%. In contrast, other algorithms exhibited accuracy rates below 90% and AUC-ROC values lower than 98%. These

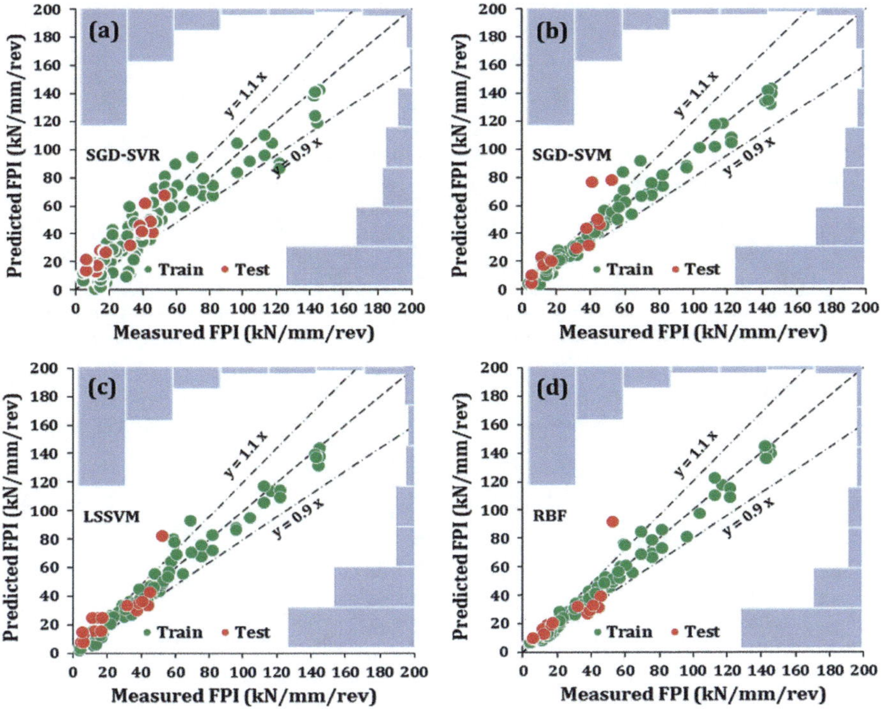

Fig. 17 The correlation between the measured and predicted results of FPI based on optimized kernel functions-algorithms

findings suggest that SCG-DNN and ESMA-KNN surpassed other models in terms of accuracy and effectiveness. The values of the AUC-ROC curve, as shown in Table 7, act as a measure of the overall performance of the algorithms utilized. This metric represents the shared area under the curve formed by the true positive rate and false positive rate. To validate and assess the accuracy of the developed models using statistical measurements, a comparison was made between the actual and predicted values of FPI (as presented in Table 8).

As a sensitivity analysis and importance ranking, Fig. 22 presents the predicted importance ranking based on the wind rose diagram for TBM performance using the employed hybrid-optimized supervised learning techniques, considering two geomechanical parameters (UCS, RQD) of excavated rock materials along projects. As can be seen, UCS has a greater impact on FPI compared to RQD, but both significantly influence FPI. SHapley Additive exPlanations (SHAP) can provide insights into model predictions and the significance of various features. By using the SHAP library with XGBoost, the SHAP values can show the impact of individual features on model predictions. The force SHAP plot for FPI analysis based on rock geomechanical factors, presented in Fig. 23, underscores the importance of UCS and RQD on FPI, which coincides with the PCA analysis.

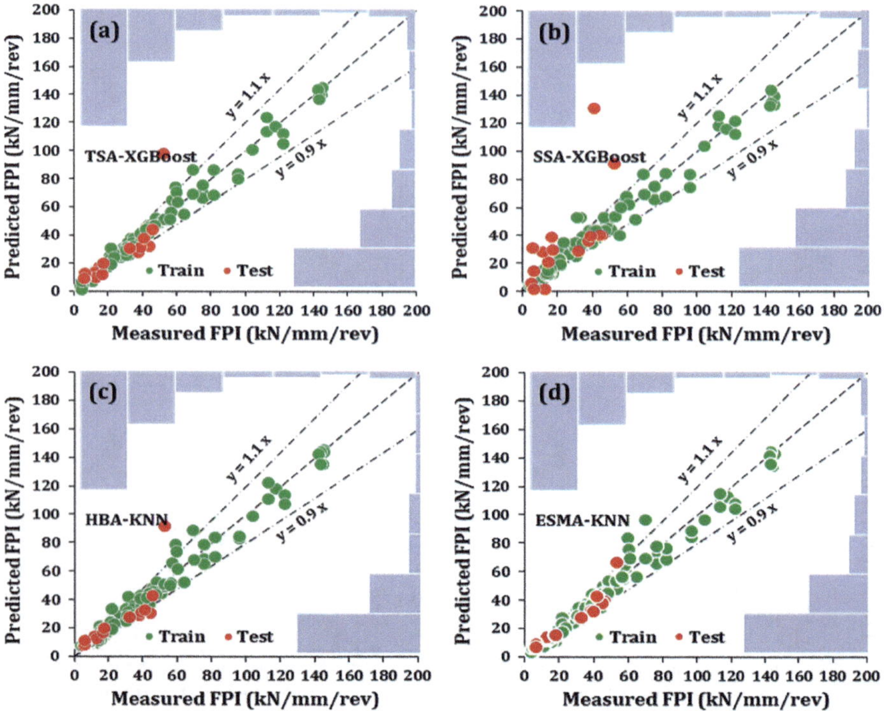

Fig. 18 The correlation between the measured and predicted results of FPI based on hybrid optimized KNN and gradient networks

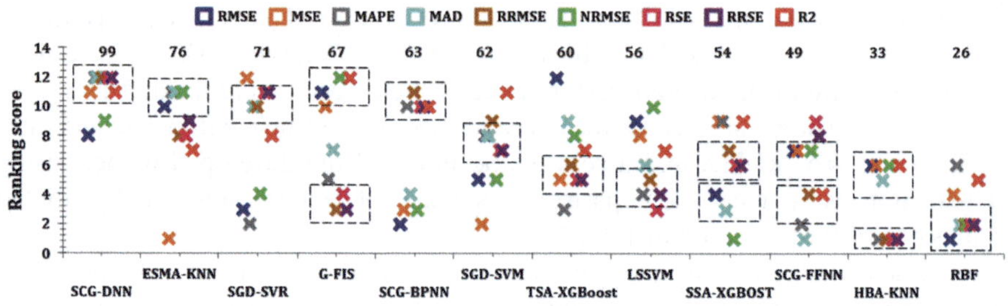

Fig. 19 Intuitive display of the comprehensive ranking of twelve hybrid supervised learning models for FPI

In this part of the analysis with explainable AI (XAI), the results of the models were classified using two different techniques. One technique involved non-numeric classification, where the database and obtained results from ML algorithms were divided into four groups based on the identified lithological types (ET-I to ET-IV) (Hassanpour et al. 2023). The other one is based on numerical classification. After analyzing

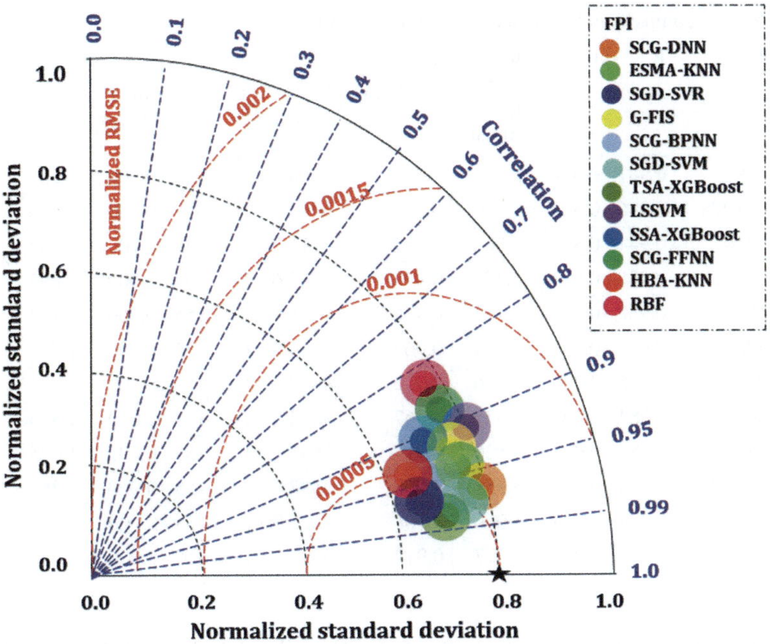

Fig. 20 Taylor chart for comparison of multiple hybrid-optimized learning algorithms to predict FPI

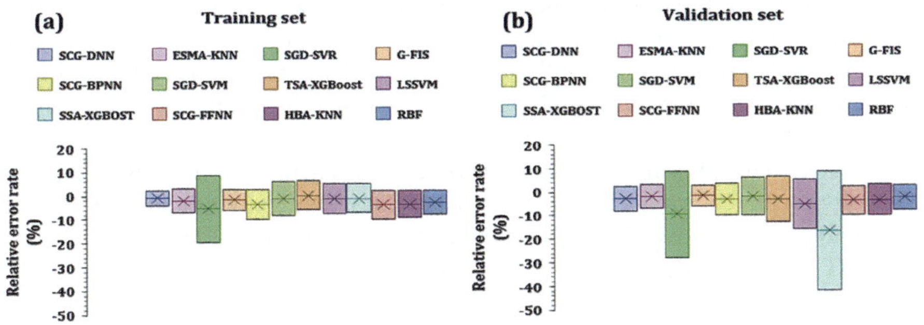

Fig. 21 The relative error rate of models in the prediction of FPI in both the training and validation sets

Table 7 Value of the area under the ROC

Algorithms and models	AUC	Algorithms and models	AUC	Algorithms and models	AUC
SCG-DNN	0.98	SCG-BPNN	0.88	SSA-XGBoost	0.95
ESMA-KNN	0.97	SGD-SVM	0.91	SCG-FFNN	0.87
SGD-SVR	0.96	TSA-XGBoost	0.83	HBA-KNN	0.85
G-FIS	0.94	LSSVM	0.86	RBF	0.82

Table 8 The comparison between models to predict FPI

Model	R	R^2	Std. Error of the estimate	Change statistics					Durbin–Watson
				R^2 change	F change	df1	df2	Sig. F change	
SCG-DNN	0.987	0.974	5.377	0.974	5744.2	1	155	0.000	1.987
ESMA-KNN	0.986	0.973	5.506	0.973	5463.1	1	155	0.000	1.876
SGD-SVR	0.944	0.892	10.98	0.892	1270.7	1	155	0.000	1.769
G-FIS	0.986	0.973	5.482	0.973	5587.6	1	155	0.000	1.895
SCG-BPNN	0.988	0.976	5.173	0.976	6328.5	1	155	0.000	1.784
SGD-SVM	0.981	0.963	6.384	0.963	3959.8	1	155	0.000	1.599
TSA-XGBoost	0.985	0.970	5.836	0.970	4949.9	1	155	0.000	1.778
LSSVM	0.984	0.967	5.932	0.967	4564.6	1	155	0.000	1.875
SSA-XGBoost	0.957	0.917	9.826	0.917	1691.6	1	155	0.000	1.685
SCG-FFNN	0.984	0.968	5.956	0.968	4588.7	1	155	0.000	1.785
HBA-KNN	0.984	0.968	6.000	0.968	4623.8	1	155	0.000	1.689
RBF	0.986	0.972	5.577	0.972	5332.3	1	155	0.000	1.652

the data, it was found that twenty-six percent of the data fell within ET-I lithological type, corresponding to pyroclastic and volcanic rocks, twenty-three percentages within ET-II lithological type, corresponding to sedimentary rocks, twenty percentages within ET-III lithological type, corresponding to low to medium grade metamorphic rocks, with the remaining data (thirty-one percentage) classified as ET-IV for intrusive rocks. An analysis of lithological types' distribution was conducted using a doughnut chart shown in Fig. 24 to compare actual distribution with predictions from models.

These visual representations show the outcomes generated by algorithms and how they align with real results based on database classification by lithological types. In the analysis of the rock classes, it is evident that for ET-I: G-FIS, SSA-XGBoost, TSA-XGBoost, ET-II: LSSVM, SCG-BPNN, SCG-DNN, ET-III: LSSVM, SCG-DNN, G-FIS, and ET-IV: SCG-DNN, SSA-XGBoost, SCG-BPNN show superior performance compared to other models in each lithological type. Also, the relative error rate in the prediction of FPI for each model in each lithological type was calculated, and the results of this analysis were presented in Fig. 25.

It can be concluded that all models show an acceptable accuracy rate in the prediction of FPI for each lithological type. As mentioned, the results of the models were classified using two different techniques, in which non-numerical analyses based on lithological types were described. The second technique used numerical classification with a confusion matrix. The results of the FPI analysis were divided into four classes for non-numerical analysis and six classes (FPI: kN/mm/rev; Min = 2.75, Max = 145.6) for numerical analysis. The non-numeric model provides an overview of the database's lithological types and data distribution among rock types. On the

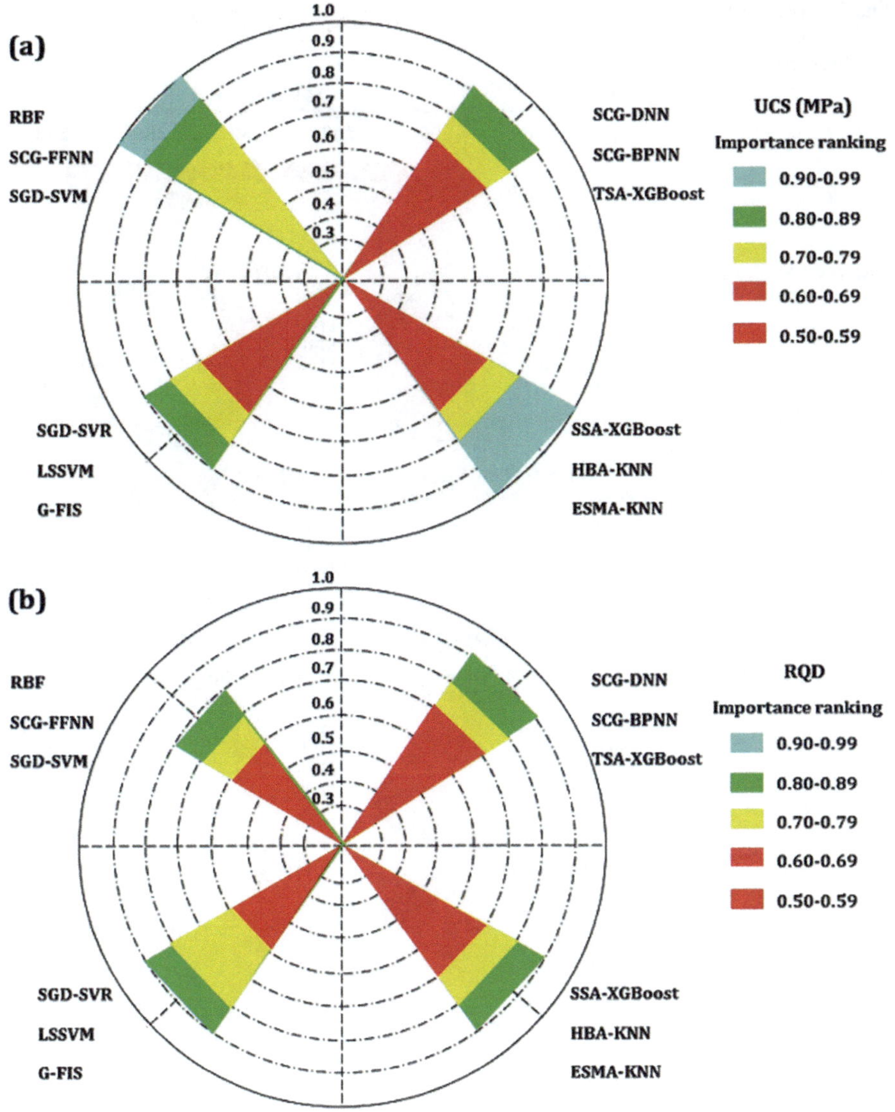

Fig. 22 The wind rose diagrams of predictor importance ranking based on two geomechanical factors

other hand, the numerical model offers a more detailed and quantitative analysis based on FPI.

Dividing FPI values into six categories allows for a detailed analysis of hybrid supervised learning algorithms' performance within each category. The confusion matrix helps evaluate the effectiveness of these algorithms across different FPI categories (Class I: 0–25, Class II: 25–50, Class III: 50–75, Class IV: 75–100, Class V:

Fig. 23 The SHAP plot for analysis of FPI using the XGBRegressor library

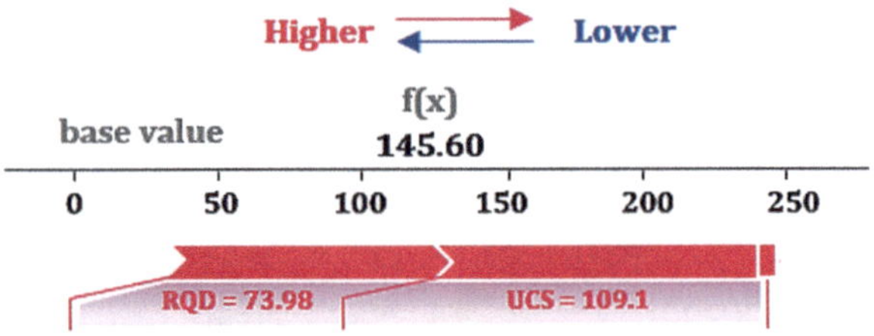

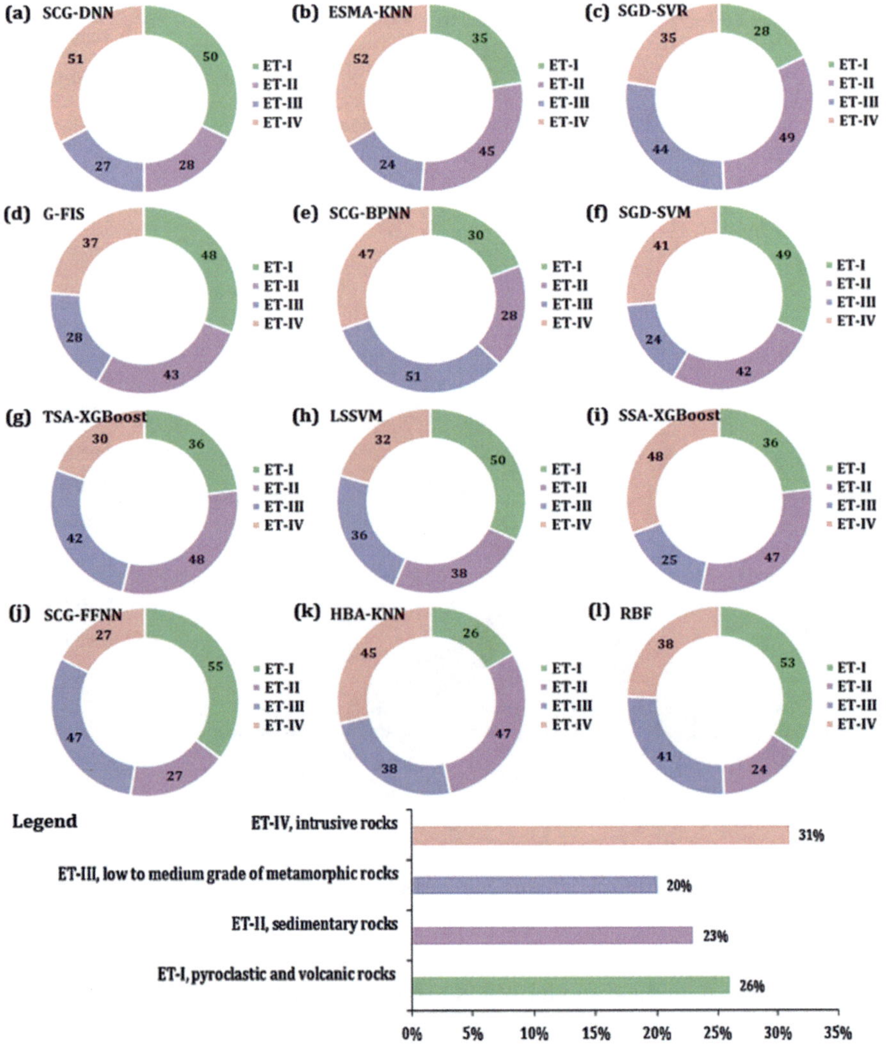

Fig. 24 The results of supervised learning algorithms considering the classification of lithological types into four ET Class No.

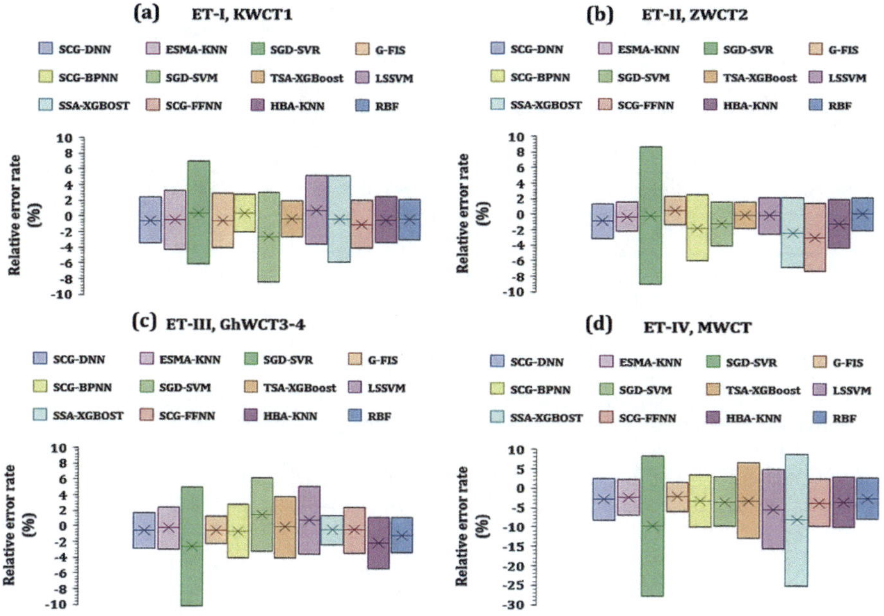

Fig. 25 The relative error rate of models in the prediction of FPI based on each lithological type

100–125, Class VI: 125–150 kN/mm/rev), showing that SCG-DNN, ESMA-KNN, and SGD-SVR have higher recall rates in identifying true positive instances within the FPI range, as depicted in Fig. 26 with over 95% accuracy.

The majority of models perform well, with high accuracy rates, particularly those with precision rates above 80%. The precision rates for the models are: SCG-DNN (0.97), ESMA-KNN (0.96), SGD-SVR (0.95), G-FIS (0.93), SCG-BPNN (0.90), SGD-SVM (0.89), TSA-XGBoost (0.87), LSSVM (0.86), SSA-XGBoost (0.85), SCG-FFNN (0.85), HBA-KNN (0.80), and RBF (0.79), indicating their accuracy and precision in predicting FPI. All of these analyses were done to introduce the best optimal models with high accuracy in predicting FPI in different lithological types. Based on the analysis, it was concluded that three models, including SCG-DNN, ESMA-KNN, and SGD-SVR, have higher accuracy and precision rate than others and thus utilized to develop a stacking-ensemble model.

A word of caution is to make sure to examine the final model for over-fitting and performance in boundary conditions. Initially, in this study optimization techniques were used to avoid over-fitting in the prediction using ML algorithms. This means that by using values of input parameters at the low and higher ends of the range, observe the output. Another approach to examine the validity of the model is to use sensitivity analysis, where the values of input parameters are set to the average of the database, and then change the average value of each parameter from − 50% to + 50% while keeping other parameters constant to see the range of changes of the output parameter. Visual presentations could be in the form of spider-grams (Fig. 27)

SCG-DNN		Predicted class						Recall
		I	II	III	IV	V	VI	
Actual class	I	119	3	0	0	0	0	97%
	II	1	29	1	0	0	0	93%
	III	0	0	18	0	0	0	1
	IV	0	0	0	2	0	0	1
	V	0	0	0	0	3	0	1
	VI	0	0	0	0	0	1	1
Precision		99%	91%	94%	1	1	1	0.97

ESMA-KNN		Predicted class						Recall
		I	II	III	IV	V	VI	
Actual class	I	118	4	0	0	0	0	96%
	II	2	29	0	0	0	0	91%
	III	0	0	18	0	0	0	1
	IV	0	0	0	2	0	0	1
	V	0	0	0	0	3	0	1
	VI	0	0	0	0	0	1	1
Precision		98%	88%	94%	1	1	1	0.96

SGD-SVR		Predicted class						Recall
		I	II	III	IV	V	VI	
Actual class	I	118	3	1	0	0	0	96%
	II	1	29	1	0	0	0	91%
	III	0	1	17	0	0	0	94%
	IV	0	0	0	2	0	0	1
	V	0	0	0	0	3	0	1
	VI	0	0	0	0	0	1	1
Precision		99%	87%	89%	1	1	1	0.95

G-FIS		Predicted class						Recall
		I	II	III	IV	V	VI	
Actual class	I	117	3	2	0	0	0	95%
	II	2	29	0	0	0	0	93%
	III	0	0	16	2	0	0	88%
	IV	0	0	0	2	0	0	1
	V	0	0	0	0	3	0	1
	VI	0	0	0	0	0	1	1
Precision		98%	90%	88%	50%	1	1	0.93

SCG-BPNN		Predicted class						Recall
		I	II	III	IV	V	VI	
Actual class	I	116	4	1	1	0	0	95%
	II	2	28	1	0	0	0	90%
	III	0	1	16	1	0	0	88%
	IV	0	0	0	2	0	0	1
	V	0	0	0	0	3	0	1
	VI	0	0	0	0	0	1	1
Precision		98%	84%	88%	50%	1	1	0.90

SGD-SVM		Predicted class						Recall
		I	II	III	IV	V	VI	
Actual class	I	116	2	2	1	0	0	95%
	II	1	28	1	1	0	0	90%
	III	0	2	15	1	0	0	83%
	IV	0	0	0	2	0	0	1
	V	0	0	0	0	3	0	1
	VI	0	0	0	0	0	1	1
Precision		99%	87%	83%	40%	1	1	0.89

Fig. 26 The results of the confusion matrix for applied models considering six classes of FPI

TSA-XGBoost	Predicted class						Recall
Actual class	I	II	III	IV	V	VI	
I	115	3	1	2	0	0	95%
II	0	27	2	2	0	0	87%
III	0	1	14	3	0	0	77%
IV	0	0	0	2	0	0	1
V	0	0	0	0	3	0	1
VI	0	0	0	0	0	1	1
Precision	1	87%	82%	28%	1	1	0.87

LSSVM	Predicted class						Recall
Actual class	I	II	III	IV	V	VI	
I	114	4	1	2	0	0	95%
II	1	26	3	1	0	0	87%
III	0	2	14	2	0	0	77%
IV	0	0	0	2	0	0	1
V	0	0	0	0	3	0	1
VI	0	0	0	0	0	1	1
Precision	99%	81%	77%	28%	1	1	0.86

SSA-XGBoost	Predicted class						Recall
Actual class	I	II	III	IV	V	VI	
I	113	5	2	1	0	0	93%
II	2	25	4	0	0	0	80%
III	0	0	14	4	0	0	77%
IV	0	0	0	2	0	0	1
V	0	0	0	0	3	0	1
VI	0	0	0	0	0	1	1
Precision	98%	83%	70%	28%	1	1	0.85

SCG-FFNN	Predicted class						Recall
Actual class	I	II	III	IV	V	VI	
I	113	5	2	1	0	0	93%
II	2	25	4	0	0	0	80%
III	0	0	14	4	0	0	77%
IV	0	0	0	2	0	0	1
V	0	0	0	0	3	0	1
VI	0	0	0	0	0	1	1
Precision	98%	83%	70%	28%	1	1	0.85

HBA-KNN	Predicted class						Recall
Actual class	I	II	III	IV	V	VI	
I	110	6	3	2	0	0	90%
II	1	24	5	1	0	0	77%
III	0	3	13	2	0	0	72%
IV	0	0	0	2	0	0	1
V	0	0	0	1	2	0	66%
VI	0	0	0	0	0	1	1
Precision	99%	72%	61%	25%	1	1	0.80

RBF	Predicted class						Recall
Actual class	I	II	III	IV	V	VI	
I	105	7	5	4	0	0	86%
II	2	23	6	0	0	0	74%
III	0	1	13	4	0	0	72%
IV	0	0	0	2	0	0	1
V	0	0	0	1	2	0	66%
VI	0	0	0	0	0	1	1
Precision	98%	74%	54%	18%	1	1	0.79

Fig. 26 (continued)

or tornado-grams (Fig. 28) to indicate the range of changes in model results when the input variables change from − 50 to + 50% of the average values.

Sharp changes indicate the sensitivity of the model to the input variables. In such cases, perhaps models with lower R^2 but more smooth changes with input variables could be preferable to avoid accumulated errors in the system that is embedded in the

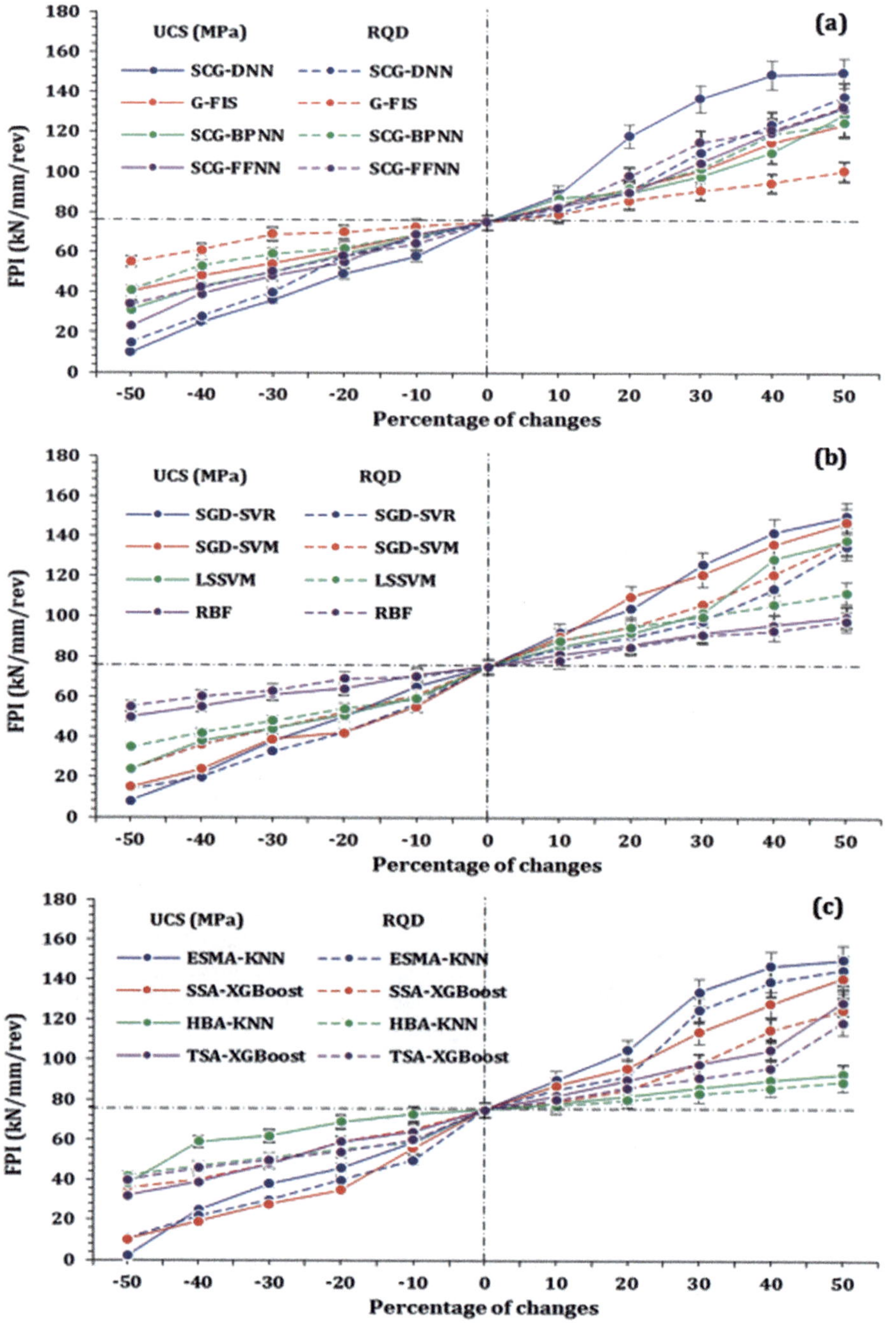

Fig. 27 Spider diagrams showing the sensitivity of FPI predictions to percentage changes in UCS and RQD for various models

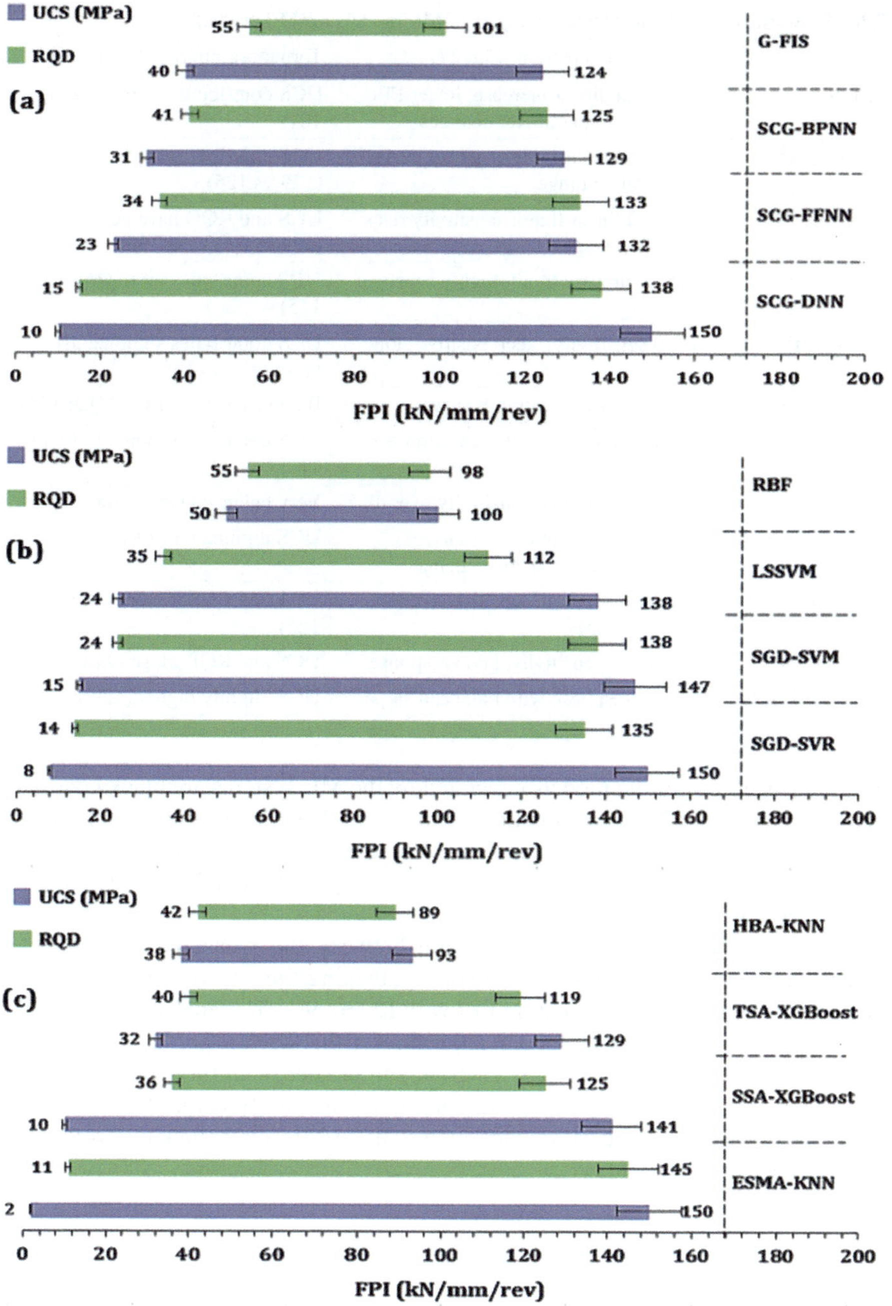

Fig. 28 Tornado diagrams illustrating the relative influence of UCS and RQD on FPI predictions across models

Table 9 Sensitivity and contribution analysis of ML models for FPI prediction

Model	Spider-gram pattern (Fig. 27)	Tornado-gram result (Fig. 28)
G-FIS	Moderate linear increase; lower FPI response to parameter changes	UCS contributes more (124 vs 101)
SCG-BPNN	Consistent sensitivity; steeper curve in the positive range	UCS contributes slightly more (129 vs 125)
SCG-FFNN	Smooth linear trend; sensitivity rises with % change	UCS and RQD have near-equal influence (132 vs 133)
SCG-DNN	Strong upward trend; highly sensitive to change	UCS contributes more (150 vs 135)
SGD-SVR	High sensitivity; steep positive slope	UCS contributes significantly more (150 vs 135)
SGD-SVM	Strong upward response curve	Balanced (UCS: 147, RQD: 138)
LSSVM	Moderate slope; consistent response	UCS contributes more (138 vs 112)
RBF	Flatter curve; low sensitivity overall	Very balanced (98 vs 100)
ESMA-KNN	Strong slope; high sensitivity, especially with positive change	UCS dominant (150 vs 145)
SSA-XGBoost	Moderate-to-high sensitivity with sharper slope	UCS contributes more (145 vs 125)
HBA-KNN	Moderate sensitivity, good response	UCS and RQD close (93 vs 89)
TSA-XGBoost	Linear response with balanced slope	UCS slightly higher (129 vs 119)

models, since most of the models generated by the ML systems are black boxes and need these explainable AI examinations for accuracy beyond the typical performance and confusion matrices. A summary of the spider-gram pattern and tornado-gram result for the contribution analysis of hybrid-optimized ML models for FPI prediction is presented in Table 9.

Figure 29 shows the strong correlation between the measured and predicted values based on stacking-ensemble model, indicating the superior performance of this ensemble network in predicting FPI with $R^2 = 99\%$. The analysis with hybrid-optimized and stacking-ensemble ML algorithms in this study to predict FPI was done in a graphical user interface (GUI) environment created with the Python programming language and developed by the authors of this study (Fig. 30). The use of a GUI can predict TBM performance with defined ML algorithms by providing a user-friendly environment.

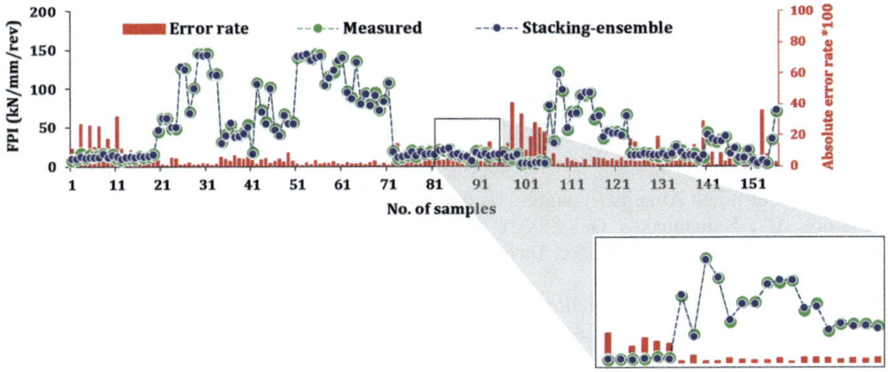

Fig. 29 Comparison between measured and predicted values of FPI based on the stacking-ensemble model

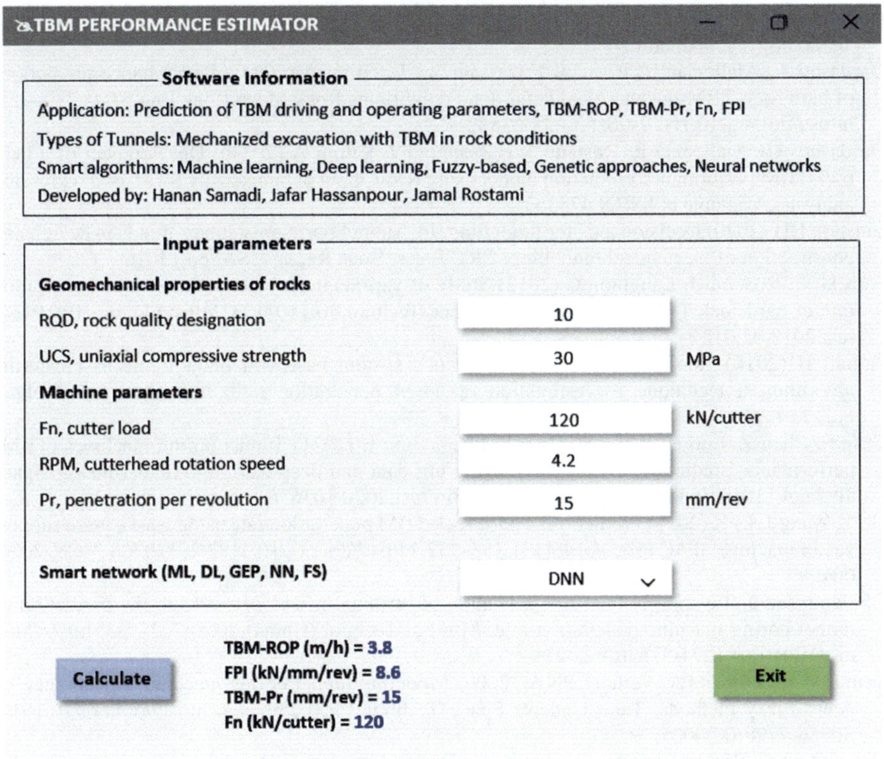

Fig. 30 GUI for predicting the FPI performance

References

Barton N (2000) TBM tunnelling in jointed and faulted rock. AA Balkema

Barton N (2009) TBM prognoses in hard rock with faults using Q_{TBM} methods. IoM3 Institute of Materials, Minerals & Mining

Barton N (2015) TBM performance, prognosis and risk caused by faulting. In: ISRM regional symposium-South American congress on rock mechanics (ISRM-SCRM). ISRM

Benardos AG, Kaliampakos DC (2004) Modelling TBM performance with artificial neural networks. Tunn Undergr Space Technol 19(6):597–605. https://doi.org/10.1016/j.tust.2004.02.128

Benardos A (2008) Artificial intelligence in underground development: a study of TBM performance. Undergr Spaces des Eng Environ Asp 102:121. https://doi.org/10.2495/US080031

Bieniawski ZT, Celada Tamames B, Galera Fernández JM, Álvarez Hernández M (2006) Rock mass excavability indicator: new way to selecting the optimum tunnel construction method. Tunn Undergr Space Technol 21:237. https://doi.org/10.1016/j.tust.2005.12.016

Bonaccorso G (2018) Machine learning algorithms: popular algorithms for data science and machine learning. Packt Publishing Ltd., Birmingham, UK

Bruland A (1998) Hard rock tunnel boring. Ph.D. Thesis, Norwegian University of Science and Technology, Trondheim

Dardashti A, Ajalloeian R, Rostami J, Hassanpour J, Salimi A (2024a) Performance predictions of hard rock TBM in subcritical cutter load conditions. Rock Mech Rock Eng 57(1):739–755. https://doi.org/10.1007/s00603-023-03582-y

Dardashti AF, Ajalloeian R, Rostami J, Hassanpour J, Salimi A (2024b) Development of a hard rock TBM performance prediction model using RMR input parameters by ML-based regression analysis. Available at SSRN 4733658

Einstein HH (2020) Decision aids for tunnelling. In: Tunnel boring machines: trends in design and construction of mechanical tunnelling. CRC Press, Boca Raton, USA, pp 13–26

Farrokh E, Rostami J, Laughton C (2012) Study of various models for estimation of penetration rate of hard rock TBMs. Tunn Undergr Space Technol 30:110–123. https://doi.org/10.1016/j.tust.2012.02.012

Fattahi H (2016) Adaptive neuro fuzzy inference system based on fuzzy c–means clustering algorithm, a technique for estimation of TBM penetration rate. Iran Univ Sci Technol 6(2):159–171

Feng S, Chen Z, Luo H, Wang S, Zhao Y, Liu L, Jing L (2021) Tunnel boring machines (TBM) performance prediction: a case study using big data and deep learning. Tunn Undergr Space Technol 110:103636. https://doi.org/10.1016/j.tust.2020.103636

Ge Y, Wang J, Li K (2013) Prediction of hard rock TBM penetration rate using least square support vector machine. IFAC Proc Vol 46(13):347–352. https://doi.org/10.3182/20130708-3-CN-2036.00105

Gholamnejad J, Tayarani N (2010) Application of artificial neural networks to the prediction of tunnel boring machine penetration rate. Min Sci Technol (China) 20(5):727–733. https://doi.org/10.1016/S1674-5264(09)60271-4

Grima MA, Bruines PA, Verhoef PNW (2000) Modeling tunnel boring machine performance by neuro-fuzzy methods. Tunn Undergr Space Technol 15(3):259–269. https://doi.org/10.1016/S0886-7798(00)00055-9

Hassanpour J, Rostami J, Khamehchiyan M, Bruland A (2009) Developing new equations for TBM performance prediction in carbonate-argillaceous rocks: a case history of Nowsood water conveyance tunnel. Geomech Geoeng Int J 4(4):287–297. https://doi.org/10.1080/17486020903174303

Hassanpour J, Rostami J, Khamehchiyan M, Bruland A, Tavakoli HR (2010) TBM performance analysis in pyroclastic rocks: a case history of Karaj water conveyance tunnel. Rock Mech Rock Eng 43:427–445. https://doi.org/10.1007/s00603-009-0060-2

Hassanpour J, Rostam J, Zhao J (2011) A new hard rock TBM performance prediction model for project planning. Tunn Undergr Space Technol 26(5):595–603. https://doi.org/10.1016/j.tust.2011.04.004

Hassanpour J, Rostami J, Zhao J, Tarigh Azali S (2015) TBM performance and disc cutter wear prediction based on ten years experience of TBM tunnelling in Iran. Geomech Tunn 8(3):239–247. https://doi.org/10.1002/geot.201500005

Hassanpour J, Ghaedi Vanani AA, Rostami J, Cheshomi A (2016) Evaluation of common TBM performance prediction models based on field data from the second lot of Zagros water conveyance tunnel (ZWCT2). Tunn Undergr Space Technol 52:147–156. https://doi.org/10.1016/j.tust.2015.12.006

Hassanpour J, Esmaeili Vardanjani S, Cheshomi A, Rostami J (2019) Engineering geological studies used for redesigning and employing a hard rock TBM in soft rock formations of Chamshir water conveyance tunnel. Geopersia 9(1):1–20. https://doi.org/10.22059/geope.2018.253479.648374

Hassanpour J, Firouzei Y, Hajipour G (2021) Actual performance analysis of a double shield TBM through sedimentary and low to medium grade metamorphic rocks of Ghomrood water conveyance tunnel project (lots 3 and 4). Bull Eng Geol Env 80:1419–1432. https://doi.org/10.1007/s10064-020-01947-z

Hassanpour J, Rostami J, Khamehchiyan M (2023) Engineering geology and tunnelling (Evaluation of TBM Performance in Rock Tunnels). University of Tehran Press

Hassanpour J, Kazemi C, Rostami J (2024) Introduction of a modified Q_{TBM} model for predicting TBM penetration rate in rock, based on data from mechanized tunneling projects in Iran. Bull Eng Geol Env 83(5):1–17. https://doi.org/10.1007/s10064-024-03659-0

Hoek E, Carranza-Torres C, Diederichs M, Corkum B (2008) The 2008 Kersten Lecture Integration of geotechnical and structural design in tunneling. In 56th Annual geotechnical engineering conference, pp 1–53

Huang S, Dastpak P, Esmaeilpour M, Liu K, Sousa RL (2023) Comparison between machine learning algorithms for TBM advance rate prediction. In Expanding underground-knowledge and passion to make a positive impact on the world. CRC Press, Boca Raton, USA, pp 2710–2716

Japkowicz N, Shah M (2011) Evaluating learning algorithms: a classification perspective. Cambridge University Press

Jakobsen PD, Babendererde T (2017) Preinvestigations for TBM tunneling. In: Proceedings of the world tunnel congress–surface challenges–underground solutions. Bergen, Norway

Kullarkar SD, Thote NR (2025) TBM performance prediction using statistical and machine learning algorithms: case study of Slurry TBM used in Mumbai Metro Rail Tunnel Line–3 (UGC-03), India. J Geol Soc India 101(2):149–162. https://doi.org/10.17491/jgsi/2025/174077

Latif K, Sharafat A, Seo J (2023) Digital twin-driven framework for TBM performance prediction, visualization, and monitoring through machine learning. Appl Sci 13(20):11435. https://doi.org/10.3390/app132011435

Ling F, Jingcheng W, Yang G, Chuang L, Langwen Z (2013) Prediction of TBM penetration rate based on the model of PLS-FNN. In: 2013 25th Chinese control and decision conference (CCDC). IEEE, pp 1295–1299. https://doi.org/10.1109/CCDC.2013.6561125

Liu B, Wang R, Guan Z, Li J, Xu Z, Guo X, Wang Y (2019) Improved support vector regression models for predicting rock mass parameters using tunnel boring machine driving data. Tunn Undergr Space Technol 91:102958. https://doi.org/10.1016/j.tust.2019.04.014

Ma T, Jin Y, Liu Z, Prasad YK (2022) Research on prediction of TBM performance of deep-buried tunnel based on machine learning. Appl Sci 12(13):6599. https://doi.org/10.3390/app12136599

Mahmoodzadeh A, Nejati HR, Ibrahim HH, Ali HFH, Mohammed AH, Rashidi S, Majeed MK (2022) Several models for tunnel boring machine performance prediction based on machine learning. Geomech Eng 30(1):75–91

Martins FF, Miranda TF (2013) Prediction of hard rock TBM penetration rate based on data mining techniques. https://hdl.handle.net/1822/25941

Mikaeil R, Naghadehi MZ, Sereshki F (2009) Multifactorial fuzzy approach to the penetrability classification of TBM in hard rock conditions. Tunn Undergr Space Technol 24(5):500–505. https://doi.org/10.1016/j.tust.2008.12.007

Minh VT, Katushin D, Antonov M, Veinthal R (2017) Regression models and fuzzy logic prediction of TBM penetration rate. Open Eng 7(1):60–68. https://doi.org/10.1515/eng-2017-0012

Monthanopparat N, Tanchaisawat T (2024) Advancing tunnel boring machine performance prediction in massive and highly fractured granite: integrating innovative deep learning and block model techniques. Geotech Eng J SEAGS AGSSEA 55(3):24–34. https://doi.org/10.14456/seagj.2024.17

Oraee K, Khorami MT, Hosseini N (2012) Prediction of the penetration rate of TBM using adaptive neuro fuzzy inference system (ANFIS). In: Proceedings of SME annual meeting and exhibit, pp 297–302

Pourhashemi SM, Ahangari K, Hassanpour J, Eftekhari SM (2022) TBM performance analysis in very strong and massive rocks; case study: Kerman water conveyance tunnel project, Iran. Geomech Geoeng 17(4):1110–1122. https://doi.org/10.1080/17486025.2021.1912410

Rostami J, Ozdemir L (1993) A new model for performance prediction of hard rock TBMs. In: Proceedings of the rapid excavation and tunneling conference

Rostami J (1997) Development of a force estimation model for rock fragmentation with disc cutters through theoretical modeling and physical measurement of crushed zone pressure, vol 38. Colorado School of Mines, Golden, CO, USA, pp 56–64

Rostami J (2008) Hard rock TBM cutterhead modeling for design and performance prediction. Geomechanik und Tunnelbau: Geomechanik und Tunnelbau 1(1):18–28. https://doi.org/10.1002/geot.200800002

Rostami J, Sepehrmanesh M, Gharahbagh EA, Mojtabai N (2013) Planning level tunnel cost estimation based on statistical analysis of historical data. Tunn Undergr Space Technol 33:22–33. https://doi.org/10.1016/j.tust.2012.08.002

Rostami J (2016) Performance prediction of hard rock tunnel boring machines (TBMs) in difficult ground. Tunn Undergr Space Technol 57:173–182. https://doi.org/10.1016/j.tust.2016.01.009

Salimi A, Esmaeili M (2013) Utilising of linear and non-linear prediction tools for evaluation of penetration rate of tunnel boring machine in hard rock condition. Int J Min Mineral Eng 4(3):249–264. https://doi.org/10.1504/IJMME.2013.053172

Salimi A, Moormann C, Singh T, Jain P (2015) TBM performance prediction in rock tunneling using various artificial intelligence algorithms. In: 11th Iranian and 2nd regional tunnelling conference-tunnels and the future

Salimi A, Rostami J, Moormann C, Delisio A (2016) Application of non-linear regression analysis and artificial intelligence algorithms for performance prediction of hard rock TBMs. Tunn Undergr Space Technol 58:236–246. https://doi.org/10.1016/j.tust.2016.05.009

Salimi A, Rostami J, Moormann C (2019) Application of rock mass classification systems for performance estimation of rock TBMs using regression tree and artificial intelligence algorithms. Tunn Undergr Space Technol 92:103046. https://doi.org/10.1016/j.tust.2019.103046

Samadi H, Hassanpour J (2022) Developing the empirical models for predicting the EPB operating parameters in strong limestones. Sci Q J Iran Assoc Eng Geol 15(1):29–41

Samadi H, Hassanpour J (2024) EPB-TBM cutterhead torque and thrust modelling in rock tunnels through an analytical method and TSFS model. Heliyon 10(11). https://doi.org/10.1016/j.heliyon.2024.e31773

Samadi H, Karimi H, Khamehchiyan M, Hassanpour J (2024a) Prediction of engineering characteristics of rock masses using actual TBM performance data with supervised and unsupervised learning algorithms (A case study in strong to very strong igneous and pyroclastic rocks). Rock Mech Rock Eng 1–30. https://doi.org/10.1007/s00603-024-03903-9

Samadi H, Hassanpour J, Rostami J (2024b) Assessment of the relationship between rock mass classification systems using hybrid XGBoost algorithms: insights from the mechanized tunnelling project database. Rock Mech Rock Eng 1–35. https://doi.org/10.1007/s00603-024-04227-4

Samadi H, Hassanpour J (2025) Application of the stacking-ensemble machine learning algorithm to investigate the rock mass cuttability index in different lithological units using tunnel boring machine performance data. Rock Mech Rock Eng 1–31. https://doi.org/10.1007/s00603-025-04501-z

Sanio HP (1985) Prediction of the performance of disc cutters in anisotropic rock. Int J Rock Mech Min Sci Geomech Abstr 22(3):153–161). https://doi.org/10.1016/0148-9062(85)93229-2

Sarker IH (2021) Machine learning: algorithms, real-world applications and research directions. SN Comput Sci 2(3):160. https://doi.org/10.1007/s42979-021-00592-x

Shao C, Li X, Su H (2013) Performance prediction of hard rock TBM based on extreme learning machine. In: Intelligent robotics and applications: 6th International conference, ICIRA 2013, Busan, South Korea, September 25–28, 2013, Proceedings, Part II 6. Springer Berlin Heidelberg, pp 409–416. https://doi.org/10.1007/978-3-642-40849-6_40

Simoes MG, Kim T (2006) Fuzzy modeling approaches for the prediction of machine utilization in hard rock tunnel boring machines. In: Conference record of the 2006 IEEE industry applications conference forty-first IAS annual meeting, vol. 2. IEEE, pp 947–954. https://doi.org/10.1109/IAS.2006.256639

Tao H, Jingcheng W, Langwen Z (2015) Prediction of hard rock TBM penetration rate using random forests. In: The 27th Chinese control and decision conference (2015 CCDC). IEEE, pp 3716–3720. https://doi.org/10.1109/CCDC.2015.7162572

Wang Y, Zhao J, Jiang K, Zhou Q, Kang Z, Chen C, Zhang H (2023) Prediction of TBM operation parameters using machine learning models based on BPSO. Adv Eng Inform 56:101955. https://doi.org/10.1016/j.aei.2023.101955

Xu C, Liu X, Wang E, Wang S (2021) Prediction of tunnel boring machine operating parameters using various machine learning algorithms. Tunn Undergr Space Technol 109:103699. https://doi.org/10.1016/j.tust.2020.103699

Xu Q, Huang X, Zhang B, Zhang Z, Wang J, Wang S (2023) TBM performance prediction using LSTM-based hybrid neural network model: case study of Baimang River tunnel project in Shenzhen, China. Undergr Space 11:130–152. https://doi.org/10.1016/j.undsp.2022.11.002

Yang H, Wang Z, Song K (2022) A new hybrid grey wolf optimizer-feature weighted-multiple kernel-support vector regression technique to predict TBM performance. Eng Comput 1–17. https://doi.org/10.1007/s00366-020-01217-2

Yu S, Zhang Z, Wang S, Huang X, Lei Q (2024) A performance-based hybrid deep learning model for predicting TBM advance rate using attention-ResNet-LSTM. J Rock Mech Geotech Eng 16(1):65–80. https://doi.org/10.1016/j.jrmge.2023.06.010

Zhang L, Guo J, Fu X, Tiong RLK, Zhang P (2024) Digital twin enabled real-time advanced control of TBM operation using deep learning methods. Autom Constr 158:105240. https://doi.org/10.1016/j.autcon.2023.105240

Disc Cutter Wear

Morteza Karami

1 Introduction

Disc cutters are fundamental components of TBMs, facilitating rock breakage by indentation into the rock surface and creating the crushed zone, which leads to inducing tensile stresses in the rock that allows formation of the rock chips. Positioned on the cutterhead, these tools bear the brunt of interactions with the rock, leading to wear of the cutter ring that is in contact with the rock. The wear of disc cutters is a pivotal aspect affecting the performance and efficiency of Tunnel Boring Machines (TBMs) in rock excavation. The operational efficiency of TBMs is closely tied to the condition of their disc cutters, which must withstand intense mechanical and thermal stresses during the excavation of abrasive and strong rock formations. Blunting of the tip due to wear will increase the cutting forces, increase the formation of fines, and increase in specific energy of cutting. A comprehensive understanding of disc cutter wear is essential for optimizing TBM performance, reducing operational costs, and ensuring project timelines are met.

Disc cutter wear directly affects the penetration rate, energy efficiency, and overall productivity of TBMs. Excessive wear leads to increased thrust and torque requirements, resulting in higher energy consumption and potential delays due to cutter replacement. In large-scale tunneling projects, cutter wear-related delays can escalate costs significantly. Records in actual projects demonstrates 30–40% of construction time could be spent on cutter inspection, maintenance, and replacement (Fu et al. 2022). Moreover, cutter consumption due to wear accounts for 30% of the total construction cost in extreme conditions when very strong and abrasive rocks are mined (Ren et al. 2018). Therefore, it is essential to clarify the wear mechanism and predict cutter service life that scholars and engineers have been working on for a long

M. Karami (✉)
Technital S.p.A, Verona, Italy
e-mail: morteza.karami@technital.it

J. Hassanpour and J. Rostami (eds.), *Performance of Tunnel Boring Machines (TBM) in Rock*, Springer Series in Geomechanics and Geoengineering,
https://doi.org/10.1007/978-3-032-20424-0_10

time. For instance, analysis of the Kerman water conveyance tunnel in Iran demonstrated that cutter wear rates varied significantly across different geological units, underscoring the importance of accurate prediction models (Karami et al. 2021a).

Estimation of the cutter wear is an integral part of TBM performance prediction, as it represents the life of the cutters and the need for inspection and changing of the cutters, as well as the need for maintenance and hence the associated cutter costs. In a given project this cost could be minimal, for example in softer limestone rocks where their cutters can offer extended life and rates of several thousand cubic meters of rock per cutter, in some cases meaning the completion of the project with one dressing of the cutterhead. This is compared to the cases where this rate drops to less than 100 cubic meters per cutter. Either way, in any projects involving hard rock TBM, cutter cost estimation needs to be conducted and provided to the pertinent party for planning purposes and cost estimations.

2 Wear Mechanisms

Disc cutter wear occurs due to the intense interaction between the cutter and rock, which involves high mechanical loads, abrasive forces, and thermal effects. However, due to the various condition of the rock, operational parameters and disc cutter material, different type of wear may be recorded during TBM tunneling. In other words, the wear mechanisms are influenced by factors such as rock abrasivity, rock strength, rock joints, cutter material, cutter geometry, TBM thrust, and rotational speed. Therefore, understanding the wear mechanisms affecting disc cutters is helps in operational planning and optimizing the performance of TBMs in rock tunneling. The main classification for disc cutter wear can be described as follows:

- Normal wear, also known as abrasive wear, is one of the most common forms of wear affecting TBM disc cutters used in hard rock TBMs. Studies have shown that normal wear contributes to approximately 70–90% of the total cutter wear observed in hard rock tunneling projects (Yang et al. 2018; Yu et al. 2021). This type of wear occurs due to continuous contact between the cutter and abrasive rock surfaces, leading to gradual material removal. This process results in a uniform decrease in the cutter's diameter over time without causing severe damage such as chipping or breakage, reducing its penetration efficiency and necessitating periodic replacements. The rate of abrasive wear depends on factors such as rock hardness, mineral composition, and the applied cutting forces. Moreover, field studies have shown that highly abrasive rock formations, like quartz-rich granites, significantly increase the rate of abrasive wear (Karami et al. 2021a). However, maintaining optimal cutter spacing, monitoring wear rates, and selecting high-quality cutter materials can help extend the lifespan of disc cutters and improve tunneling efficiency.

- Abnormal wear of disc cutters in hard rock mechanized tunneling presents significant challenges, impacting both operational efficiency and project costs. Unlike normal wear, which occurs uniformly, abnormal wear is manifested through irregular patterns such as chipping, spalling, tapering/grooving, mushrooming, blockage, and brittle fracture, often leading to premature cutter failure (Janc et al. 2024). These irregularities can result from various factors, including heterogeneous rock formations, suboptimal cutter design, or improper machine operation (using excessive thrust). Understanding and mitigating abnormal wear is crucial for enhancing TBM performance and ensuring the longevity of disc cutters (Ren et al. 2018). Abnormal wear could be due to issues with material properties of the cutter ring, for example heat treatment of a given batch of rings, or due to high loading of the cutters in harder rock and hence bearing failures.

As shown in Fig. 1, normal or abrasive wear occurs as the cutter ring gradually loses material due to consistent contact with rock surfaces, leading to uniform wear patterns. Tapering or grooving manifests as uneven wear, often resulting from misalignments or improper cutter rotation, causing the cutter ring to develop grooves or a tapered profile. Mushrooming refers to the deformation of the cutter ring edges, where the metal spreads outward, resembling a mushroom cap; this is typically due to excessive thrust forces or impacts with exceptionally hard rock formations, but also due to cutter ring material being softer than standard HRC ~ 55–56. Blockage happens when debris obstructs the bearing and hence cutter's rotation, leading to uneven wear or damage. Brittle fracture involves the sudden cracking or breaking of the cutter ring, often due to high-stress concentrations or flaws within the cutter material (i.e. very hard materials with HRC ~ 58–60), resulting in a rapid failure without significant prior deformation (Ellecosta et al. 2018).

Quantification of the abnormal wear is a very difficult task. Away from the material defects in the cutter ring that is unpredictable and incidental, most of the abnormal

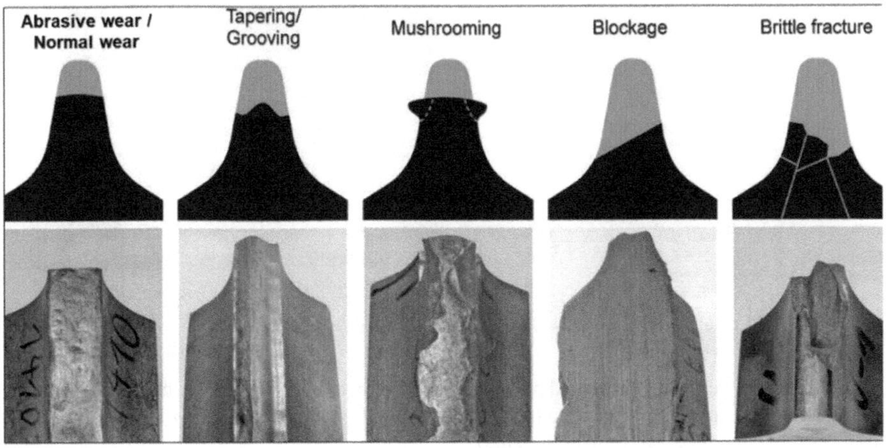

Fig. 1 Types of wear of a disc cutter due to TBM tunneling (Ellecosta et al. 2018)

wear has to do with the cutter load and machines operating thrust. Various studies have shown that the rate of abnormal wear increases at higher cutter load rating during the operation which is a function of the rock strength. The ratio of abnormal wear to normal wear can be as high as 20–30% in hard rock conditions. The estimation of the total number of cutters to be replaced in a project is the sum of normal and abnormal wear, but since the normal wear is predictable, often abnormal wear is expressed as a percentage of normal wear to allow for calculation of cutter cost. Meanwhile, to study the mechanism and quantification of abnormal wear some efforts have been made in recent years, as will be discussed here. With advancements in computational methods, numerical simulations have been employed to evaluate the rock breakage process and wear mechanism of disc cutters aimed to predict cutter life with greater precision. Researchers have attempted to study the effect of various parameters derived from geological features, rock properties, machine operational parameters, disc geometry on wear mechanism using simulating crack propagation, rock-breaking mechanisms, and disc cutter interactions with rock mass properties. These efforts contribute to a deeper understanding of TBM disc cutter behavior, aiding in the optimization of cutting efficiency and tool durability in tunneling projects. For Instance, Karami et al. (2022) simulated the dynamic process of the TBM boring in the jointed rocks using a real-scale numerical analysis using the discrete element method (Fig. 2). The stress distributions induced within the disc cutters as well as the development of the plastic zones in the rock are investigated and compared with the actual results. The numerical results indicated that the wear mechanism of disc cutters in jointed rocks is different than intact rocks so that a higher penetration rate as well as more lifetime for disc cutters can be achieved in highly fractured rocks.

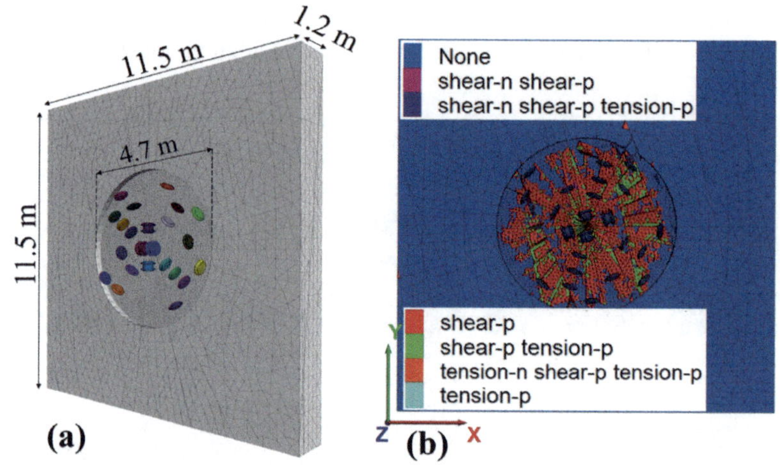

Fig. 2 **a** Model geometry generated by 3DEC for numerical analysis of rock mass cutting by disc cutters mounted on a TBM cutterhead; **b** Development of plastic zones due to TBM boring during after 10 s dynamic time (Karami et al. 2022)

This chapter focuses on the normal wear of disc cutters as the most common type of wear in TBM tunneling, and will offer some means of estimating the abnormal wear to allow for calculation of total cutter changes and hence cutter cost. As noted earlier, normal wear occurred due to the continuous interaction between the cutter and the rock, significantly affects the overall performance and efficiency of tunnel boring machines. Understanding and predicting this type of wear is crucial for optimizing cutter consumption and reducing operational costs. To address this, the chapter presents an overview of the main prediction models developed for estimating the lifespan of disc cutters. These models are based on various factors such as rock properties, machine operating conditions, and cutter design parameters. By analyzing these models, the chapter aims to provide practical insights into improving cutter life estimation and enhancing the efficiency of TBM tunneling projects. Hereafter, the term "wear" when used in a general context, specifically refers to the "normal wear" of the disc cutter and should not be mistaken for the general concept of wear.

3 Wear Measurement

Normal wear is the primary mechanism responsible for TBM disc cutter deterioration, occurring due to, among other things, the plastic deformation of steel when it comes into contact with high-strength or hard minerals. Numerous studies indicate that this type of wear is the most prevalent in TBM operations and has been documented across various tunneling projects. For instance, abrasive wear accounted for 88% of disc cutter replacements in the Qinling Tunnel in China (Lin et al. 2018), 82% in the Ghomrud Tunnel in Iran (Hassanpour 2018), 82% in the Tsuen Wan Tunnel in Hong Kong (Farrokh and Kim 2018), 85% in a metro tunnel excavation in Shenzhen in China (Su et al. 2020).

Regular visual assessments by experienced personnel can identify wear patterns and potential issues early. This method is invaluable when combined with other quantitative approaches. Accurate measurement of normal wear in TBM tunneling is crucial for maintaining operational efficiency and planning maintenance schedules. The measurements can be done directly utilizing calipers, micrometers, template, and indirectly by weighing the worn disc and correlating it to the extent of the wear. In Fig. 3, the extent of disc cutter wear is measured using a simple scaled template. In case of no wear, the template can be radially aligned with the tip of the disc cutter. As wear progresses, the gap between the template and the cutter tip increases, indicating the degree of material loss.

Additionally, an indirect measurement of disc cutter wear using the weighing method, which correlates wear to weight loss, is presented in Fig. 4 (Karami et al. 2021b).

Moreover, modern TBMs equipped with sensors can monitor parameters like thrust force and rotational speed. Anomalies in these readings may signal excessive wear in real-time (Han et al. 2023). In addition, advanced techniques such as close-range photogrammetry create detailed 3D models of cutters by processing photos,

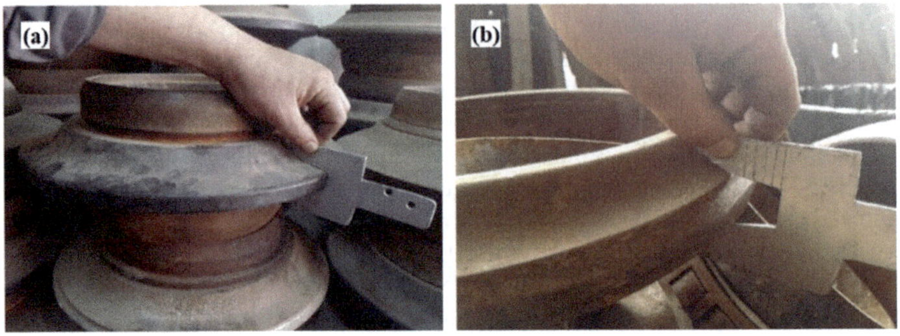

Fig. 3 Measurement of disc cutter wear using a scaled template: **a** No wear—template aligns radially with the cutter tip; **b** Worn cutter—the gap between the template and the cutter tip increases with wear (Karami et al. 2021b)

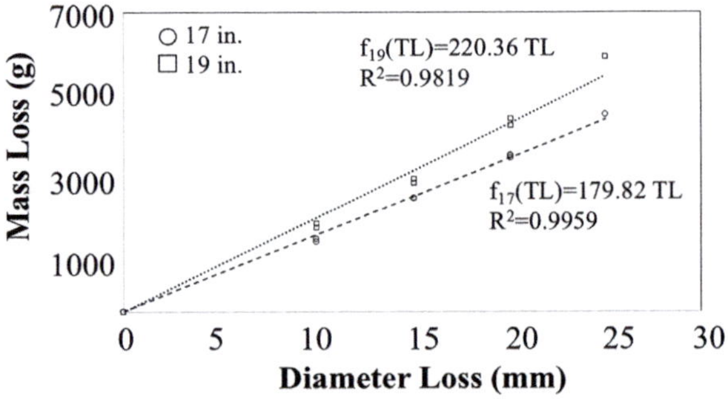

Fig. 4 The relationship between the mass loss (in gram) and tip loss (in millimeter) for 17-inch and 19-inch disc cutters (Karami et al. 2021b)

selected reference points, and control points, allowing for precise wear analysis. The resulting 3D model consisted of a numerous mesh including faces vertices, as shown in Fig. 5 (Janc et al. 2024). As the 3D scanning devices are becoming cheaper and more common in industrial applications, they can be used to scan the disc cutters at the face very quickly and offer a sub-millimeter accuracy in measurement and hence offer a quick process of measuring the ring wear. A technology involving "Disc Cutter Wear Sensor Package and TBM Monitoring System" has recently been unveiled that offers continuous monitoring of ring wear to assist the TBM operations and avoid frequent inspections (Ju et al. 2024).

4 Wear Limit

In tunnel boring machine (TBM) operations, the cutterhead is equipped with various groups of disc cutters (i.e. center group, face group, pre-gage group, and gage group), each assigned specific roles to optimize excavation efficiency. Center cutters are positioned close to the cutterhead's axis and are primarily responsible for breaking the

Fig. 5 A 3D model of a worn disc cutter created in Agisoft Metashape 1.6.3 software (Janc et al. 2024)

rock at the tunnel's centerline. Due to their location, they experience lower speeds and often utilize double-blade or center quat (first 4 cutters) designs to enhance rock-breaking efficiency in this region. Face cutters are distributed across the cutterhead's frontal surface and serve as the main tools for excavating the bulk of the tunnel face. These cutters typically feature single-blade designs and are strategically arranged to ensure uniform rock fragmentation and load distribution. Gage cutters are located along the outer perimeter of the cutterhead and define the tunnel's final diameter. They are crucial for cutting the boundary between the excavated section and the surrounding rock, ensuring the tunnel maintains its designed cross-sectional profile. Pre-gage or transition cutters are situated just before the gage cutters and function to reduce the load on the gage cutters by initiating the breakage of peripheral rock, thereby facilitating smoother and more efficient cutting at the tunnel's edges. The coordinated operation of these cutter groups ensures effective and balanced rock excavation during TBM tunneling. Figure 6 illustrates the hard rock double shield TBM featuring a cutterhead equipped with 27 disc cutters arranged in different groups. This configuration was specifically designed to address the geological challenges encountered during the excavation, optimizing cutter performance and tunnel advancement rates.

Setting wear limit of disc cutters is a crucial in hard rock TBM operations, directly impacting excavation efficiency, project costs, and machine performance. As disc cutters progressively wear due to continuous contact with rock, their cutting efficiency declines, increasing torque, thrust, and energy consumption. To maintain optimal TBM performance, it is essential to establish a wear limit—defined as the maximum allowable material loss of the ring (tip loss in radial direction) before cutter replacement becomes necessary. This limit varies depending on the cutter position on the cutterhead, with center cutters typically reaching their wear threshold faster than face, gage, and pre-gage cutters due to higher rock-breaking loads (Frenzel 2011). Properly defining and monitoring wear limits allows for timely cutter changes, minimizing downtime and reducing excessive cutter consumption. The maximum allowable wear limit for 17- and 19-inch disc cutters is 25 and 30 mm, respectively (Frenzel et al. 2008). However, irrespective of disc diameter, the typical wear limits for different cutter groups can be considered as follows:

- Center cutters: 20–25 mm

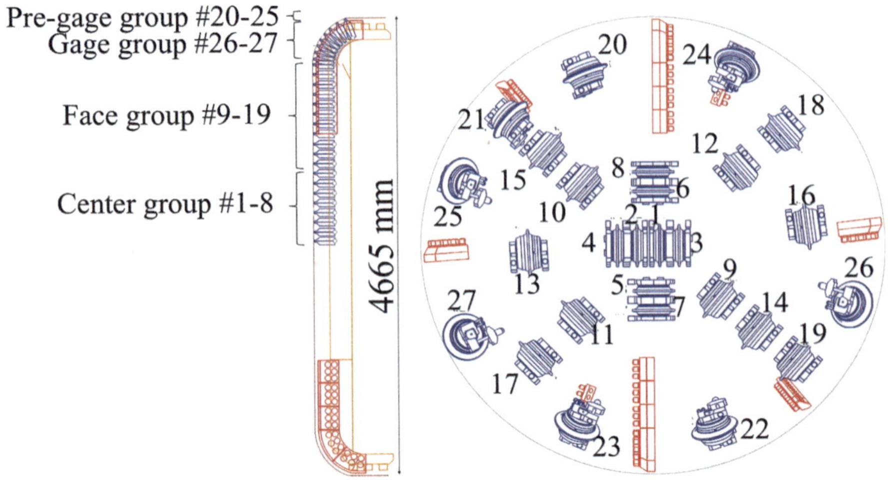

Hard Rock TBM	Center group	Face group	Pre-gage group	Gage group
Disc No.	1 to 8	9 to 19	20 to 25	26 to 27
Type	Twin	Single	Single	Single
Diameter (in)	17	19	19	19
Wear limit (mm)	25	20	15	10

Fig. 6 Groups of disc cutters mounted on a hard rock TBM cutterhead. A relative lower wear limit is defined for inner groups experience (Karami et al. 2021c)

- Face cutters: 15–20 mm
- Pre-gage cutters: 10–15 mm
- Gage cutters: 10 mm

Lower tip loss or wear limit for the gage cutters is to assure that tunnel diameter is maintained and that the cutterhead support or the open machines or front shield of the shielded machines do not get stuck by the reduced tunnel size. Another consideration is the fact that at the time of changing the cutters, the gage cutters have to start from a position where they have to penetrate the rock in front of them, hence very high penetration before they establish their track. This is not an issue for the face and center cutters. For this reason, lower limit is set for the gage cutters but it is not unusual to transpose these cutters to the face cutters to get another few hours of performance and cutting from them before they are removed to replace the cutter ring.

Table 1 Classification of CAI (Alber et al. 2014) and AVS (Dahl et al. 2012)

Classification	Mean CAI	AVS (mg)
Extremely low	0.1–0.4	≤ 1.0
Very low	0.5–0.9	1.1–3.9
Low	1.0–1.9	4.0–12.9
Medium	2.0–2.9	13.0–25.9
High	3.0–3.9	26.0–35.9
Very high	4.0–4.9	36.0–44.0
Extremely high	≥ 5	≥ 44.0

5 Rock Abrasivity Testing

For estimation of the cutter life in hard rock TBM tunneling, it is important to define and measure rock abrasion. Hamzaban et al. (2025) have provided a comprehensive review of rock abrasion measurements. In general, there are several approaches to measurement of rock abrasion. The simplest approach is to use the mineral compositions of the rock and the Vickers hardness of the constituent minerals to estimated weighted average of rock abrasion as represented by Vickers hardness (VHNR). Another approach is to use a more specific test such as Cerchar Abrasivity Index (CAI) where the contact between the cutter and rock is simulated by using a steel stylus scratching against the rock. CAI is by far the most common abrasivity testing used for application of TBMs. Recent modifications of the CAI introduced by Hamzaban et al. (2025) have shown very good results, yet the original CAI is what is mostly available in the contract documents and the modified version is still not available on the industrial scale. The other alternative abrasion measurement is the Abrasivity Value (AV) or the Abrasion Value of Steel (AVS) that is measured as part of the Norwegian (NTNU) suite of tests. The results of AVS testing can be used in conjunction with the NTNU model to estimate cutter life in their proposed TBM performance prognosis system (Macias et al. 2016). In Table 1, the abrasivity of the rock has been classified based on the CAI and AVS.

6 Cutter Life Prediction Models

Accurate prediction of disc cutter wear and lifetime is essential for optimizing TBM performance, accurate project planning, and reducing operational costs. Over the past few decades, numerous parties have developed empirical, analytical, and numerical models to estimate the wear rate and lifetime of disc cutters under varying geological and operational conditions. These models typically incorporate key rock properties such as uniaxial compressive strength (UCS), Cerchar Abrasivity Index (CAI),

Vickers Hardness Number Rock (VHNR), and rock mass characteristics. Additionally, TBM operational parameters, including thrust force, penetration depth, and cutter spacing, play a significant role in determining cutter consumption.

Some researchers have focused to correlate the wear of the disc cutters to key effective parameters as a quick solution. Such correlations can be considered as a guide or classification in which the wear or the life of the disc cutter is easily and fast estimated. In Table 2, a wear classification is presented based on the range of certain intact rock properties (Hassanpour et al. 2015; Farrokh and Kim 2018). According to Table 2, an approximate cutter life (H_f) can be predicted for TBM disc cutters operating under various strength and abrasiveness conditions.

Early empirical models were based on field data and established relationships between cutter wear and rock abrasivity. For instance, Gehring (1995) introduced a model linking the CAI to the mass loss of disc cutters, while Bruland (1998) developed the NTNU TBM prognosis model, which employs a Cutter Life Index (CLI) to predict cutter lifetime. Rostami et al. (2005) further refined empirical approaches by incorporating additional rock properties to estimate cutter wear rates in various geological conditions. Several studies have focused on statistical and regression-based approaches to predict cutter consumption. Frenzel (2011) analyzed a database of multiple tunneling projects to develop a statistical wear prediction model, while Hassanpour et al. (2014, 2018) proposed models specifically tailored to pyroclastic and mafic igneous rocks, incorporating UCS and VHNR for enhanced accuracy.

Additionally, Bieniawski et al. (2009) explored the relationship between rock mass excavability (RME), CAI, and cutter consumption, emphasizing the influence of rock mass properties. While many models primarily rely on intact rock properties, recent studies have emphasized the role of rock mass characteristics, such as joint frequency and orientation on cutter wear. Naghadehi and Ramezanzadeh (2017) and Koopialipoor et al. (2019) highlighted that joints facilitate rock breakage, reducing overall cutter consumption compared to intact rock conditions. Furthermore, Karami et al. (2021b) further investigated the impact of rock mass strength on cutter life,

Table 2 Approximate abrasivity, wear and cutter life (H_f) for different indices (Hassanpour et al. 2015; Farrokh and Kim 2018)

Abrasivity class	Wear class	H_f (m^3/cutter)	UCS (MPa)	CAI	EQC (%)	CLI
Extremely abrasive	Extremely high	<100	200–250	4–6	75–100	35–75
Very abrasive	Very high	100–500	150–200	2–4	35–75	15–34
Medium abrasive	High	500–1000	100–150	1–2	20–35	8–14.9
Slightly abrasive	Normal	1000–2000	50–100	0.5–1	10–20	6–7.9
Not very abrasive	Low	2000–2500	25–50	0.3–0.5	5–10	5–5.9
Not abrasive	No wear	>2500	<25	0–0.3	0–5	0–5

H_f: Average cutter life in m^3/cutter; UCS: Intact rock compressive strength; CAI: Cerchar abrasivity index; EQC: Equivalent quartz content; CLI: Cutter life index

demonstrating that models based solely on intact rock properties tend to overestimate cutter wear, leading to conservative predictions.

The following section addressed the most significant prediction models for estimating the life of TBM disc cutters, along with their corresponding equations. These models and their parameters are presented in a way that enables the reader to easily estimate cutter life for a given project, provided that the relevant input data is available.

6.1 Gehring (1995) Model

In 1995, Gehring developed a simple equation based on data analysis from various tunneling projects, establishing a correlation with a coefficient of 0.7 between the Cerchar Abrasivity Index (CAI) and the weight loss of TBM disc cutters due to abrasive wear, expressed in milligrams of steel per meter of rolling distance (vs). This model is presented in Eq. 1. Later, Maidl et al. (2008) modified this model by adjusting the coefficient from 0.74 to 0.65 to account for improvements in steel alloy properties (Eq. 2). In Gehring (1995) model, the normal life of the disc cutter is determined using Eq. 3 (Ko et al. 2014). As a point, ΔG representing Maximum permissible mass loss of the disc cutter, should be determined based on the diameter of the disc cutter, cross section and density of the disc cutter, and maximum allowable wear. As a suggestion, for a 17" Disc cutter with a maximum allowable wear 30 mm, ΔG can be assumed to be 3700 g.

$$\mathrm{v_s} = 0.74\mathrm{CAI}^{1.93} \tag{1}$$

$$\mathrm{v_s} = 0.65\mathrm{CAI}^{1.93} \tag{2}$$

$$H_f = \frac{\Delta \mathrm{G.P.D}^2}{4\overline{\mathrm{D}}.\mathrm{v}_s.\mathrm{N}} \tag{3}$$

Where, H_f is disc cutter life (m^3/cutter), v_s is specific cutter ring wear in milligram per roll meter (mg/Rm)), ΔG is maximum permissible mass loss of the disc cutter (g), P is cutterhead penetration (mm/rev), D is TBM diameter (m), $\overline{D}$ is average rolling diameter of the disc cutter which is $\overline{D}$ =0.6D, and N is number of disc cutters on the cutterhead.

6.2 CSM Model

The CSM model, developed by Rostami (1997) at the Colorado School of Mines, estimates the cutting forces acting on a disc cutter to provide machine penetration rate for a given size TBM with specified Thrust, Torque, and Power. Within this model, a separate module was introduced to estimate lifespan of disc cutters using the laboratory CAI index and the average rolling distance of a cutter which was explained in more details in Rostami et al. (2005). Additionally, it is used to estimate other wear-related parameters, such as the total number of cutters required, cutter costs per unit volume of excavated rock, total cutter costs, cutter replacement time, and associated operational delays. The main equations of this model are defined in Eqs. 4 to 8 (Ko et al. 2014).

$$\mathrm{CR} = \frac{2054}{\mathrm{CAI}}\frac{D_c}{432} \tag{4}$$

$$\overline{s_r} = \frac{0.6\mathrm{D}\pi}{1000}[\mathrm{km}] \tag{5}$$

$$U_c = \frac{\mathrm{CL}.\overline{\mathrm{k}}}{\overline{s_r}} \tag{6}$$

$$\overline{k} = \sum \mathrm{gk}_k k_v \tag{7}$$

$$H_f = \frac{\pi \mathrm{D}^2}{4}.\mathrm{U}_c.\frac{P}{1000}.\frac{2\mathrm{S}}{1000\mathrm{D}} \tag{8}$$

Where, H_f is disc cutter life (m^3/cutter), CR is maximum linear rolling distance of a disc cutter before wear-out and replacement (km), CAI is Cerchar abrasivity index, D_c is disc cutter diameter (m), $\overline{s_r}$ is average rolling distance of disc cutters (km), D is TBM diameter (m), U_c is maximum number of cutter rotations around its axis before replacement, P is cutterhead penetration (mm/rev), S is spacing between disc cutters (mm), and D/2S is approximately equal to the total number of disc cutters mounted on the cutterhead.

The initial coefficient of 2057 has recently been increased to 3000 to account for better quality of steel rings and more advanced alloys used by various manufacturers, effectively increasing the cutter life by nearly 45% compared to the cutters used in the original data base used for development of the model introduced by Rostami et al. (2005). Additionally, $\overline{\mathrm{k}}$ is the cutter wear correction factor, which depends on its installation zone on the cutterhead, while g represents the weight ratio of center, face, and gage cutters ($\sum g = 1$). It should be noted that kk and kv are determined based on the data provided in Table 3. If no information is available regarding the number of disc cutters, the correction factor is assumed to be 0.67 (Ko et al. 2014).

Table 3 Correction factor for disc CUTTERS (Ko et al. 2014)

Installation zone	k_k	k_v
Center	0.56	1.0
Face	0.76	1.0
Gage	0.92	0.54

6.3 NTNU Model

The NTNU model was initially proposed in a report published at Norwegian University of Science and Technology (NTH 1988) and later updated by Bruland (1998). Compared to the CSM model, the NTNU model has a more extensive field data and operational history and is based on a large dataset collected from various tunneling projects. This model utilizes specific charts and simplified equations to predict disc cutter lifetime and associated costs. The primary input parameter of this model is the Cutter Life Index (CLI) (Bruland 1998) which is typically presented for some rocks in Fig. 7. The life of disc cutters is determined using Eq. 9 (Ko et al. 2014).

$$H_f = \frac{\pi D^2}{4} . \frac{H_0 k_D k_Q k_{\mathrm{RPM}} k_N}{N} .\mathrm{P}.\mathrm{RPM}. \frac{60}{1000} \tag{9}$$

Where, H_f is disc cutter life (m^3/cutter), D is TBM diameter (m), H_0 is basic cutter ring life (h), as illustrated in Fig. 8. P is cutter penetration (mm/rev), RPM is cutterhead rotation speed (rev/min), D is tunnel or TBM diameter (m), N is the number of disc cutters mounted on the cutterhead and N_0 is normal number of cutters, as illustrated in Fig. 9. Correction factors include k_N for cutter numbers equals to N/N_0, k_{RPM} for RPM equals to (50/D)/RPM, k_D for TBM diameter (Fig. 10), and k_Q for quartz content (Fig. 11).

6.4 Maidl et al. (2008) Model

Maidl et al. (2008) demonstrated that the normal wear of cutting tools is correlated with the Cerchar Abrasivity Index (CAI) and the Uniaxial Compressive Strength (UCS) of rock. According to this model, the disc cutters' wear can be minimal in some weak rocks with a CAI value of approximately 2 and extremely high in strong granites with a CAI of around 6. Moreover, the Specific Disc Cutter Wear Rate (SDCWR) expressed as cubic meters of excavation per cutter (m^3/cutter) is determined by the intact rock strength (σ_c) in MPa and the Cerchar Abrasivity Index (CAI), as defined by Eqs. 10 to 14 (Maidl et al. 2008).

$$\mathrm{CAI} = 2 : \mathrm{SDCWR} \cong 10^{\left(0.000020\sigma_c^2 - 0.0088\sigma_c + 3.9944\right)} \tag{10}$$

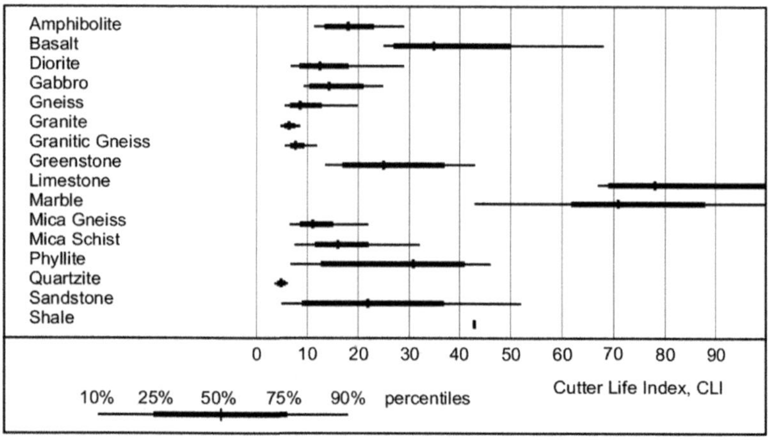

Fig. 7 Typical Cutter Life Index (CLI) for some rock types (Bruland 1998)

Fig. 8 Basic cutter ring life (H0) in hour for different disc size (Bruland 1998)

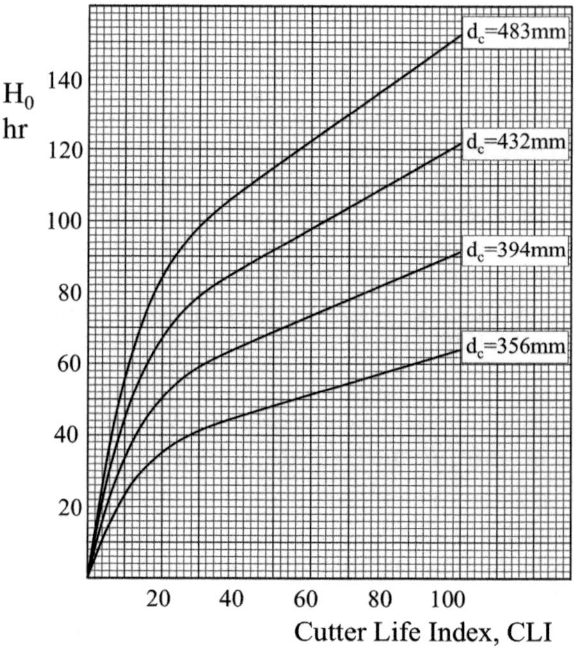

$$\text{CAI} = 3 : \text{SDCWR} \cong 10^{(0.000007\sigma_c^2 - 0.0056\sigma_c + 3.1889)} \tag{11}$$

$$\text{CAI} = 4 : \text{SDCWR} \cong 10^{(0.000010\sigma_c^2 - 0.0037\sigma_c + 2.8387)} \tag{12}$$

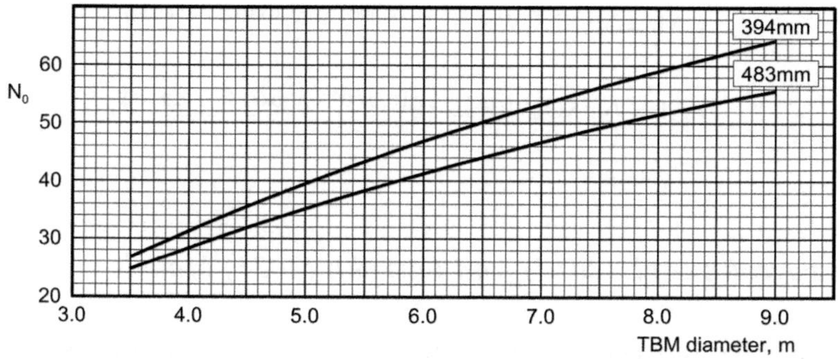

Fig. 9 Normal number of cutters based on TBM diameter and disc size (Bruland 1998)

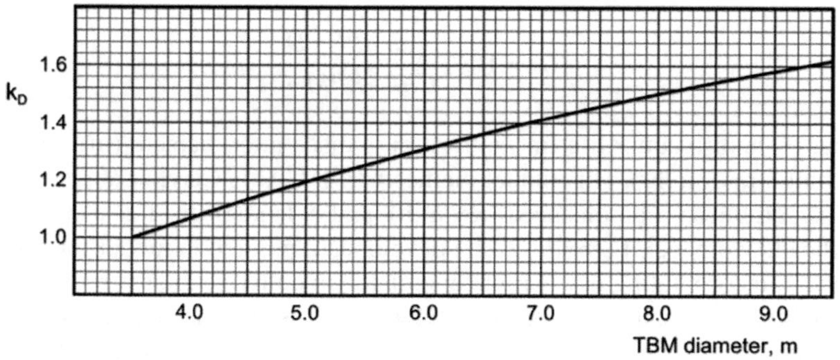

Fig. 10 Correction factor for TBM diameter (Bruland 1998)

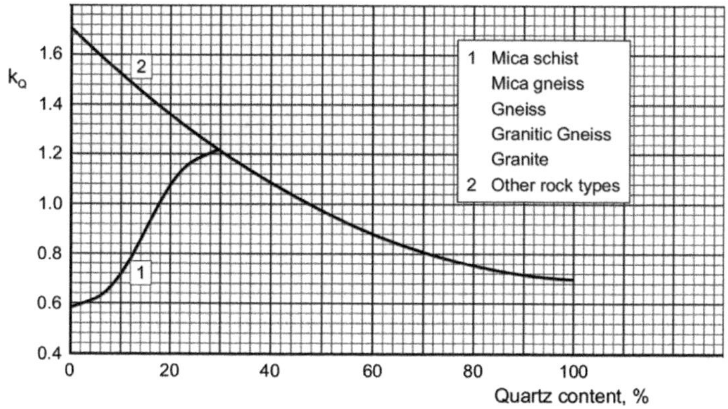

Fig. 11 Correction factor for quartz content (Bruland 1998). It should be noted if quartz content is less than 30%, the user should be aware of rock type to select the correct line (i.e. line 1 or 2)

$$\text{CAI} = 5 : \text{SDCWR} \cong 10^{(0.000010\sigma_c^2 - 0.0037\sigma_c + 2.4669)} \tag{13}$$

$$\text{CAI} = 6 : \text{SDCWR} \cong 10^{(0.000005\sigma_c^2 - 0.0037\sigma_c + 2..2371)} \tag{14}$$

6.5 Farrokh et al. (2013) And Farrokh (2021) Models

Based on factors such as the Cerchar Abrasivity Index (CAI), and operational parameters of the TBM, Farrokh et al. (2013) introduced an empirical model to predict disc cutter life as Eq. 15.

$$H_f = 1577 \frac{d_i \cdot PR \cdot D}{CAI \cdot RPM \cdot N} \tag{15}$$

Where, H_f is disc cutter life (m^3/cutter), di is disc cutter diameter(in), PR is TBM penetration rate (m/hr), D is tunnel diameter (m), CAI is Cerchar abrasivity index, RPM is cutterhead rotation speed (rev/min), and N is number of disc cutters on the cutterhead. Moreover, Farrokh (2021) developed a prediction model for estimating cutter life, incorporating key parameters related to cutter geometry, rock properties, and machine operation. The specific cutter weight loss (Scwl), measured in milligram per roll meter, is defined as Eq. 16. It should be noted that Scwl is closely related to the geometrical characteristics of the disc cutter as shown in Fig. 12. Accordingly, the cutter life can be predicted using Eq. 17.

$$Scwl = \frac{8\pi \cdot d \cdot TL \cdot T}{R} \tag{16}$$

$$H_f = 2\pi \cdot r \cdot S \cdot N \cdot p = R \cdot S \cdot p = \frac{8\pi \cdot d \cdot TL \cdot T \cdot S \cdot p}{Scwl} \tag{17}$$

In the above equations, Scwl is specific cutter weight loss (mg/Rm), D is disc diameter, TL is tip loss of disc cutter, T is tip width of disc cutter, R is cutter rolling distance, H_f is cutter life (m^3/cutter), r is radial position of the cutter on the cutterhead, S is spacing between disc cutters, N is number of cutterhead revolutions, and P is cutterhead penetration.

6.6 Hassanpour et al. (2014) and (2018) Model

Hassanpour et al. (2014) developed a model for predicting normal wear of disc cutters based on geological conditions. The database for this model includes actual disc cutter wear and geological data from various projects completed between 2005–2014. In

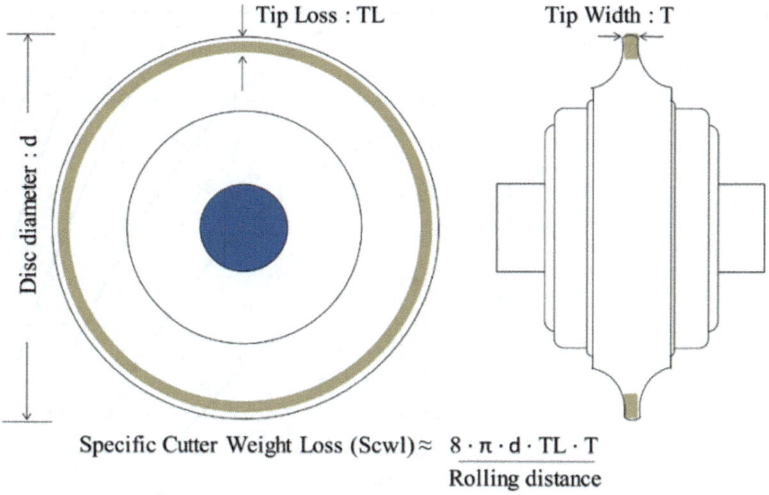

Fig. 12 Specific cutter weight loss and geometrical characteristic of a disc cutter (Farrokh 2021)

his model, the Abrasivity Index (ABI) as a new parameter, formulated using the Vickers Hardness Number of Rock (VHNR) and the Uniaxial Compressive Strength (UCS), as defined in Eq. 18. Using simple and multivariate regression analyses, empirical prediction models for disc cutter life were developed for pyroclastic and mafic rocks, represented by Eqs. 19 and 20, with regression coefficients of 0.683 and 0.693, respectively. Figure 13 illustrates a chart for predicting disc cutter life based on Eq. (10.20) (Hassanpour et al. 2014). Later, Hassanpour et al. (2018) offered an empirical prediction model for the foliated and schistose metamorphic rocks as Eq. 21.

$$ABI = VHNR\left(\frac{UCS}{100}\right) \tag{18}$$

$$H_f = -940.1\ \mathrm{Ln(ABI)} + 6939.4 \tag{19}$$

$$H_f = -2.669\ \mathrm{VHNR} - 7.891\ \mathrm{UCS} + 3430.955 \tag{20}$$

$$H_f = -756\ \mathrm{Ln\ (ABI)} + 5605.6 \tag{21}$$

In the above equations, ABI is Abrasivity Index, VHNR is Vickers Hardness Number of Rock, UCS is uniaxial compressive strength (MPa), and H_f is cutter life (m^3/cutter).

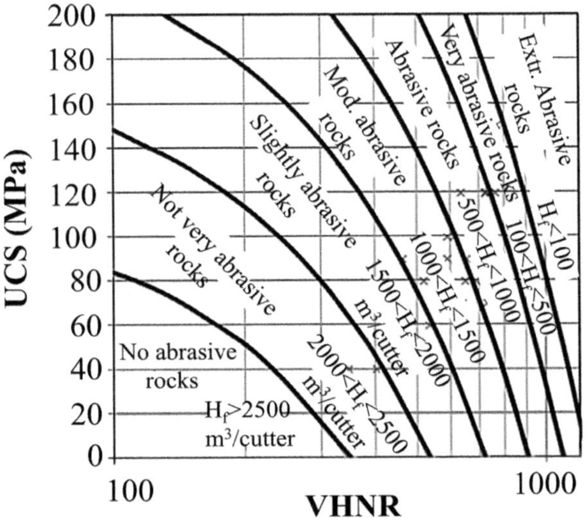

Fig. 13 Prediction chart for disc cutter life in igneous and mafic rocks (Hassanpour et al. 2014)

6.7 *Karami et al. (2021) Model*

As earlier discussed, empirical models are widely used in tunneling to predict TBM disc cutter life during design and construction. However, these models, based on different databases and input parameters, often yield varying predictions that may not align with actual project conditions. In this regard, Karami et al. (2021a) analyzed various empirical models under different rock abrasivity conditions and found that selecting an appropriate model is critical, especially for slightly abrasive rocks, due to the wide range of predicted lifetimes. Figure 14 shows how the various empirical prediction models can lead to different results of disc cutter life under same abrasivity class.

Comparison of the predicted life and measured life of disc cutters recorded in a hard rock TBM tunneling project, Karami et al. (2021b) found that predicted cutter life are generally lower than actual ones due to the influence of joints and fractures in rock masses. Since empirical models often overestimate wear by focusing on intact rock properties, they recommended to incorporate rock mass parameters such as RMR, Q, and GSI to enhance prediction accuracy. According to the analysis of field data from various projects, Karami et al. (2021b) proposed a new empirical prediction model using nonlinear multivariable regression based on CAI as a rock material parameter and RQD (Rock Quality Designation) as a rock mass parameter to estimate the cumulative volumetric mass loss of disc cutters. The modified version of this model is presented in Eq. 22.

Figure 15 is a chart introduced to ease of using the model expressed in Eq. 22. Knowing ΔG as the maximum permissible mass loss of the disc cutter (g) can lead to estimate the cutter life as presented in Eq. 23. It should be noted that ΔG depends on the cross section of a disc cutter, the maximum allowable wear, and the cutter density.

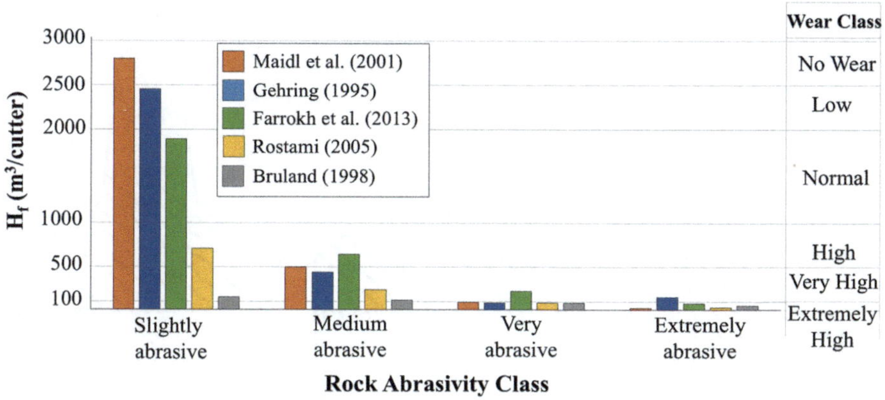

Fig. 14 Comparison of cutter life predicted by different empirical models for the different rock abrasivity class (Karami et al. 2021a)

In Fig. 16, the cross section of cutter rings with different diameter and calculation of maximum wear volume has been illustrated (Log et al. 2018). By considering an average radius for the maximum wear area respect to the disc axis and knowing the density of steel of the cutter ring, the maximum permissible mass loss can be easily calculated.

$$\mathrm{MV}_{Cumulative} = 3.042\text{e-}7\ \mathrm{CAI}^{0.586} RQD^{3.865} \tag{22}$$

$$H_f = \Delta G / MV_{Cumulative} \tag{23}$$

Where, $\mathrm{MV}_{\mathrm{cumulative}}$ is cumulative volumetric mass loss of disc cutters (g/m^3), CAI is Cerchar abrasivity index, RQD is rock quality designation, $\mathrm{H_f}$ is cutter life (m^3/cutter), and ΔG is maximum permissible mass loss of the disc cutter (g).

Karami et al. (2021c) introduced an updated empirical model to predict the life of disc cutters in jointed rock masses. The model incorporates both intact rock and

Fig. 15 Prediction chart for estimating $\mathrm{MV}_{\mathrm{cumulative}}$ based on CAI and RQD (Karami et al. 2021b)

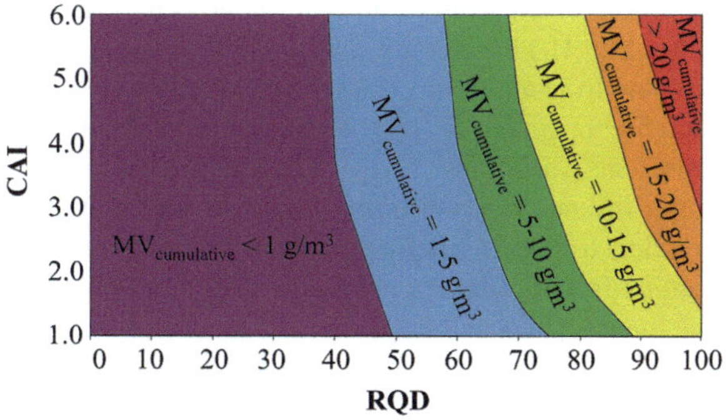

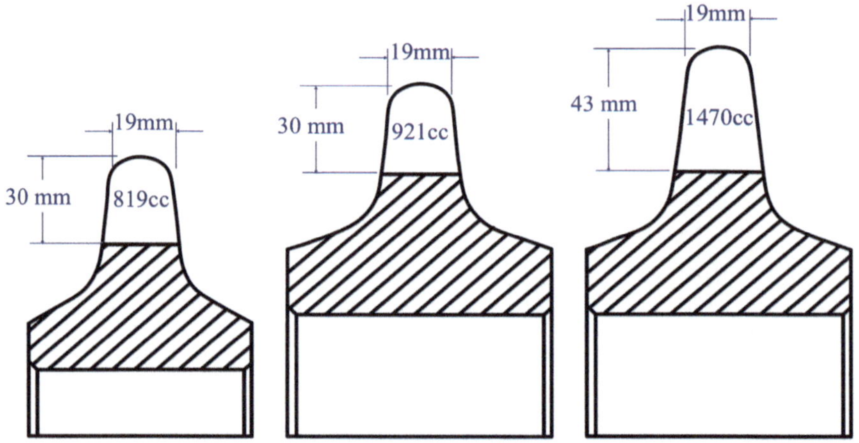

Fig. 16 Cross section of cutter rings and calculation of wear volume (Log et al. 2018)

rock mass properties, aiming to improve the limitations of pre-existing models that primarily consider intact rock characteristics. The key parameters involved in this model include the total fracturing factor from NTNU model (ks-tot), which accounts joint orientation and spacing, the rock mass strength (UCSrm), and the Cerchar abrasivity index (CAI), representing the abrasivity of the intact rock. Based on Eq. 24, The proposed model was developed using multivariable regression analysis and demonstrated improved prediction accuracy compared to existing empirical models. Validation against field data from another tunneling project confirmed the model's reliability in estimating cutter wear in jointed rock conditions.

$$\mathrm{H}_f = 963 \frac{k_{s-tot}^{0.32}}{UCS_{rm}^{0.15} \cdot CAI^{0.2}} \tag{24}$$

Where H_f is cutter life (m^3/cutter), $k_{s\text{-}tot}$ is total fracturing factor, UCS_{rm} is rock mass strength (MPa), and CAI is Cerchar abrasivity index.

Figure 17 illustrates four color-coded charts prepared to offer and estimate of disc cutter life (H_f) based on Eq. 24 for a range of CAI of 3 to 6. These charts use a color gradient to visually illustrate the relationships between rock mass strength (UCS_{rm}), total fracturing factor ($k_{s\text{-}tot}$), and Cerchar abrasivity index (CAI). The color spectrum represents variations in H_f (m^3/cutter), with cooler colors (blue) indicating longer cutter life and warmer colors indicating shorter cutter life. These charts serve as a practical tool for tunneling engineers, providing a quick and reliable way to estimate cutter life without requiring equations. In following, the relations and process to calculate UCS_{rm} and $k_{s\text{-}tot}$ are briefly mentioned.

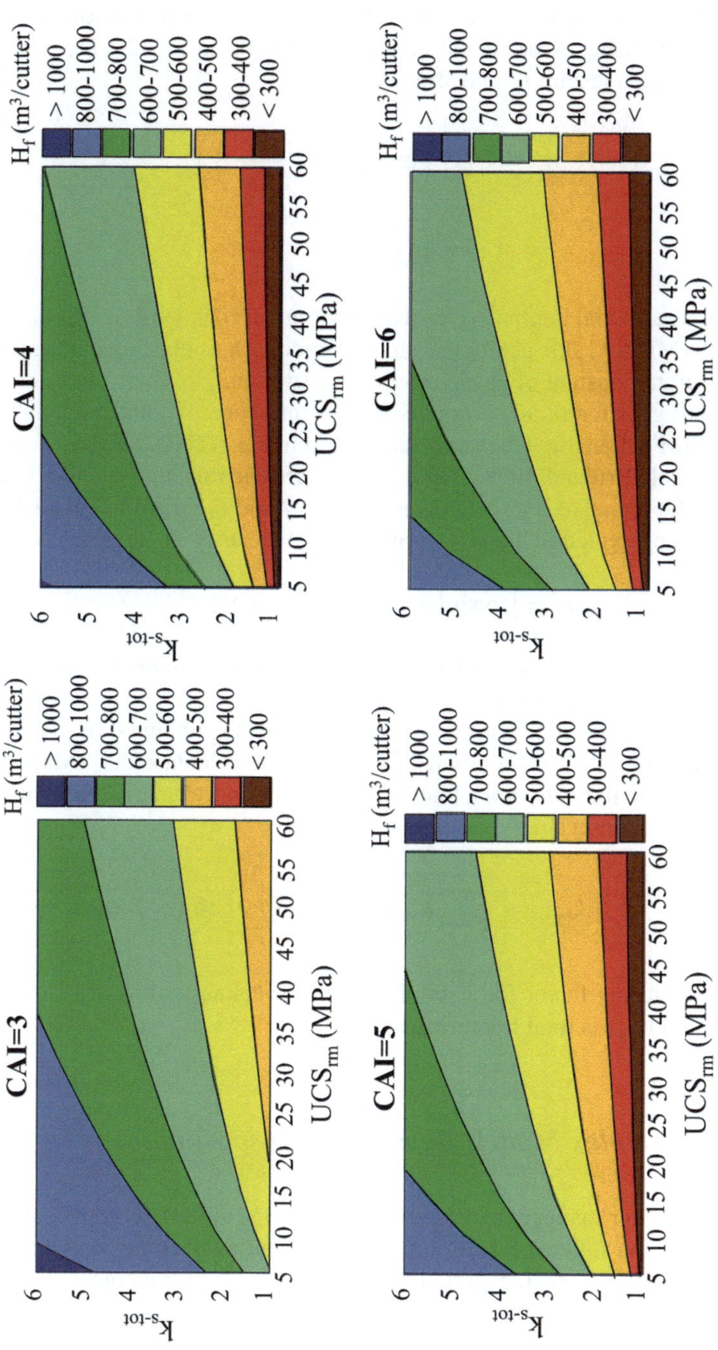

Fig. 17 Prediction charts plotted based on empirical prediction model as Eq. 24 (Karami et al. 2021c)

To determine the rock mass strength, Hoek and Brown (2019) presented an empirical relationship which utilize parameters such as GSI, UCS and empirical coefficients. This relationship is presented in Eqs. 25 and 26. In Eq. 25, D_b is blasting damage factor which is zero for TBM tunneling.

$$UCS_{rm} = UCS_i\left(\exp\left(\frac{GSI - 100}{9 - 3D_b}\right)\right)^a \tag{25}$$

$$a = \frac{1}{2} + \frac{1}{6}\left(e^{-\frac{GSI}{15}} - e^{-\frac{20}{3}}\right) \tag{26}$$

Where, UCS_i is uniaxial compressive strength of intact rock (MPa), UCS_{rm} is rock mass strength (MPa), GSI is geological strength index, D_b is blasting damage factor, and A is curvature constant of Hoek and Brown criterion.

The fracturing factor ($k_{s\text{-}tot}$) accounts for the influence of joint orientation and spacing on TBM cutter life which was introduced in the NTNU TBM Performance Prediction Model (Bruland 1998, 2000). It is determined using the angle between joint sets and the tunnel axis as calculated by Eq. 27 and the fracture class based on joint spacing as presented in Table 4 (Yagiz 2003; Macias et al. 2014).

$$\alpha = \left|\arcsin\left(\sin\alpha_f . \sin(\alpha_t - \alpha_s)\right)\right| \tag{27}$$

Where, α_s and α_f are strike and dip angles of the weakness plane, respectively and α_t is the azimuth of the tunnel axis.

The individual fracturing factor (k_s) for each joint set is obtained from empirical charts as shown in Fig. 18 and combined using Eq. 28 to calculate $k_{s\text{-}tot}$ (Macias et al. 2014). Higher $k_{s\text{-}tot}$ indicate more fractured rock, leading to increased penetration rates and extended cutter life.

$$k_{s-tot} = \left(\sum_{i=1}^{n} k_{si}\right) - (n - 1) \times 0.36 \tag{28}$$

where k_{si} is fracturing factor for a set of planes of weakness of i_{th}, n is number of fracture sets and $k_{s\text{-}tot}$ is total fracturing factor.

6.8 More Complex Models Based on Transposition

The process of cutter management on a hard rock TBM is fairly complex and often involves tracking the cutters per position and implementing the transposition to minimize the number of cutters transported to the face and out of the tunnel. This refers to the reuse of gage cutter disc where they have smaller wear limits at the face, so that they can wear to lowest allowable level at the face, before they are pulled out for replacement and shipped to cutter shop to remove the worn ring and install a new

Table 4 Fracture class as defined by the spacing between planes of weakness (Yagiz 2003; Macias et al. 2014)

Fracture class (joints/fissures)	Average spacing between planes of weakness a_f (cm)	Range of spacing (cm)	Discription
0 0+	∞ 190	240- ∞ 160–240	Massive rock interval with few joints or fissures. Seldom found in complexly deformed terranes except for granoblastic metamorphic rocks and equiangular, crosscutting igneous rocks. Fracture spacing must be greater than 160 cm
0-I	140	110–160	Massive rock interval with fracture spacing of 110–160 cm
I-	90	60–110	Relatively massive rock interval with fracture spacing of 60–110 cm
I I+ I-II II-	40 35 30 25	37.5–60 32.5–37.5 27.5–32.5 22.5–27.5	Fractured rock
II	20	17.5–22.5	Well fractured rock mass
II-III	15	12.5–17.5	
III	10	8.75–12.5	Highly fractured rock mass
III-IV	7.5	6.25–8.75	
IV	5	4–6.25	Highly brecciated with closely spaced anastomosing fractures completelt associated with zones of stress relief, fault breccia, and fault gouge

ring. This requires careful tracking of cutters on the face, as the TBM goes from one formation to another. In other words, the cutters installed at a certain position may bore through a certain rock mass and transition to another formation along the tunnel, thus the wear in each formation has to be calculated for a given cutter position to see when the cutter should be replaced. This means that the wear of each individual cutters should be monitored to determine when they should be changed, also to see if a cutter at the gage is going to be mounted at the face, where the replacement cutter does not start at zero wear, rather at certain level of wear when it is installed. A more comprehensive model for tracking of disc cutters across the face and across different formations has been offered by Rajati et al. (2023) and additional details are being provided in a publication that is under review. This level of details are warranted if a contractor wants to optimize the supply chain of their replacement parts to avoid front loading the cost of cutters, also to manage resources and have the proper stock of parts for uninterrupted operation at the site.

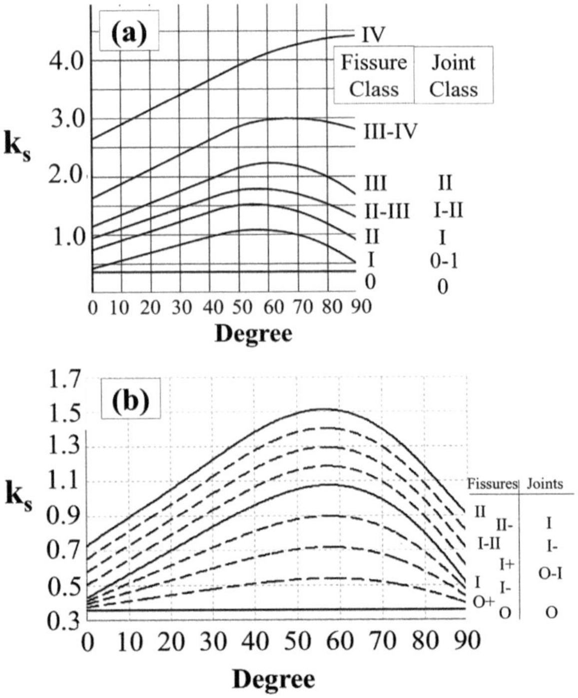

Fig. 18 **a** Calculation of fracturing factor (k_S) based on the angle between the tunnel axis and the plane of weakness and fracture class; **b** Interpolated curves for low fracture classes (i.e. joints: 0-I, and fissures: 0-II) (Macias et al. 2014)

7 Summary and Recommendations

Importance of the accurate estimation of cutter life as part of the performance prediction for hard rock TBM is very high, as the cutter wear can impact TBM efficiency and rate of penetration, as well as machine utilization, hence heavily influencing daily advance rate. Furthermore, given the cost of cutters and the time it takes to inspect, replace, and maintain the disc cutters in hard rock TBM operations, accurate estimate of the cutter life allows for proper planning of the cutter management and hence project success.

While different cutter life estimation models are available, it is highly recommended that an expert be engaged to look at the test results, TBM design and specifications, and overall project ground conditions to provide a reasonable estimate. For rough initial estimates, models such as Hassanpour et al. (2014) could often suffice. However, for more detailed analysis and planning of the cutter changes, more delicate models such as Karami et al. (2021c) could be used. Ultimately, for more detailed analysis the method presented by Rajati et al. (2023) or CSM model in conjunction with the performance prediction models offered by CSM can be considered. The impacts of rock mass and joints cannot be overlooked and should be incorporated into the estimates to adjust the estimated cutter wear rate and number of cutters for completion of the project. It is highly recommended that multiple models to be used, for example NTNU+ Karami+CSM to be able to see the potential range of cutter life that can be expected.

It is also important to reiterate that the models listed above are for normal wear. In very hard and abrasive rocks, the number of abnormal cutter wear should be estimated and added to the normal wear to arrive at total number of cutters for completion of the project. As a rule of thumb, the abnormal wear can be calculated as follows:

- If the anticipated cutter load in a given project is less than 60% of the nominal cutter load capacity, Abnormal wear $\leq 5\%$. For example, if the anticipated cutter load for a 17-inch disc cutter with nominal capacity of 250 kN is less than 150 kN per cutter (low thrust conditions in soft to medium strength rock), around or less than 5% abnormal wear can be expected.
- If the applied cutter load is 60% < FN/CL < 80%, roughly 10–15% abnormal wear can be expected.
- If the applied cutter load is 80% < FN/CL, 25–30% abnormal wear can be anticipated using Eq. 29:

$$\%Abnormal\ Wear = 30 \times \left(\frac{F_N}{CL}\right)^3 \tag{29}$$

where F_N is the applied cutter load in kN which can be calculated as (thrust / $N_{cutters}$), and CL is the nominal cutter load capacity per manufacturer's recommendation/ specification (i.e. 250 kN for 17-inch disc cutters).

References

Alber M et al (2014) ISRM suggested method for determining the abrasivity of rock by the CERCHAR abrasivity test. In: The ISRM suggested methods for rock characterization, testing and monitoring: 2007-2014, pp 101–106

Bieniawski ZT, Celada B, Galera JM, Tardáguila I (2009) Prediction of cutter wear using RME. In: Proceedings of ITA Congress, Budapest

Bruland A (1998) Hard rock tunnel boring. Dissertation, Norwegian University of Science and Technology (NTNU)

Bruland A (2000) Hard rock tunnel boring. Fakultet for ingeniørvitenskap og teknologi

Dahl F, Bruland A, Jakobsen PD, Nilsen B, Grøv E (2012) Classifications of properties influencing the drillability of rocks, based on the NTNU/SINTEF test method. Tunn Undergr Space Technol 28:150–158

Ellecosta P, Kasling H, Thuro K (2018) Tool wear in TBM hard rock drilling—backgrounds and special phenomena. Geomech Tunn 11:142–148

Farrokh E, Rostami J, Askilsrud OG (2013) A discussion on TBM cutter change time and cutter life. In: SME annual meeting, Denver, pp 13–105

Farrokh E, Kim DY (2018) A discussion on hard rock TBM cutter wear and cutterhead intervention interval length evaluation. Tunn Undergr Space Technol 81:336–357

Farrokh E (2021) Cutter change time and cutter consumption for rock TBMs. Tunn Undergr Space Technol 114:104000

Frenzel C (2011) Disc cutter wear phenomenology and their implications on disc cutter consumption for TBM. In: ARMA US rock mechanics/Geomechanics symposium. ARMA, p ARMA–11

Frenzel C, Käsling H, Thuro K (2008) Factors influencing disc cutter wear. Geomech Und TunnBau 1(1):55–60

Fu J, Zeng G, Zhang H, Zhang L, Yin X, Xia Y (2022) Experimental investigation on wear behaviors of TBM disc cutter ring with different cooling methods. Eng Fail Anal 134:106076

Gehring K (1995) Prognosis of advance rates and wear for underground mechanized excavations (in German). Felsbau 13(6):439–448

Hamzaban MT, Rostami J, Farrokhzadeh S, Faridazad M, Rasouli F (2025) A critique on the correlations between rock abrasiveness, hardness, and mechanical characteristics: implication of LCPC abrasiveness and breakability coefficients. Rock Mech Rock Eng 58(2):2487–2529

Han J, Xiang H, Feng Q, He J, Li R, Zhao W (2023) Detection for disc cutter Wear of TBM using magnetic force. Machines 11(3):388. https://doi.org/10.3390/machines11030388

Hassanpour J (2018) Development of an empirical model to estimate disc cutter wear for sedimentary and low to medium grade metamorphic rocks. Tunn Undergr Space Technol 75:90–99

Hassanpour J, Rostami J, Tarigh Azali S, Zhao J (2014) Introduction of an empirical TBM cutter wear prediction model for pyroclastic and mafic igneous rocks; a case history of Karajwater conveyance tunnel, Iran. TunnIng Undergr Space Technol 43:222–231

Hassanpour J, Rostami J, Zhao J, Tarigh Azali S (2015) TBM performance and disc cutter wear prediction based on ten years' experience of TBM tunnelling in Iran. Geomech TunnLing 8(3):239–247

Hoek E, Brown ET (2019) The Hoek–Brown failure criterion and GSI–2018 edition. J Rock Mech Geotech Eng 11(3):445–463

Janc B, Vižintin G, Pal A (2024) Investigation of disc cutter wear in tunnel-boring machines (TBMs): integration of photogrammetry, measurement with a caliper, weighing, and macroscopic visual inspection. Appl Sci 14(6):2443

Ju MS, Park MS, Kim JJ, Song SW, Do SC, Jeong H (2024) Development and performance evaluation of real-time Wear measurement system of TBM disc cutter. Tunn Undergr Space 34(2):154–168

Karami M, Zare S, Rostami J (2021a) Study of common wear prediction models for hard rock TBM disc cutters and comparison with field observation in Kerman water conveyance tunnel. Bull Eng Geol Environ 80:1467–1476

Karami M, Zare S, Rostami J (2021b) Tracking of disc cutter wear in TBM tunneling: a case study of Kerman water conveyance tunnel. Bull Eng Geol Environ 80:201–219

Karami M, Zare S, Rostami J (2021c) Introducing an empirical model for prediction of disc cutter life for TBM application in jointed rocks: case study, Kerman water conveyance tunnel. Bull Eng Geol Environ 80(5):3853–3870

Karami M, Zare S, Rostami J (2022) Real-scale numerical analyzing dynamic process of TBM boring in jointed rock; a case study: Kerman water conveyance tunnel in Iran. J Min Environ 13(3):643–666

Ko TY, Yoon HJ, Son YJ (2014) A comparative study on the TBM disc cutter wear prediction model. J Korean TunnLing Undergr Space Assoc 16(6):533–542

Koopialipoor M, Nikouei SS, Marto A, Fahimifar A, Armaghani DJ, Mohamad ET (2019) Predicting tunnel boring machine performance through a new model based on the group method of data handling. Bull Eng Geol Environ 78(5):3799–3813

Lin L, Xia Y, Mao Q, Zhang X (2018) Experimental study on wear behaviors of TBM disc cutter ring in hard rock conditions. Tribol Trans 61(5):920–929

Log S, Smading S, Mosavat K (2018) Recent cutter technology advances to efficiently bore through extremely challenging conditions in hard rock. White Pap Robbins Co

Macias FJ, Jakobsen PD, Seo Y, Bruland A (2014) Influence of rock mass fracturing on the net penetration rates of hard rock TBMs. Tunn Undergr Space Technol 44:108–120

Macias FJ, Dahl F, Bruland A (2016) New rock abrasivity test method for tool life assessments on hard rock tunnel boring: the rolling indentation abrasion test (RIAT). Rock Mech Rock Eng 49(5):1679–1693

Maidl B, Schmid L, Ritz W, Herrenknecht M (2008) Hard rock tunnel boring machines. Ernst & Sohn
Naghadehi MZ, Ramezanzadeh A (2017) Models for estimation of TBM performance in granitic and mica gneiss hard rocks in a hydropower tunnel. Bull Eng Geol Environ 76(4):1627–1641
NTH (1988) Hard Rock Tunnel Boring., Project Report 1-88. Norwegian Institute Technology, Div. of Construction Engineering, Trondheim, Norway
Rajati MH, Rostami J, Memarian H, Hamzaban MT (2023) A study on predicting the wear of TBM disc cutters using Cerchar testing. Tunn Undergr Space Technol 140:105290
Ren DJ, Shen SL, Arulrajah A, Cheng WC (2018) Prediction model of TBM disc cutter wear during tunnelling in heterogeneous ground. Rock Mech Rock Eng 51:3599–3611
Rostami J (1997) Development of a force estimation model for rock fragmentation with disc cutters through theoretical modeling and physical measurement of crushed zone pressure, vol 38. Colorado School of Mines, Golden, pp 56–64
Rostami J, Ozdemir L, Bruland A, Dahl F (2005) Review of issues related to Cerchar abrasivity testing and their implications on geotechnical investigations and cutter cost estimates. Proc RETC:738–751
Su W, Li X, Jin D, Yang Y, Qin R, Wang X (2020) Analysis and prediction of TBM disc cutter wear when tunneling in hard rock strata: a case study of a metro tunnel excavation in Shenzhen, China. Wear 446:203190
Yagiz S (2003) A model for prediction of TBM performance in hard rock condition. 56th Geol Congr Turk
Yang Y, Hong K, Sun Z, Chen K, Li F, Zhou J, Zhang B (2018) The derivation and validation of TBM disc cutter wear prediction model. Geotech Geol Eng 36(6):3391–3398
Yu H, Tao J, Huang S, Qin C, Xiao D, Liu C (2021) A field parameters-based method for real-time wear estimation of disc cutter on TBM cutterhead. Autom Constr 124:103603

Index

J. Hassanpour and J. Rostami (eds.), *Performance of Tunnel Boring Machines (TBM) in Rock*, Springer Series in Geomechanics and Geoengineering,
https://doi.org/10.1007/978-3-032-20424-0

GPSR Compliance

The European Union's (EU) General Product Safety Regulation (GPSR) is a set of rules that requires consumer products to be safe and our obligations to ensure this.

If you have any concerns about our products, you can contact us on ProductSafety@springernature.com

In case Publisher is established outside the EU, the EU authorized representative is:

Springer Nature Customer Service Center GmbH
Europaplatz 3
69115 Heidelberg, Germany

Batch number: 10755329

Printed by Printforce, the Netherlands